MACMILLAN WORK OUT SERIES

Electronics

G. Waterworth

First published 1988 by
MACMILLAN PRESS LTD
Houndmills, Basingstoke, Hampshire RG21 2XS
and London
Companies and representatives
throughout the world

ISBN 0–333–45871–0

A catalogue record for this book is available
from the British Library.

13 12 11 10 9 8 7 6 5
03 02 01 00 99 98 97 96 95

Printed in Malaysia

Contents

Acknowledgements

The author and publishers wish to thank the following who have kindly given permission for the use of their examination questions:

The Institution of Electrical Engineers (IEE)
The Council of Engineering Institutions (CEI)
The Engineering Council (EC)

The author also wishes to thank all those who have helped in the preparation of this book, both for their helpful advice and for their encouragement. In particular, he would like to thank Fred Robson who, as a colleague at Leeds Polytechnic, has provided support at all stages throughout the preparation of the book.

The examination boards accept no responsibility whatsoever for the accuracy or method in the answers given to questions set by them. The answers are the entire responsibility of the author.

Introduction

How to Use this Book

This book is intended primarily as a text to help electrical engineering students, who are taking a first course in electronics, prepare for an examination. It will also be useful to students on some second-year courses and to students of mechanical engineering, physics and other disciplines for which a first course in electronics is included.

The questions vary in their level of difficulty and the book should be found useful at various stages of both degree and diploma courses.

Each chapter contains:

(a) A brief summary of the major facts and concepts;
(b) Worked examples of representative university, polytechnic and IEE/CEI/EC examination questions;
(c) Unworked problems of a similar standard with answers at the back of the book.

The topics are arranged in the traditional order of semiconductor devices (diodes, BJTs, FETs, op-amps), followed by their applications in a variety of electronic subsystems such as amplifiers, oscillators, non-linear circuits, power amplifiers, regulated power supplies, power electronics systems, combinational and sequential logic circuits. This should enable the reader to use these worked examples alongside any of the standard textbooks in electronics, or as a guide in working through past examination papers and tutorial sheets provided by the appropriate examining body.

The text at the beginning of each chapter is deliberately brief, but a comprehensive range of techniques and equations is developed from basic principles, allowing the reader to use this both as a companion guide during initial studies and as a revision book towards the end of the course.

The author has attempted to give an engineering approach to the subject by including questions that have a sound application and a realistic solution. Examples are included both on discrete devices and on integrated-circuit devices.

The book uses up-to-date techniques and devices and includes questions on the current mirror, bootstrapping, the constant current source, the Miller effect, the cascode circuit, feedback amplifiers, differential amplifiers, instrumentation amplifiers, i.c. regulators, switched mode regulators, chopper control, programmable logic arrays, synchronous counter design, and logic hazard detection.

Extensive use has been made of device models in analysing circuit operation, and in the case of the bipolar junction transistor, both the h-parameter and the model based on the Ebers–Moll equation have been included.

Examination Technique

No student can expect to pass examinations without a sound knowledge of the subject material to be examined, but it is possible to improve one's chances by attention to the following.

The Candidate's Presentation

When a candidate is faced with an examination paper, he or she may avoid common errors (which can seriously prejudice the performance) and present the work most favourably by attention to the following advice.

(a) *Choice*
Take time to read through the whole paper, decide on the questions to be answered, and mark them in order of preference; note any rubric concerning compulsory questions or restricted choice from a sectional paper.

(b) *Timing*
Do not spend too much time on a question (e.g. with 5 questions to answer in 3 hours not more than 30–40 minutes should be spent on each). It is unwise to persist with complicated and laborious work that seems to be leading nowhere: the answer sought by the examiners is unlikely to involve several pages of analysis.

(c) *Interpretation*
Carefully read each question before starting to answer it: answer the question set, not some alternative of your own; do not just 'describe' when the question states, 'compare'; do not omit an essential part of the question.

(d) *Presentation*
Either read each answer carefully after writing it, or allow adequate time towards the end of the examination period, to revise and correct the work.

(e) *Formula*
Do not attempt to memorise extensive lists of formula. A candidate is expected to know (by common usage and familiarity rather than by a conscious effort of memory) a few simple expressions directly based on fundamental laws and principles. Complex formulae will normally be quoted in the question, or the candidate asked to develop them from 'first principles'.

The Examiner's Requirements

In assessing a candidate's ability, the examiner looks particularly for the following:

(a) Sound knowledge of the appropriate fundamental principles; facility in the analysis of problems based thereon; and a clear and explicit understanding of the methods used and approximations legitimately made.

(b) Facility in numerical and graphical work in the solution of (in the examination sense) 'practical' engineering problems; logical layout of steps in the working; and an appreciation of practical orders of size and of numerical accuracy.

(c) Collective marshalling of logic and critical arguments for and against the choice of methods, equipment and lines of action.

(d) Some knowledge, derived from wider reading and professional motivation, of the general trend of modern developments.

(e) Ability to present descriptive answers legibly, concisely and grammatically; the use, where appropriate, of clear and neat freehand sketches.

1 Diode Circuits

1.1 Diode Operation

A semiconductor diode is a junction between a *p*-type and an *n*-type semiconductor (a *pn* junction). A *pn* junction is essentially a device which allows current to flow in one direction but restrains the flow in the other direction, except for a small leakage current.

The diode symbol is shown in Fig. 1.1; current will flow in the direction of the arrow.

Figure 1.1 The diode symbol

Semiconductor diodes are used in a wide variety of applications, some of the more common being clippers, comparators, diode gates, rectifiers and voltage doublers.

1.2 Diode Characteristic

A forward characteristic for a silicon diode is shown in Fig. 1.2. For forward voltages below about 0.6 V the current is very small. In the reverse bias condition

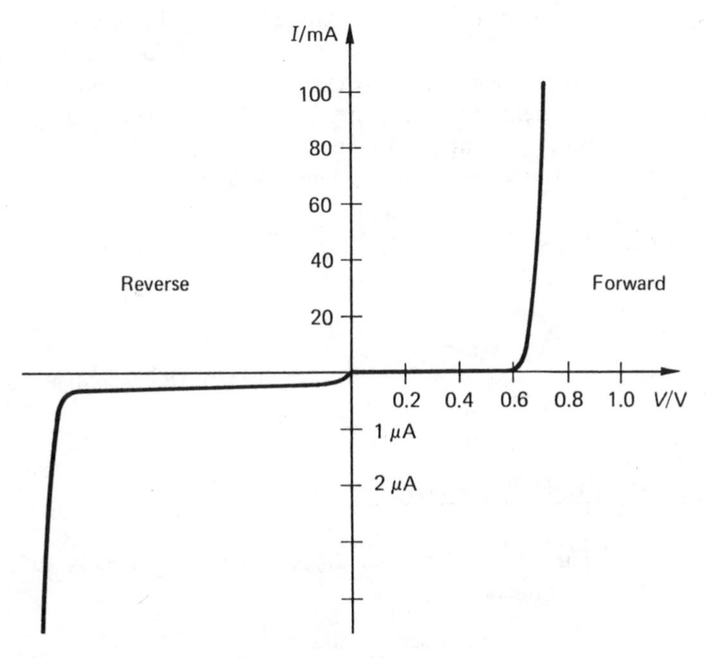

Figure 1.2 The characteristic for a silicon diode

the reverse saturation current (leakage current) is of the order of several nA to several μA. Leakage current is very temperature-dependent and it is found that the reverse saturation current approximately doubles for every 10 °C rise in temperature.

At the reverse breakdown voltage (or peak inverse voltage, PIV) the diode will break down, and unless the current is limited by a resistor the device will be damaged. Diodes are not used on this part of the characteristic except in the case of zener diodes.

In the forward bias condition the change in forward voltage due to temperature is typically –2.5 mV/°C, the current being held constant. Usually the forward voltage drop of about 0.6 V to 0.8 V is of little concern, and the diode can be approximated to an ideal device.

1.3 Diode Parameters

The important diode parameters usually supplied by the manufacturer are as follows:

$V_{R(max)}$ or PIV or V_{RRM} The voltage reverse maximum or peak inverse voltage, this being the maximum reverse voltage the diode can withstand.

$I_{R(max)}$ The reverse leakage current at a specified reverse voltage (usually $\leqslant$ several microamps).

V_F The forward voltage drop at a specified forward current, I_F.

$I_{av(max)}$ The maximum value of full-cycle average current that the diode can safely conduct without becoming overheated.

1.4 Zener Diodes

A diode biased in the reverse direction will break down at some specified voltage, as shown in Fig. 1.2. Since the breakdown characteristic is so abrupt, the diode may be used as a voltage reference device. Such diodes are referred to as avalanche or zener diodes, depending on the breakdown mechanism. The term zener is commonly used for all breakdown diodes. In low-voltage zeners (below 6 V), the breakdown is due to the zener effect, which has a negative temperature coefficient. In higher-voltage zeners, the breakdown is due to the avalanche effect with a positive temperature coefficient.

The symbol for the zener diode is shown in Fig. 1.3.

Figure 1.3 The zener diode symbol

1.5 Zener Diode Parameters

The important zener diode parameters usually supplied by the manufacturer are as follows:

V_z Nominal zener voltage.

r_z Dynamic resistance (reciprocal of slope of V–I characteristic in the operating range).

P Maximum power dissipation.

α_z Temperature coefficient (percentage change in reference voltage per °C change in zener diode temperature).

1.6 Rectifiers

Most electronic circuits require a source of d.c. power. Rectifiers provide a means of converting an alternating voltage into a unidirectional voltage and thus form the basis of a power supply. The a.c. voltage is usually provided by transforming down the mains supply voltage.

(a) Half-wave Rectifier

A half-wave rectifier circuit and its associated waveforms are shown in Fig. 1.4.

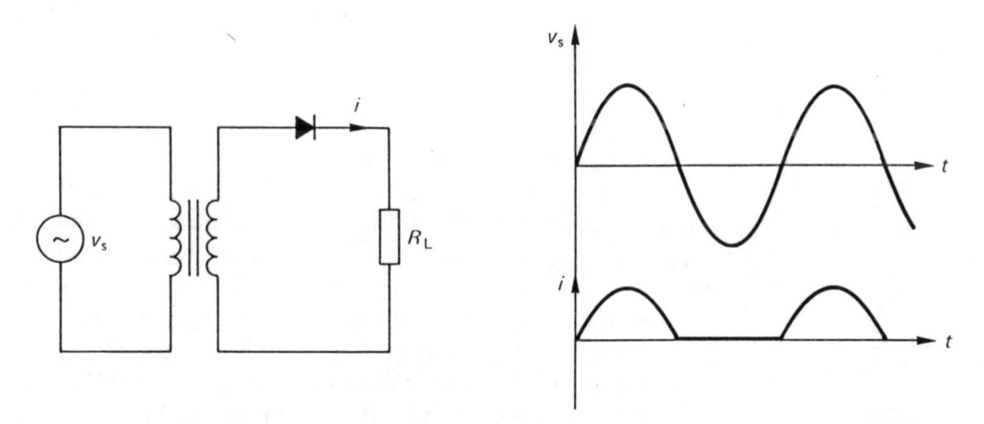

Figure 1.4 Half-wave rectifier circuit

(b) Full-wave Rectifier

One method of full-wave rectification using two diodes and a centre-tapped transformer is shown in Fig. 1.5(a), together with the associated waveforms in (b).

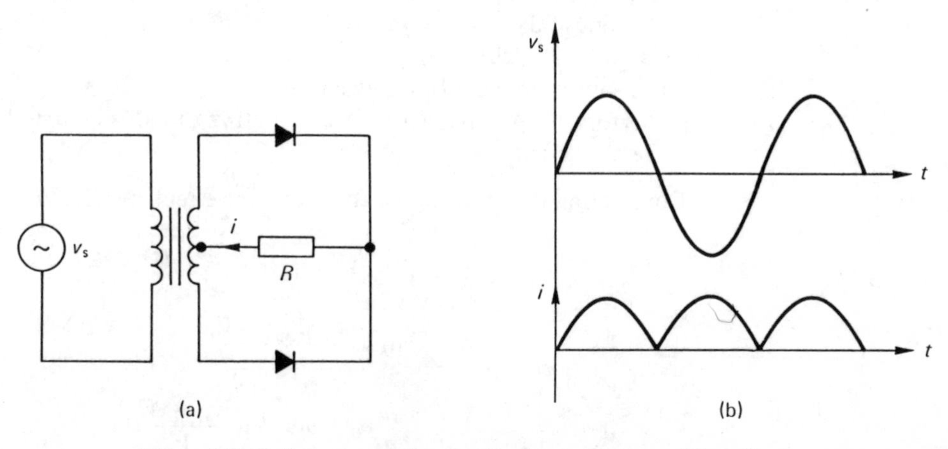

Figure 1.5 Full-wave rectifier circuit using centre-tapped transformer

An alternative full-wave rectifier circuit is shown in Fig. 1.6. It is referred to as a bridge rectifier. Only two diodes conduct at any one time. The transformer requires no centre tap, and each diode has only the transformer voltage across it on the inverse cycle, making it more suitable for high-voltage applications.

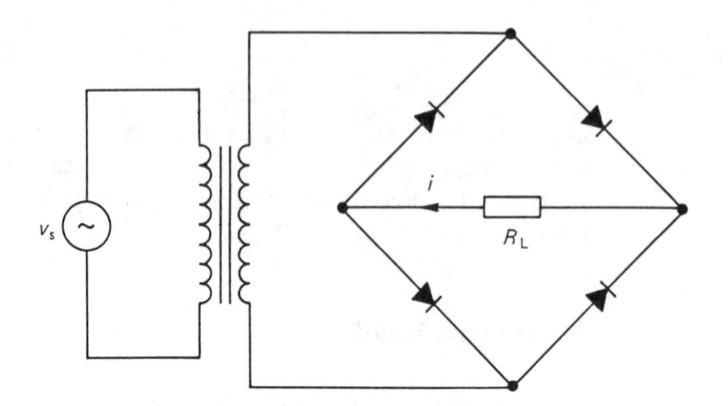

Figure 1.6 Bridge rectifier circuit

(c) Rectifier with Capacitor Filter

A half- or full-wave rectified-voltage waveform may be smoothed out to provide an approximate d.c. voltage using a capacitor shunted across the load to act as a filter. The capacitor stores energy during the conducting period and delivers this energy to the load during the non-conducting period. The deviation of the load voltage from its average or d.c. value is referred to as the ripple voltage.

Referring to Fig. 1.7, which shows the half-wave rectifier circuit with capacitor smoothing, and the associated output voltage and diode current waveforms, the following simplifying approximations are made:

(a) The load-time constant CR_L is large compared with the period of the rectifier output waveform, so that the charging interval θ_c is small compared with the cycle time T.
(b) The diode current i_d is assumed to be triangular.
(c) The diode switching time α_2 when the capacitor is fully charged occurs at the peak of the supply voltage (i.e. at $\omega t = \pi/2$).
(d) The sinusoidal and exponential parts of the output voltage can be approximated by straight lines.
(e) The source impedance is negligible.
(f) The forward voltage across a conducting diode is assumed constant, irrespective of current.

Considering the waveforms, for a half-wave rectifier shown in Fig. 1.7,

$$V_{dc} \approx V_m - \frac{V_r}{2}$$

and

$$\sin \alpha_1 = \frac{V_m - V_r}{V_m},$$

$$\therefore \quad \alpha_1 = \sin^{-1}\left(\frac{V_m - V_r}{V_m}\right)$$

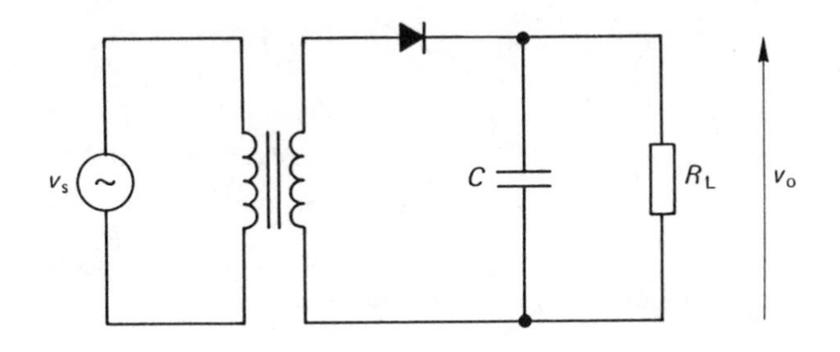

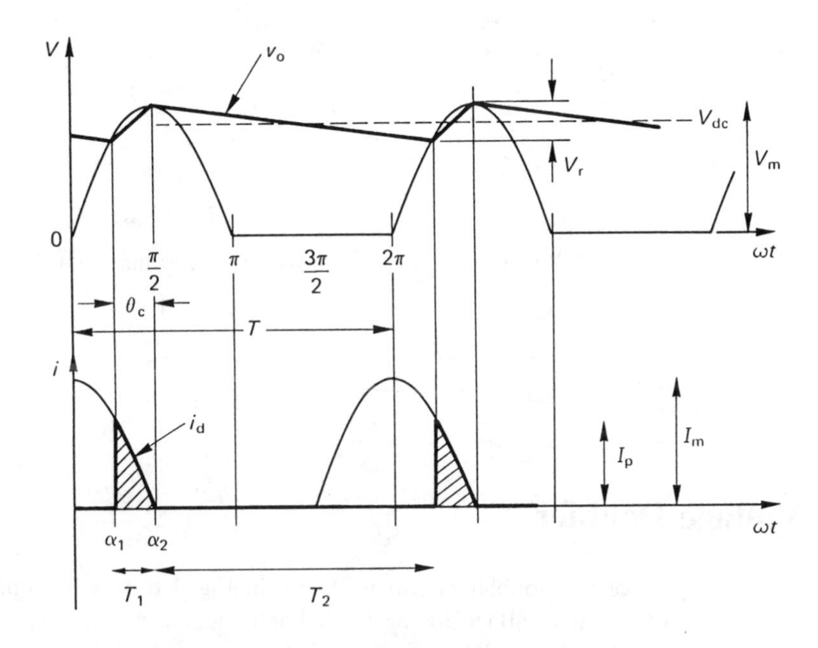

Figure 1.7 Half-wave rectifier with capacitor smoothing

But
$$\alpha_2 = \tfrac{1}{2}\pi,$$
$$\therefore\ \theta_c = \alpha_2 - \alpha_1 = \tfrac{1}{2}\pi - \sin^{-1}\left(\frac{V_m - V_r}{V_m}\right),$$

or
$$\cos\theta_c = \frac{V_m - V_r}{V_m}.$$

During the charging period, the charge accumulated in the capacitor C is given by
$$Q = CV_r.$$

This charge is transferred to the load during discharge, the charge lost by the capacitor being given approximately by:
$$Q = I_{dc}T,$$

since $T_2 \approx T$.
$$\therefore\ I_{dc}T = CV_r,$$
$$\therefore\ V_r = \frac{I_{dc}T}{C}.$$

But $$T = \frac{1}{f},$$

$$\therefore \quad V_r = \frac{I_{dc}}{fC}$$

and $$V_{dc} = I_{dc}R_L = fCR_L V_r.$$

The ripple factor γ is given by

$$\gamma = \frac{\text{r.m.s. value of all a.c. components}}{\text{d.c. component}}.$$

Since we have assumed that the ripple has a triangular waveform of peak value $V_r/2$, its r.m.s. value can be shown to be

$$\frac{V_r}{2} \times \frac{1}{\sqrt{3}}.$$

$$\therefore \gamma = \frac{V_r}{2\sqrt{3}V_{dc}}$$

For the full-wave case, the waveform period is half that for the half wave,

$$\therefore T_2 = \frac{1}{2f}$$

and $$V_r \approx \frac{I_{dc}}{2fC}.$$

1.7 Voltage Doubler

A voltage doubler circuit is shown in Fig. 1.8. It will supply approximately twice the supply voltage to the load. Each capacitor is alternately charged to the transformer peak voltage V_m, current being continually drained from the capacitors through the load. The PIV across the diode during the non-conducting period is twice the transformer peak voltage.

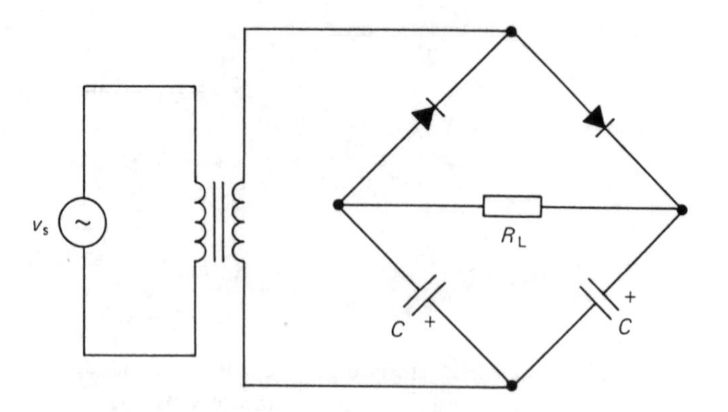

Figure 1.8 Voltage doubler circuit

1.8 Worked Examples

Example 1.1

Explain the term 'clipping' as referred to diode circuits.

Given a sinusoidal input, show how a diode may be used to

(a) clamp the signal to allow positive variations only;
(b) clip the signal to provide variations only above a reference voltage V_r.

An RS232 serial interface signal has signal levels of ± 12 V and is to be clipped to provide a 0 V to +5 V signal. Show how this may be done, using

(c) two catching diodes;
(d) a zener diode.

Solution 1.1

'Clipping' is a means of selecting the parts of a voltage waveform that lie above or below some reference level. Clipping circuits are also referred to as 'voltage limiters' or 'catching diodes'.

Clamping and clipping are shown in diagrams (a) and (b).

RS232-voltage level conversion to TTL voltage level is shown in diagrams (c) and (d).

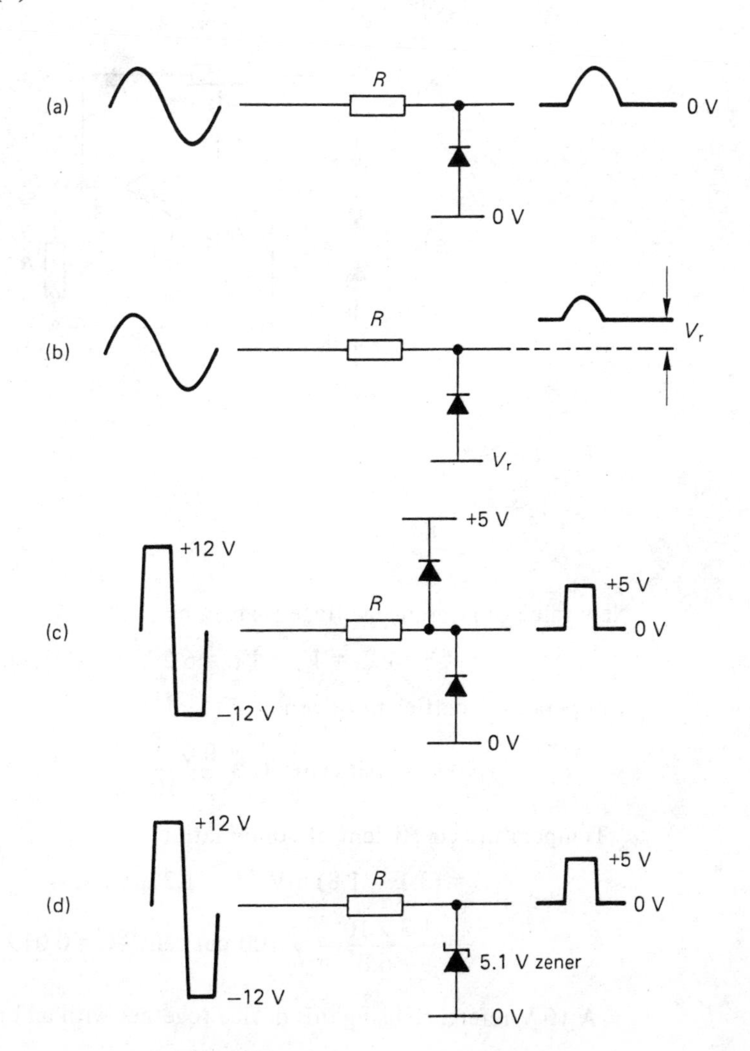

Example 1.2

A 6.2 V zener diode has a temperature coefficient of +0.05 per cent/°C at 25 °C and a zener current of 5 mA. If a silicon diode with a forward voltage of 0.6 V at this current and a temperature coefficient of −1.8 mV/°C is connected in series with the zener diode, calculate the new reference voltage and the temperature coefficient of the combination. Show how this combined device could be used together with an op-amp to provide a voltage reference of 10 V.

Solution 1.2

The arrangement is shown in diagram (a).

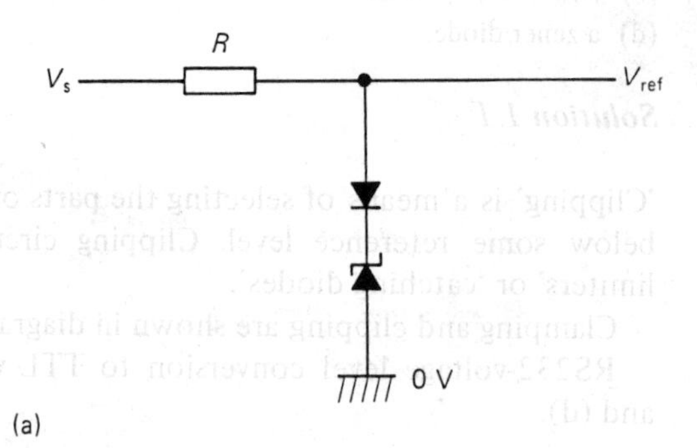

(a)

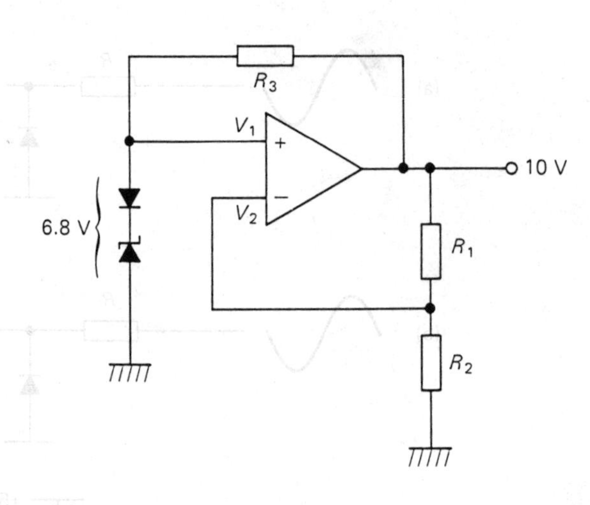

(b)

New value of reference voltage is given by

$$V_{ref} = V_z + V_F = 6.2\text{ V} + 0.6\text{ V} = \underline{6.8\text{ V}}.$$

Temperature coefficient of zener

$$= 0.05\text{ per cent/°C} = \frac{0.05 \times 6.2}{100}\text{ V/°C} = 3.1\text{ mV/°C}.$$

∴ Temperature coefficient of combination

$$= (3.1 - 1.8)\text{ mV/°C} = 1.3\text{ mV/°C}$$

$$= \frac{1.3 \times 10^{-3}}{6.8} \times 100\text{ per cent/°C} = \underline{0.019\text{ per cent/°C}}.$$

A 10 V reference using this device together with an op-amp is shown in diagram (b), where R_1 and R_2 are chosen such that when $v_o = 10$ V then

$$V_1 = V_2 = 6.8 \text{ V}.$$

$$\therefore \frac{R_2}{R_1 + R_2} = \frac{6.8}{10},$$

with, say, $R_1 = 10$ kΩ and $R_2 = 21$ kΩ.
Choose R_3 to give 5 mA in zener diode;

$$\therefore R_3 = \frac{v_o - v_i}{5 \text{ mA}} = \frac{10 \text{ V} - 6.8 \text{ V}}{5 \times 10^{-3} \text{ A}} = 640 \, \Omega.$$

Example 1.3

A silicon diode has an ideal forward and reverse characteristic and is used in a half-wave rectifier circuit to supply a resistive load of 500 Ω from a sinusoidal supply of 250 V. Calculate:

(a) the peak diode current;
(b) the mean diode current;
(c) the r.m.s. diode current;
(d) the power dissipated in the load.

Solution 1.3

(a) $$I_p = \frac{V_p}{500 \, \Omega} = \frac{250\sqrt{2} \text{ V}}{500 \, \Omega} = \underline{707 \text{ mA}}.$$

(b) The current is half-wave rectified and averaged over a full cycle.

$$\therefore I_m = \frac{1}{2\pi} \int_0^{\pi} I_p \sin \theta \, d\theta$$

$$= \frac{1}{2\pi} \times I_p \left[-\cos \theta \right]_0^{\pi}$$

$$= \frac{707 \times 2}{2\pi} = \underline{225 \text{ mA}}.$$

(c) $$I_{rms} = \sqrt{\frac{1}{2\pi} \int_0^{\pi} I_p^2 \sin^2 \theta \, d\theta}$$

$$= I_p \sqrt{\frac{1}{2\pi} \int_0^{\pi} \frac{1 - \cos 2\theta}{2} \, d\theta}$$

$$= I_p \sqrt{\frac{1}{4\pi} \left[\theta - \frac{\sin 2\theta}{2} \right]_0^{\pi}}$$

$$= I_p \sqrt{\frac{1}{4\pi} [\pi]}$$

$$= \frac{I_p}{2} = \frac{708 \text{ mA}}{2} = \underline{354 \text{ mA}}.$$

(d) Power dissipated in load is given by

$$P_D = I_{rms}^2 R_L$$

$$= (354 \times 10^{-3})^2 \times 500$$

$$= \underline{62.7 \text{ W}}.$$

Example 1.4

A full-wave rectifier circuit uses two silicon diodes that may be considered to have a constant forward resistance R_F of 2 Ω and infinite reverse resistance. The circuit is fed from a 200-0-200 volt r.m.s. secondary winding of a transformer, giving a mean current in the resistive load R_L of 10 A. Calculate:

(a) the value of the load resistance;
(b) the maximum value of the voltage that appears across the diodes (PIV);
(c) the circuit efficiency.

Solution 1.4

(a)

$$I_{mean} = \frac{1}{\pi}\int_0^{\pi} I_p \sin\theta \, d\theta$$

$$= \frac{I_p}{\pi}\Big[-\cos\theta\Big]_0^{\pi}$$

$$= \frac{2I_p}{\pi}.$$

But

$$I_p = \frac{V_p}{R_F + R_L} = \frac{V_p}{R},$$

where $R = R_F + R_L$,

$$\therefore \; I_p = \frac{200\sqrt{2}}{R},$$

$$\therefore \; I_{mean} = \frac{2}{\pi} \times \frac{200\sqrt{2}}{R} = 10 \text{ A},$$

$$\therefore \; R = \frac{2}{\pi} \times \frac{200\sqrt{2}}{10} = 18 \, \Omega.$$

$$\therefore \; R_L = R - R_F = 18 - 2 = \underline{16 \, \Omega}.$$

(b) The maximum voltage across the diodes is when the anode is at $-V_p$ and the cathode at $+V_p$ (neglecting small drop across diode).

$$\therefore \text{ Max. voltage } = 2V_p = \underline{566 \text{ V}}.$$

(c)

$$\text{Efficiency } \eta = \frac{\text{power in load}}{\text{power delivered}}$$

$$= \frac{I_{mean}^2 R_L}{I_{rms}^2 R} = \frac{\left(\frac{2I_p}{\pi}\right)^2 R_L}{\left(\frac{I_p}{\sqrt{2}}\right)^2 R}$$

$$= \frac{4 \times 2 \times 16}{\pi^2 \times 18} = \underline{72 \text{ per cent.}}$$

Example 1.5

The output waveform of a 50 Hz half-wave rectifier is smoothed using a capacitor filter circuit with $C = 50\ \mu F$. The circuit supplies a direct current of 50 mA to a 10 kΩ load resistor. If the transformer has a 'step-up' ratio of 1 : 2, calculate:

(a) the voltage applied to the primary winding of the transformer;
(b) the ripple factor;
(c) the diode conduction angle;
(d) the peak diode current.

Solution 1.5

$$V_{dc} \simeq V_m - \frac{I_{dc}}{2fC}.$$

But

$$V_{dc} = I_{dc}R_L$$

$$= 50 \times 10^{-3} \times 10 \times 10^3 = 500\ \text{V}.$$

$$\therefore\ V_m = V_{dc} + \frac{I_{dc}}{2fC}$$

$$= 500 + \frac{50 \times 10^{-3}}{2 \times 50 \times 50 \times 10^{-6}} = 510\ \text{V}.$$

For the transformer, $\dfrac{V_{1m}}{V_{2m}} = \dfrac{1}{2}.$

$$\therefore\ V_{1m} = \frac{V_{2m}}{2} = \frac{510}{2} = 255\ \text{V},$$

$$\therefore\ V_{1rms} = \frac{255}{\sqrt{2}} = \underline{180\ \text{V.}}$$

(b)

$$\text{Ripple factor } \gamma = \frac{V_r}{2\sqrt{3}\,V_{dc}}.$$

But

$$V_{r(p-p)} = \frac{I_{dc}}{fC} = \frac{50 \times 10^{-3}}{50 \times 50 \times 10^{-6}} = 20\ \text{V},$$

$$\therefore\ \gamma = \frac{20}{2\sqrt{3} \times 500} = \underline{1.15\ \text{per cent.}}$$

(c) Conduction angle found from

$$\cos\theta_c = \frac{V_m - V_r}{V_m} = \frac{510 - 20}{510}$$

$$= 0.96,$$

$$\therefore\ \theta_c = \underline{16.1^\circ}.$$

(d) Diode current is found from

$$i = I_m \sin(\omega t + \phi),$$

where $I_m = V_m \sqrt{\dfrac{1}{R^2} + \omega^2 C^2}$

$$= 510 \sqrt{\frac{1}{(10 \times 10^3)^2} + (2\pi \times 50 \times 50 \times 10^{-6})^2}$$

$$= 8.00\ \text{A},$$

and ϕ = arctan ωCR_L

$$= \arctan (2\pi \times 50 \times 50 \times 10^{-6} \times 10 \times 10^3)$$

$$= 89.6°.$$

In this case the conduction commences at

$$\text{angle } \alpha_1 = 90° - 16.1° = 73.9°,$$

$$\therefore i = I_m \sin (73.9° + 89.6°)$$

$$= 8 \times 0.284 = \underline{2.27 \text{ A}}.$$

Example 1.6

A power supply delivers 200 mA at 24 V to a resistive load. The supply is 240 V 50 Hz and a bridge rectifier is used with a 2000 μF smoothing capacitor connected across the resistive load. Making any necessary assumptions, estimate:

(a) the ripple voltage;
(b) the transformer secondary voltage,
(c) the peak inverse voltage across a diode;
(d) the peak current in a diode;
(e) the power dissipated in a diode.

Solution 1.6

(a) Ripple voltage is given by

$$V_r = \frac{I_{dc} T}{2C} = \frac{I_{dc}}{2fC}$$

$$= \frac{200 \times 10^{-3}}{2 \times 50 \times 2000 \times 10^{-6}} = \underline{1 \text{ V (peak to peak)}}$$

(b) The transformer secondary voltage is found from

$$V_m = V_{dc} + \frac{V_r}{2} = 24 + 0.5 = 24.5 \text{ V}.$$

If we allow for a diode forward voltage drop of, say, 0.7 V then the peak transformer secondary voltage is given by

$$24.5 + 1.4 = 25.9 \text{ V (peak)}$$

$$= \underline{18.3 \text{ V (r.m.s.)}}.$$

(c) Diode PIV is simply $\underline{24.5 \text{ V}}$.
(d) The diode peak repetitive current is found as follows:

The charge passed by the diode during the conduction time must equal that supplied to the load during the discharge time.

But the charge passed by the diode is given by the area under the triangular current waveform.

$$\therefore \tfrac{1}{2} I_p T_1 = I_{dc} \frac{T}{2}.$$

But the conduction time is given by

$$T_1 = \frac{\theta_c}{\pi} \times \frac{T}{2}\,.$$

$$\therefore \tfrac{1}{2} \times \frac{I_p\,\theta_c}{\pi} \times \frac{T}{2} = I_{dc}\,\frac{T}{2}\,,$$

$$\therefore I_p = \frac{2\pi\, I_{dc}}{\theta_c}\,.$$

Now, from the text,

$$\theta_c = \arccos\left(\frac{V_m - V_r}{V_m}\right)$$

$$= \arccos\left(\frac{24.5 - 1}{24.5}\right) = 16.4\,°,$$

$$\therefore I_p = \frac{360°}{16.4°} \times 0.2\ \text{A} = \underline{4.4\ \text{A}}.$$

(e) The mean power dissipation for each diode in the bridge is given by

$$P = \frac{1}{T}\int v_d i_d\, dt = \frac{1}{2\pi}\int_{\alpha_1}^{\alpha_2} v_d i_d\, d\theta,$$

where v_d and i_d are the instantaneous values of diode voltage and current. Assuming v_d is constant at 0.7 V when conducting, then

$$P = \frac{0.7}{2\pi}\int_{\alpha_1}^{\alpha_2} i_d\, d\theta = 0.7 \times I_{av}$$

$$\approx 0.35 I_{dc} = \underline{70\ \text{mW}}$$

(since the average current per diode is half the average load current).

1.9 Unworked Problems

Problem 1.1

What are the important features of a diode for use in a power rectifier circuit? List the parameters to be considered in selecting a diode for this application.

Define the percentage regulation of a smoothed power supply and explain briefly how it may be reduced.

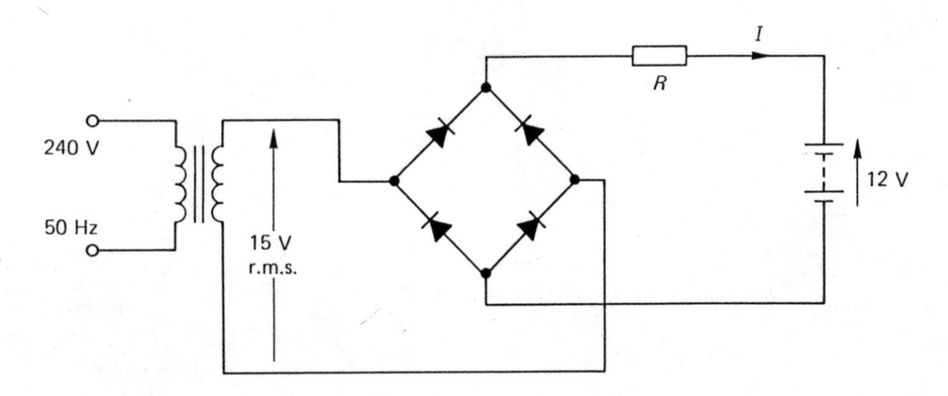

The battery charger circuit shown gives a peak current of 10 A. Calculate the mean current. Assume that the voltage across the battery remains constant and the voltage drop across the diode in conduction is 0.7 V. (EC Part 1)

Problem 1.2

A bridge rectifier circuit is made up using silicon diodes that may be considered to have a constant forward resistance R_F of 0.1 Ω and infinite reverse resistance. If the circuit supplies a mean current of 8 A to a resistive load R_L from an a.c. supply of 50 V, calculate:

(a) the load resistance;
(b) the circuit efficiency;
(c) the power dissipated in each diode.

Problem 1.3

Design a 50 Hz full-wave rectifier circuit with a capacitor filter to supply 200 V at 50 mA with a ripple factor not greater than 2 per cent and specify:

(a) the required capacitance;
(b) the transformer secondary voltage;
(c) the peak diode current expected.

2 Bipolar Junction Transistors

2.1 Transistor Operation

The bipolar junction transistor is a three-terminal 'active' device that may be used to control the flow of current through a load resistor from an external power source (such as a power supply) by means of a small power signal applied to its control terminal. It may thus be used as a power amplifier. The bipolar junction transistor (BJT) is so called because both polarities of current carrier are involved in the transport of current. It is normally referred to simply as a transistor. The transistor may be used in a wide variety of electronic circuit applications, such as amplifiers, switches, oscillators, filters, power supplies and many more.

The transistor symbol is shown in Fig. 2.1, from which it may be seen that there are two types, these being the *n p n* and the *p n p*. The device has three terminals, the emitter, the base and the collector. The flow of 'majority' carriers is between emitter and collector, and is controlled by means of the voltage applied between the base and emitter. The arrow on the emitter terminal indicates the direction in which conventional current would flow in normal operation. In the case of the *n p n*, conduction is due mainly to flow of electrons. In the *p n p* the conduction is mainly due to holes (electron vacancies).

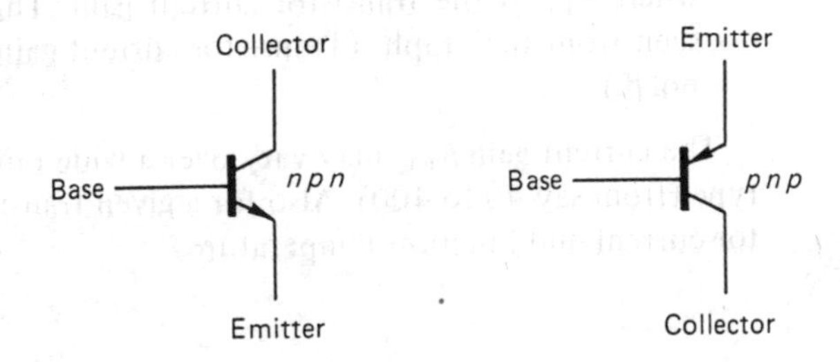

Figure 2.1 The transistor symbol

2.2 Transistor Bias

In the case of both the *n p n* and the *p n p*, to ensure that the device behaves as a transistor for amplifying signals it is necessary to:

(a) forward bias the base–emitter junction,
(b) reverse bias the collector-base junction.

Methods of achieving these bias conditions are shown for an *n p n* transistor in Fig. 2.2, where R_B is the base bias resistor. The arrangement in Fig. 2.2(c) is generally the preferred choice. The purpose of the other components will be explained shortly.

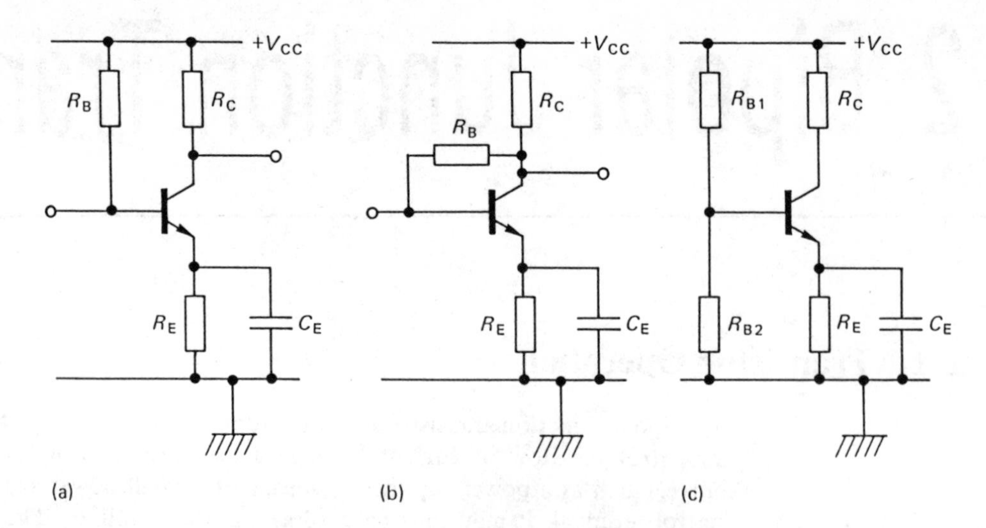

Figure 2.2 Biasing methods

Under normal transistor action,

(a) the base–emitter junction acts like a forward-biased diode with a forward voltage drop of about 0.6 V, i.e. $V_{BE} \approx 0.6$ V;
(b) the currents in the emitter, base and collector are related by the equation $I_E = I_C + I_B$;
(c) the collector current (I_C) is proportional to the base current (I_B), the relationship being shown by the graph of Fig. 2.3. The equation which relates I_C and I_B is

$$I_C = h_{FE} I_B,$$

where h_{FE} is the transistor current gain. This relationship is linear as may be seen from the graph. (Transistor current gain is also sometimes given the symbol β.)

The current gain h_{FE} may vary over a wide range of values for a given transistor type (from say 40 to 400). Also for a given transistor, h_{FE} varies with both collector current and junction temperature.

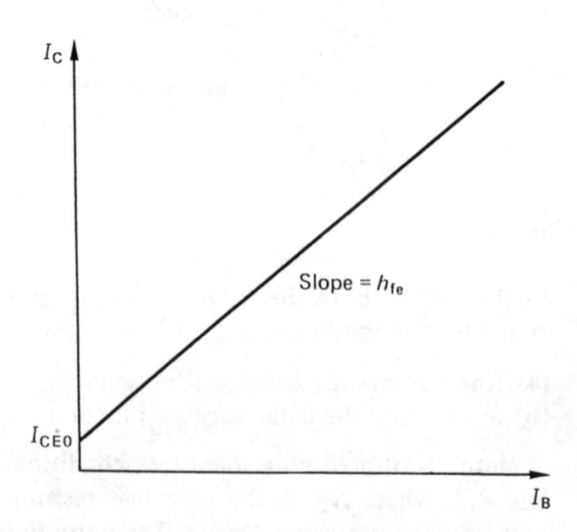

Figure 2.3 Relationship between I_C and I_B where I_{CEO} is the leakage current

For this reason it is necessary to avoid using h_{FE} as a 'design' parameter, and transistor circuits should be designed in such a way as to reduce as far as possible the effect of variations in this parameter. This is the purpose of the stabilising circuit shown in later amplifier circuits.

2.3 The Transistor Amplifier

One of the most common applications of the transistor is as a voltage amplifier. A common-emitter, single-stage a.c. amplifier using an *npn* transistor is shown in Fig. 2.4. The common-emitter amplifier is so called because the emitter side is common to both the input signal and the output signal.

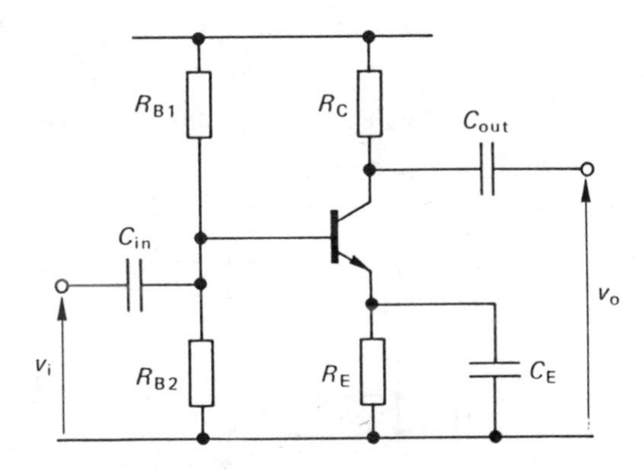

Figure 2.4 The CE single-stage amplifier

In the single-stage amplifier of Fig. 2.4, the resistors R_{B1} and R_{B2} form a potential divider to provide the bias arrangement for the base. The resistor R_C acts as the collector resistor to convert the collector current variations into voltage variations, thus providing an output signal voltage. The resistor R_E acts as a stabilising circuit to stabilise the d.c. output voltage level against variations in h_{FE} due to device parameter spread and temperature. (In fact, it provides series negative feedback to minimise any changes in the d.c. levels at the output. See section 6.3.) This circuit provides stabilisation against changes in h_{FE}, I_{CO}, and V_{BE} due to changes of temperature. The capacitor C_E acts as an a.c. bypass capacitor to bypass the resistor R_E for a.c. signals. The capcitors C_{in} and C_{out} act as d.c. level decoupling between the input signal stage and the output stage.

If an a.c. signal voltage is applied at the input, then an amplified version appears at the output. Notice that the output signal is 180° out of phase with the input signal.

The voltage gain of such an amplifier is of the order of several hundred. One way of reducing the voltage gain of the amplifier is to insert some unbypassed emitter resistance as shown in Fig. 2.5. It is thus possible to design the circuit to have a chosen voltage gain.

The transistor may also be used in two other configurations as shown in Fig. 2.6. These are referred to as the common-collector amplifier (CC) and the common-base amplifier (CB). Notice that the d.c. bias circuit remains basically the same.

In the case of the common-collector amplifier the collector is the common terminal, the base is the input, and the emitter is the output. This arrangement is

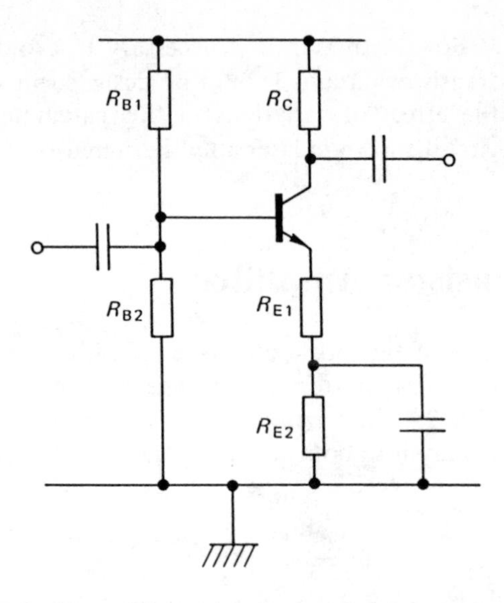

Figure 2.5 CE amplifier with unbypassed resistor in the emitter

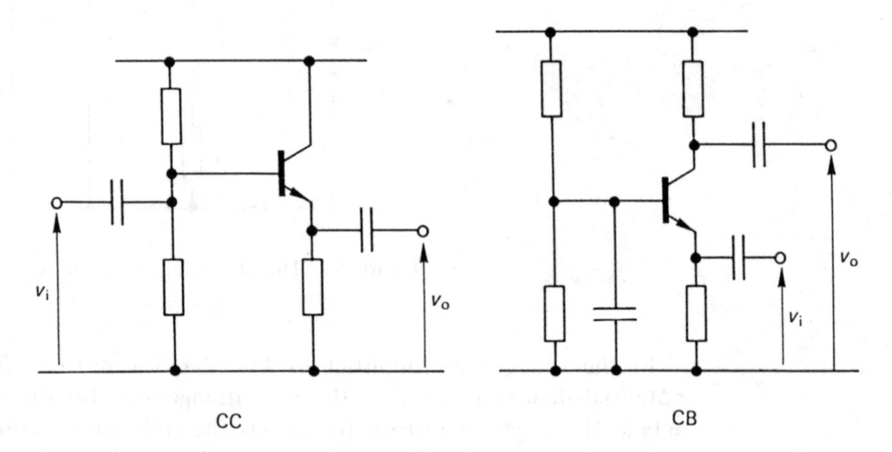

Figure 2.6 Common-collector and common-base amplifiers

often referred to as an emitter follower, since the output voltage at the emitter almost exactly follows the input voltage at the base.

In the case of the common-base amplifier the base is decoupled to ground as the common terminal, the emitter is the input, and the collector is the output.

The features of these three amplifier configurations are shown in Table 2.1 with typical values given.

Notice that the common-emitter arrangement provides good voltage, current and power gain. The common-collector arrangement (emitter follower) has a voltage gain of approximately unity together with a high input resistance and a low output resistance, thus making it useful as a buffer stage.

2.4 Transistor Characteristics

The input and output characteristics of the transistor in the common-emitter configuration are shown in Fig. 2.7.

The input characteristic shows the relationship between the input current I_B and the input voltage V_{BE}.

Table 2.1

	CE	CC	CB
voltage gain	high (300)	low (0.99)	high (300)
current gain	high (100)	high (100)	low (0.99)
power gain	very high (30 000)	high (100)	high (300)
phase relation between v_o and v_i	180° out of phase	in phase	in phase
R_{in}	medium (1 kΩ)	high (100 kΩ)	low (20 Ω)
R_{out}	high (10 kΩ)	low (20 Ω)	high (10 kΩ)

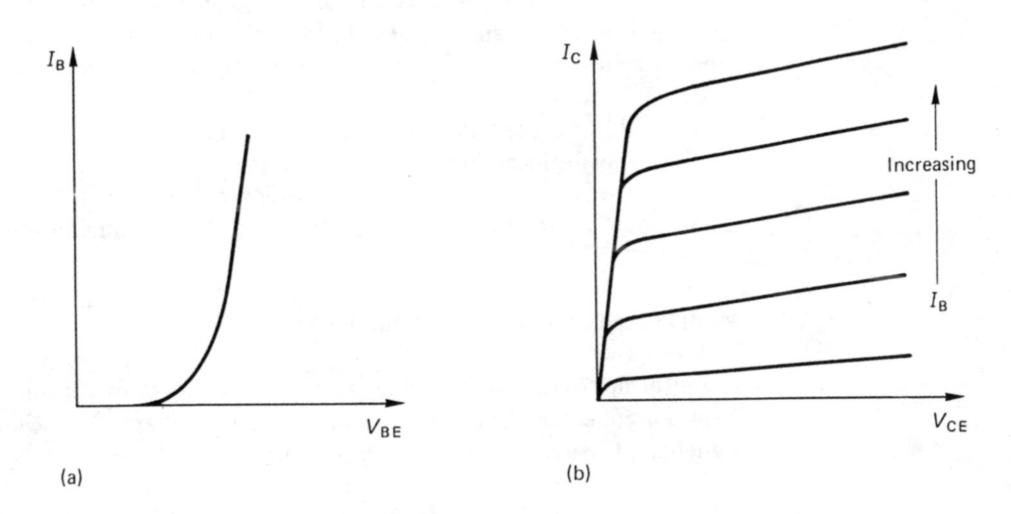

Figure 2.7 Transistor characteristics:
(a) input characteristic, (b) output characteristics

The output characteristic shows the relationship between the output current I_C and the output voltage V_{CE}.

Notice that the input characteristic is similar to that of a forward-biased diode.

The output characteristic for any given value of I_B has a collector current which rises rapidly with V_{CE} and then shows a pronounced 'knee' beyond which the current is almost constant. This part of the characteristic allows the device to be used as a constant current 'source' which is considered later.

For any given transistor type there are maximum specified levels of collector current I_C, base current I_B, collector-emitter voltage V_{CE} and reverse base-emitter voltage V_{EB}, which, if they are exceeded, may cause the transistor to be damaged.

2.5 Transistor Models

In analysing transistor circuits it is often useful to use a transistor model. For small-signal, low-frequency operation a convenient four-terminal model is the one shown in Fig. 2.8.

The model is considered from the point of view of the transistor in common emitter. It takes into account the input resistance r_i and the fact that the output

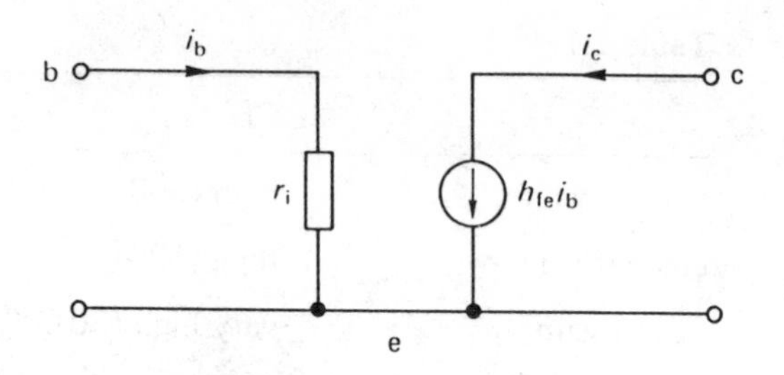

Figure 2.8 The transistor model

appears as if there were a current generator between the collector and the emitter whose current is given by $i_C = h_{fe}i_b$.

Notice that h_{fe} represents the small-signal current gain (small-signal, short-circuit, forward-current transfer ratio to be precise). The term h_{FE} is used to represent the static current gain, which may often have a somewhat different value.

It should be noted that the transistor model is only approximate and that it applies to small-signal, low-frequency variations.

There are two commonly used variations based on this model, these being the model based on the Ebers–Moll equation and the h-parameter model.

Model Based on Ebers–Moll Equation

A useful approach to analysing transistor circuits is to consider the relationship between collector current I_C and base-emitter voltage V_{BE}. Under normal small-signal amplifier conditions the relationship is given by

$$I_C = I_S \exp\left(\frac{V_{BE}}{V_T} - 1\right),$$

where $V_T = 25$ mV, provided that the junction is at room temperature.

The equation is known as the Ebers–Moll equation. At room temperature (20 °C) it may be approximated to

$$I_C = I_S \exp(40\, V_{BE}).$$

This equation quite closely describes the shape of graph of Fig. 2.7(a).

The transconductance g_m of the transistor in the common-emitter configuration is by definition given by

$$g_m = \frac{dI_C}{dV_{BE}} = 40 I_S \exp(40 V_{BE}) = 40 I_C,$$

where I_C is in amps.

The small-signal collector current is thus given by

$$i_c = g_m\, v_{be} = 40 I_C\, v_{be}.$$

The input resistance r_i in the common-emitter configuration is given by

$$r_i = \frac{dV_{BE}}{dI_B} = \frac{h_{fe}\, dV_{BE}}{dI_C} = \frac{h_{fe}}{g_m} = \frac{h_{fe}}{40 I_C}.$$

These two equations for g_m and r_i are very useful in transistor circuit analysis and design using a model.

The transistor model based on the Ebers–Moll equation is shown in Fig. 2.9. It

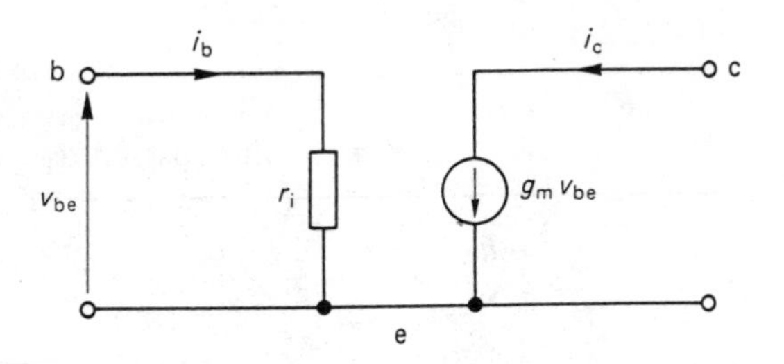

Figure 2.9 Model based on Ebers-Moll equation

has input resistance r_i between base and emitter and provides a collector current proportional to v_{be} given by $g_m v_{be}$.

Notice that g_m is directly proportional to collector current and inversely proportional to temperature. This is because V_T in the Ebers-Moll equation is temperature-dependent.

Notice also that r_i is inversely proportional to current and directly proportional to temperature.

A summary of the approximate equations for the transistor at low frequency derived from the simplified model based on the Ebers-Moll equation is given in Table 2.2.

In Tables 2.2 and 2.3, R_o is the transistor output resistance, and R'_o is the output resistance of the transistor and associated circuitry (including the effects of R_C or R_E). R_S is the source resistance (i.e. the output resistance of the signal generator).

Table 2.2

	CE	CE with unbypassed R_E	CC	CB
A_i	$-h_{fe}$	$-h_{fe}$	$1 + h_{fe}$	$\frac{h_{fe}}{1 + h_{fe}}$
R_i	$\frac{h_{fe}}{g_m}$	$\frac{h_{fe}}{g_m}(1 + g_m R_E)$	$\frac{h_{fe}}{g_m} + h_{fe} R_E$	$\frac{1}{g_m}$
A_v	$-g_m R_C$	$-\frac{g_m R_C}{1 + g_m R_E}$	$\frac{g_m R_E}{1 + g_m R_E}$	$g_m R_C$
R_o	infinite	infinite	$\frac{R_S}{h_{fe}} + \frac{1}{g_m}$	infinite
R'_o	R_C	R_C	$R_o \parallel R_E$	R_C

Note: $g_m = 40 I_C$ S, where I_C is measured in amps.

Table 2.3

	CE	CE with unbypassed R_E	CC	CB
A_i	$-h_{fe}$	$-h_{fe}$	$h_{fc} = 1 + h_{fe}$	$-h_{fb} = \frac{h_{fe}}{1 + h_{fe}}$
R_i	h_{ie}	$h_{ie} + h_{fe}R_E$	$h_{ie} + h_{fe}R_E$	$\frac{h_{ie}}{h_{fe}}$
A_v	$-\frac{h_{fe}R_C}{h_{ie}}$	$-\frac{h_{fe}R_C}{R_i}$	$1 - \frac{h_{ie}}{R_i}$	$\frac{h_{fe}R_C}{h_{ie}}$
R_o	infinite	infinite	$\frac{R_S + h_{ie}}{h_{fe}}$	infinite
R'_o	R_C	R_C	$R_o \parallel R_E$	R_C

Model Based on *h*-Parameters

The approximate equivalent *h*-parameter model is shown in Fig. 2.10. This is a model showing how the output current relates to the input current. The input resistance in the *h*-parameter model is referred to as h_{ie}. Manufacturers' data sheets generally provide typical values for *h*-parameters at specified conditions.

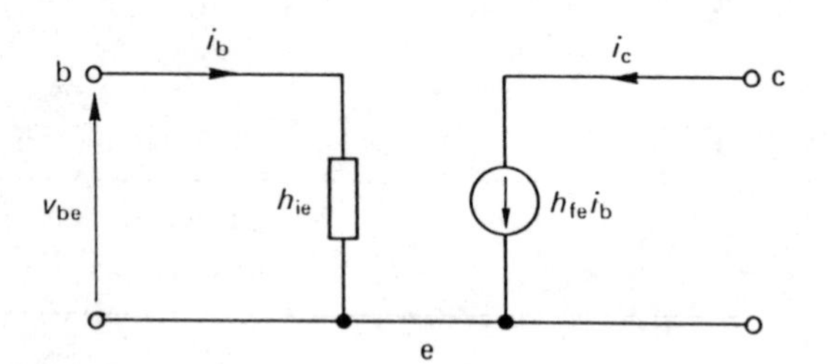

Figure 2.10 *h*-parameter model

Both of these models are useful for analysing transistor circuits. In both cases a more sophisticated model would include the following:

(a) The transistor output resistance $1/h_{oe}$ between collector and emitter, where h_{oe} is the output admittance. This is large enough to neglect in most cases, since the transistor in normal operation is almost a constant current source for a given I_B as shown in Fig. 2.7(b). Errors will be less than 10 per cent, provided $h_{oe}R_C$ is less than 0.1.
(b) The Early effect, which considers the variation in I_B due to V_{CE}, which again may be neglected in most cases. The full *h*-parameter model takes this into account by inserting a voltage generator $h_{re}v_{ce}$ in series with the input.

A summary of the approximate equations for the transistor at low frequency derived from the simplified *h*-parameter model is given in Table 2.3.

The full *h*-parameter model includes the effects of h_{oe} and h_{re}. From this model can be derived a more accurate set of equations for the various gains and impedances of an amplifier.

The designer should be wary of assuming accuracy by using the full equations, since h-parameters vary significantly with collector current. This means that the calculations are valid only for one value of collector current.

In practice, the full h-parameter model is often too cumbersome and many practising electronics engineers use the simplified model based on the Ebers–Moll equations. The advantage of using the Ebers–Moll model is that it takes into account the variation of g_m with I_C, whereas the h-parameter does not.

2.6 High Frequency Effects

At signal frequencies greater than a few kilohertz, the interelectrode capacitance of the transistor must be considered.

The Ebers–Moll model becomes modified to that shown in Fig. 2.11. This is referred to as the hybrid-π model, the one shown here being an approximation to the full hybrid-π.

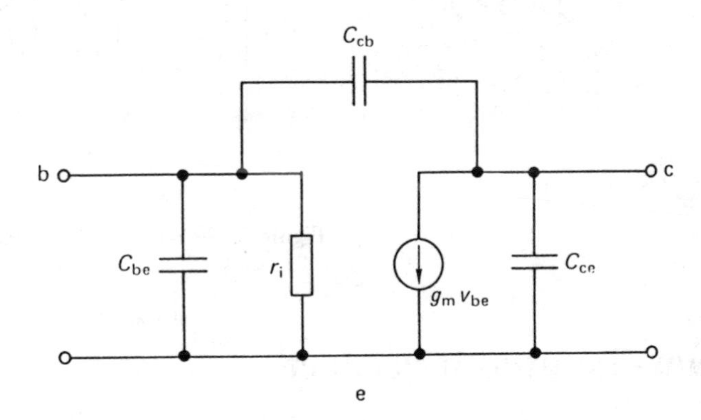

Figure 2.11 Hybrid-π model

Typical values for C_{be} and C_{cb} are 100 pF and 3 pF respectively. The effect of C_{ce} can usually be neglected.

Although C_{cb} is less than C_{be} it has the more pronounced effect on the performance of the transistor in amplifier circuits owing to the Miller effect.

Miller's theorem allows the effect of C_{cb} to be seen more clearly. The theorem is illustrated in Fig. 2.12 where the two circuits are equivalent. A proof is given in Example 2.5.

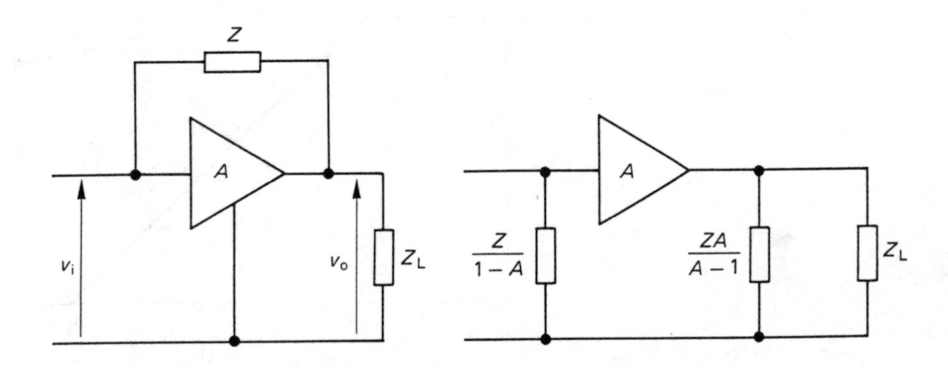

Figure 2.12 Miller's theorem

The voltage gain of a common emitter stage is given by

$$A_v = \frac{v_o}{v_i} = -g_m Z_L$$

The effect of C_{cb} is thus equivalent to that of a capacitor $(1 + g_m Z_L) C_{cb}$ across the input (and an additional capacitance of approximately C_{cb} across the output whose effect can normally be neglected).

The hybrid-π model when used in common emitter configuration may thus be simplified to that shown in Fig. 2.13, where

$$C = C_{be} + (1 + g_m R_C) C_{cb}.$$

The 'Miller effect' multiplication of C_{cb} prevents an amplifier from achieving both high gain and wide bandwidth simultaneously.

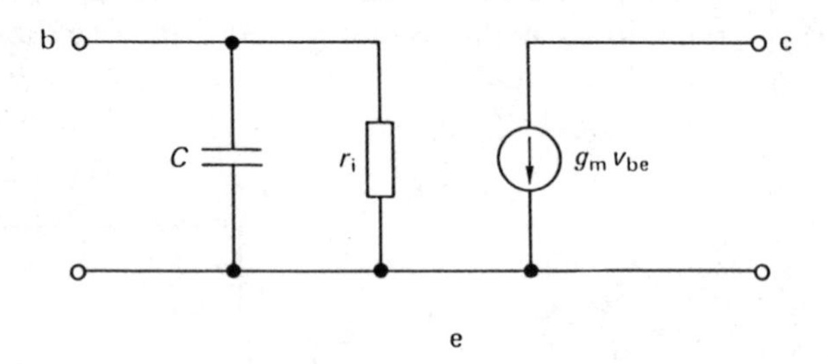

Figure 2.13 Simplified hybrid-π model

2.7 Gain–Bandwidth Product

The parameter f_T is an important high-frequency characteristic of the transistor. It is defined as the frequency at which the short-circuit common-emitter current gain becomes zero. Figure 2.14 shows this parameter in decibels against the logarithm of frequency where h_{fe} is the low-frequency value of current gain. f_T is shown as the point when the curve crosses the 0 dB axis. Also shown is the frequency f_B, where the frequency up to this point is referred to as the bandwidth. This is the point at which the gain has fallen to $1/\sqrt{2}$ of its low-frequency

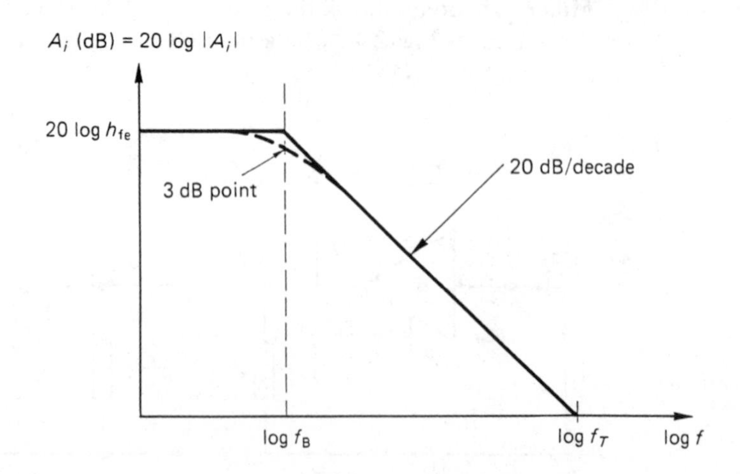

Figure 2.14 Current gain vs. frequency

value. Since $f_T \approx h_{fe} f_B$, then another interpretation of f_T is as the gain–bandwidth product where we are considering the short-circuit current gain.

2.8 Calculation of Output Impedance

The Thévenin representation of a circuit, as shown in Fig. 2.15, may conveniently be used to calculate the output impedance of a network. The Thévenin representation shows that the open-circuit voltage across the terminals would give the generator voltage V_{oc}. Alternatively, if a short circuit is applied between the terminals, then the short-circuit current will be given by $I_{sc} = \frac{V_{oc}}{Z}$. The output impedance may thus be found from $Z = \frac{V_{oc}}{I_{sc}}$.

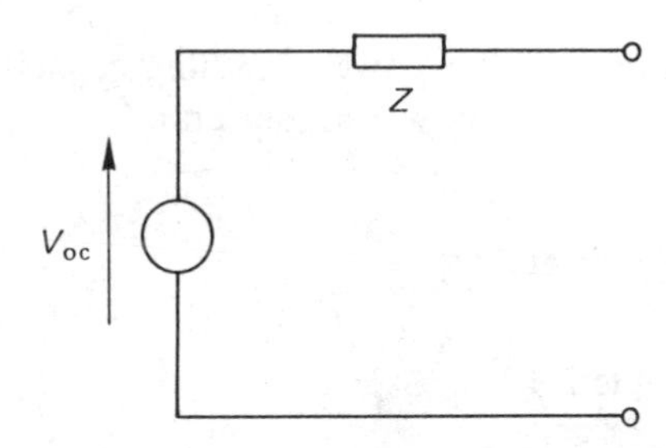

Figure 2.15 Thévenin's representation of a network

2.9 Advantages and Disadvantages of Bipolar Junction Transistors

Bipolar junction transistors, when compared with field effect transistors, have the advantages of possessing good current gain h_{fe} and good transconductance g_m. They have the disadvantages of poor stability of operating point with temperature, high noise, poor distortion, high intermodulation products, low input impedance when used in common-emitter or common-base configurations, and a risk of thermal runaway.

2.10 Transistor as a Switch

In many applications the transistor is used simply as a switch, as shown in Fig. 2.16. A small base-current may be used to control the larger current flowing in the collector. The large signal-current gain is given the symbol h_{FE} and is somewhat less than the small-signal current gain h_{fe}. To ensure that the transistor saturates (switches hard on), it is best to use a base current that is about a twentieth of the chosen collector current. A transistor in saturation has a $V_{CE(sat)}$ of about 0.05 to 0.2 V. In cases where the collector current is near to the full rated value, it is important to ensure that sufficient base current is provided to saturate the device, otherwise a significant voltage will exist across the collector–emitter

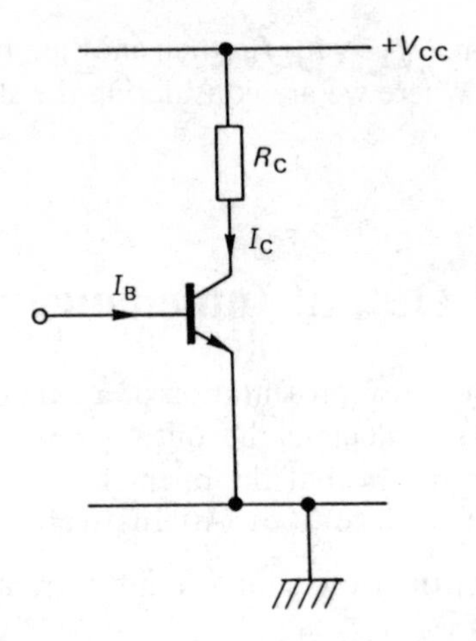

Figure 2.16 A transistor as a switch

junction and the power handling capacity of the transistor may be exceeded (see Chapter 10 on power amplifiers).

2.11 Worked Examples

Example 2.1

(a) Use the simplified model based on the Ebers-Moll equation to derive the equations for the voltage gain (A_v), input impedance (R_{in}), output impedance (R_{out}), and current gain (A_i) for the common-emitter transistor amplifier shown in diagram (a).

(b) If the circuit is now modified to that shown in diagram (b), show that the new voltage gain is given approximately by R_C/R_E and the input impedance by $h_{fe}R_E$.

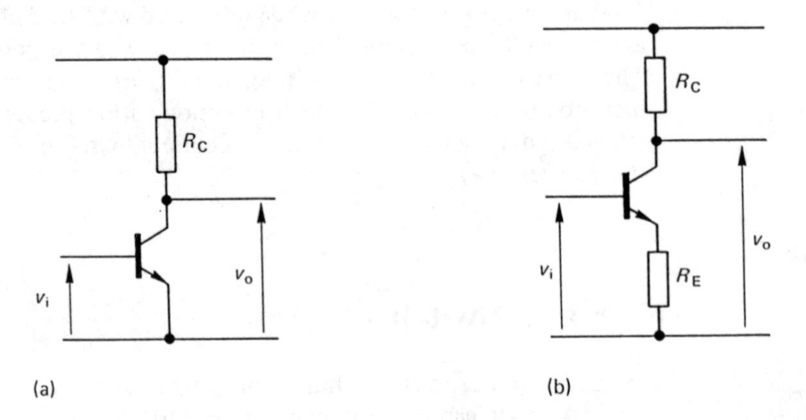

Solution 2.1

(a) Using the equivalent circuit of diagram (c),

$$\text{voltage gain } A_v = \frac{v_o}{v_i} = -\frac{g_m\, v_{be} R_C}{v_i}.$$

But $v_{be} = v_i$,

$$\therefore A_v = -g_m R_C = \underline{-40 I_C R_C}$$

$$\text{Input resistance } R_{in} = \frac{v_{be}}{i_b} = r_i = \frac{h_{fe}}{g_m} = \underline{\frac{h_{fe}}{40 I_C}}.$$

$$\text{Output resistance } R_{out} = \frac{v_{o\,(oc)}}{i_{o\,(sc)}}$$

$$= \frac{g_m v_{be} R_C}{g_m v_{be}} = \underline{R_C}.$$

$$\text{Current gain } A_i = \frac{g_m v_{be}}{i_b} = \frac{g_m v_{be} r_i}{v_{be}}$$

$$= \frac{g_m h_{fe}}{g_m} = \underline{h_{fe}}.$$

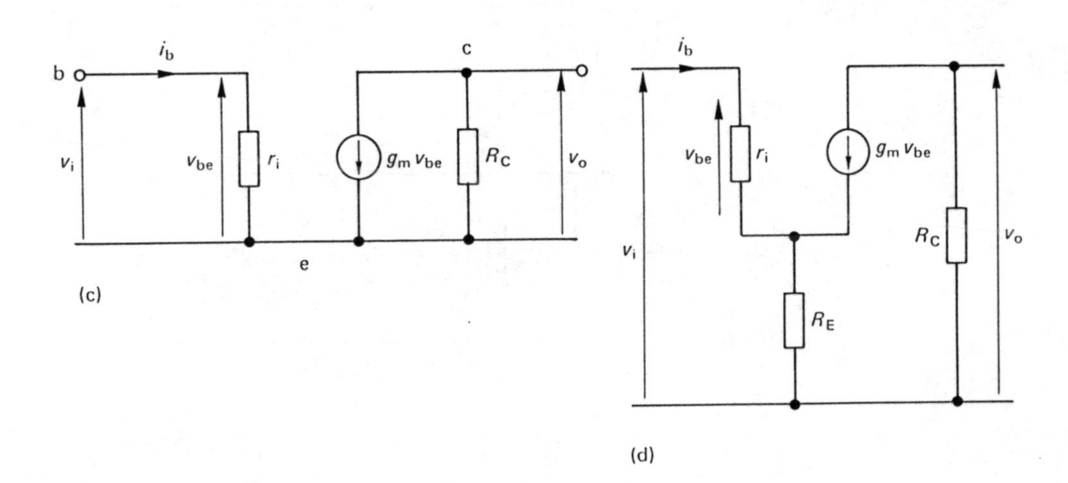

(b) Using the equivalent circuit of diagram (d),

$$A_v = \frac{v_o}{v_i} = -\frac{g_m v_{be} R_C}{v_i}$$

$$\approx -\frac{g_m v_{be} R_C}{v_{be} + g_m v_{be} R_E}$$

$$= -\frac{g_m R_C}{1 + g_m R_E} \approx \underline{\frac{R_C}{R_E}};$$

$$R_{in} = \frac{v_i}{i_b} = \frac{i_b r_i + (i_b + g_m v_{be}) R_E}{i_b}$$

$$\approx r_i + g_m r_i R_E$$

$$= \frac{h_{fe}}{g_m}(1 + g_m R_E)$$

$$\approx \underline{h_{fe} R_E}.$$

Example 2.2

Use the simplified model based on the Ebers–Moll equation to derive the equations for the voltage gain, A_v, input impedance, R_{in}, output impedance, R_{out}, and current gain, A_i, for the common-base transistor amplifier shown in diagram (a).

State the advantages and limitations of the CB configuration.

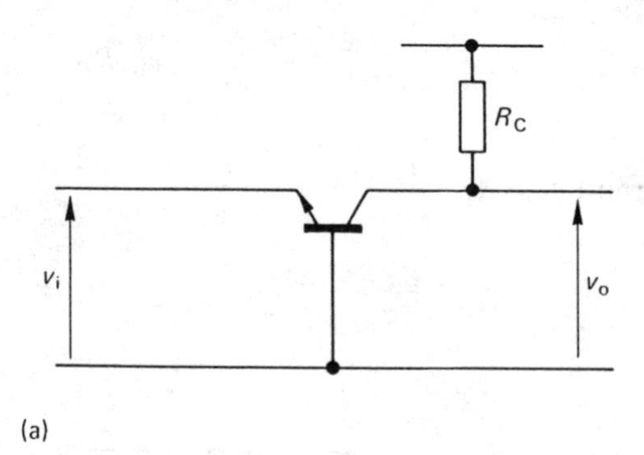

(a)

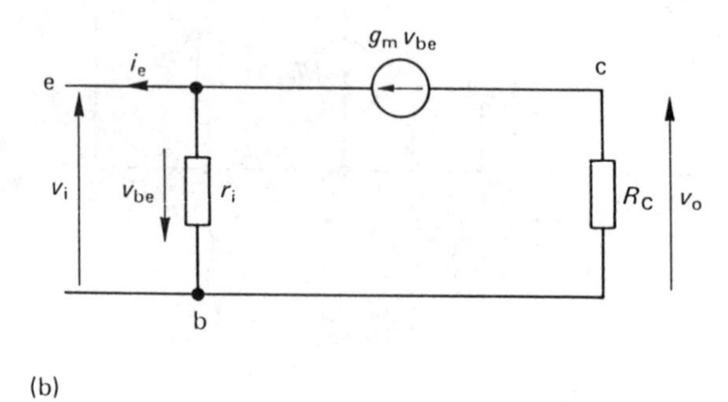

(b)

Solution 2.2

The equivalent circuit is shown in diagram (b).

$$A_v = \frac{v_o}{v_i} = -\frac{g_m v_{be} R_C}{v_i}.$$

But $v_i = -v_{be}$,

$$\therefore A_v = g_m R_C = \underline{40 I_C R_C}.$$

To find input resistance,

$$v_{be} = (i_e - g_m v_{be})\, r_i,$$

$$\therefore \frac{v_{be}}{i_e} = \frac{r_i}{1 + g_m r_i} \approx \frac{1}{g_m},$$

$$\therefore R_{in} \approx \frac{1}{g_m} = \underline{\frac{1}{40 I_C}}.$$

$$\text{Output resistance } R_{out} = \frac{v_{o(oc)}}{i_{o(sc)}} \approx \frac{g_m v_{be} R_C}{g_m v_{be}}$$

$$= \underline{R_C}.$$

The advantages of the CB connection are good voltage gain and good frequency response, which are due to there being no Miller feedback with no phase inversion.

The limitations of CB are low input impedance and high output impedance, although this is sometimes useful in impedance matching.

Example 2.3

Use the simplified h-parameter model to derive equations for the current gain A_i, the input impedance R_{in}, the voltage gain A_v and the output impedance R_{out} for the emitter follower circuit shown in diagram (a).

Calculate the values of these parameters, assuming:

$h_{ie} = 1100\ \Omega$, $R_E = 10\ k\Omega$, $h_{fe} = 50$ and $R_S = 1\ k\Omega$.

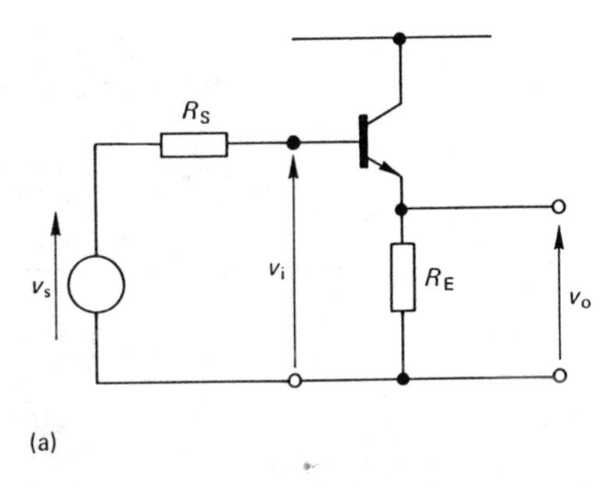

(a)

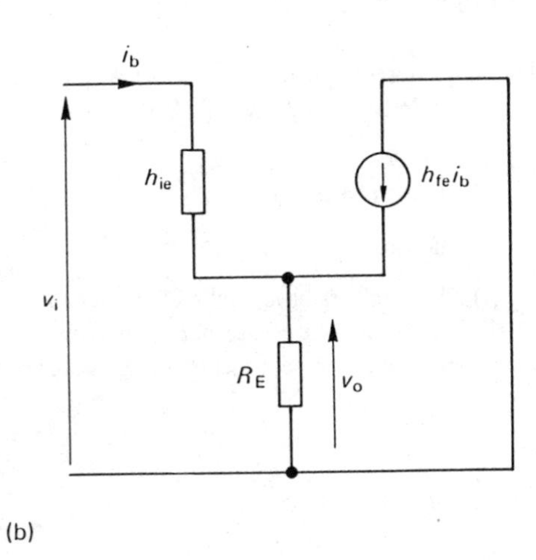

(b)

Solution 2.3

From the equivalent circuit of diagram (b),

$$A_v = \frac{v_o}{v_i} \approx \frac{h_{fe}\, i_b R_E}{h_{ie}\, i_b + R_E h_{fe}\, i_b}$$

$$= \frac{h_{fe} R_E}{h_{ie} + h_{fe} R_E} \quad \left(= 1 - \frac{h_{ie}}{R_{in}}\right),$$

$$R_{in} = \frac{v_i}{i_b} \approx \frac{h_{ie} i_b + h_{fe} i_b R_E}{i_b}$$

$$= h_{ie} + h_{fe} R_E \approx \underline{h_{fe} R_E},$$

$$R_{out} = \frac{v_{o(oc)}}{i_{o(sc)}} = \frac{i_b (R_S + h_{ie})}{h_{fe} i_b} \quad \text{(remove } R_E \text{ to find } v_{o(oc)})$$

$$= \underline{\frac{R_S + h_{ie}}{h_{fe}}};$$

$\therefore \underline{R'_{out} = R_{out} \| R_E}.$

$$\underline{A_i = 1 + h_{fe}}$$

Substituting values,

$$A_i \approx \underline{50},$$

$$R_{in} = 1.1 \text{ k}\Omega + 50 \times 10 \text{ k}\Omega \approx \underline{500 \text{ k}\Omega},$$

$$A_v = 1 - \frac{1100}{50 \times 10^3} = \underline{0.98},$$

$$R_{out} = \frac{1 \text{ k}\Omega + 1.1 \text{ k}\Omega}{50} = \underline{42 \ \Omega}$$

$$R'_{out} = 10 \text{ k}\Omega \| 42 \ \Omega \approx \underline{42 \ \Omega}.$$

Example 2.4

The amplifier shown in diagram (a) is required to provide a voltage gain $v_o/v_i = -5$, with a d.c. quiescent collector voltage of 10 V. Draw a small-signal equivalent circuit model and determine

(a) R_1 and R_2;
(b) the input resistance.

Assume that

(i) all capacitors have negligible impedance;
(ii) the transistor operation is defined by $I_C = I_S \exp(40 V_{BE})$, the current gain $\beta = h_{fe} = 200$, and the d.c. base-emitter voltage is approximately 0.6 V.

Solution 2.4

Ignoring the effects of I_B,

$$\therefore V_B = \frac{5}{15} \times 15 \text{ V} = 5 \text{ V},$$

$$\therefore V_E = 5 \text{ V} - 0.6 \text{ V} = 4.4 \text{ V}.$$

If the quiescent collector voltage is 10 V then

$$I_C = \frac{(15 - 10) \text{ V}}{1 \text{ k}\Omega} = 5 \text{ mA},$$

$$I_E \approx I_C = 5 \text{ mA},$$

$$\therefore R_1 + R_2 = \frac{V_E}{I_E} = \frac{4.4 \text{ V}}{5 \text{ mA}} = 880 \ \Omega.$$

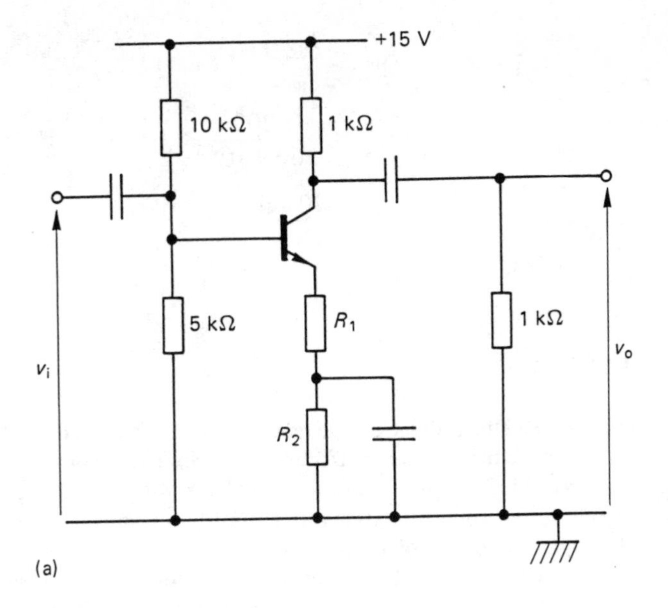

(a)

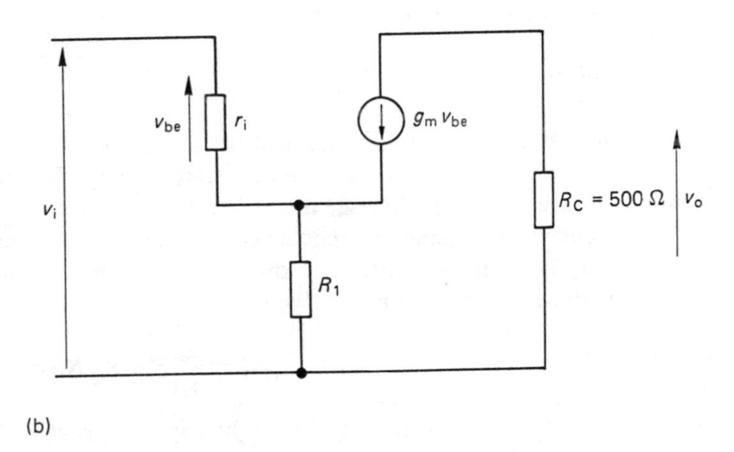

(b)

From the equivalent circuit of diagram (b),

$$v_o = -g_m v_{be} R_C,$$

$$v_i = v_{be} + g_m v_{be} R_1 \quad \text{(neglecting } i_b\text{)},$$

$$\therefore v_{be} = \frac{v_i}{1 + g_m R_1},$$

$$\therefore A_v = \frac{v_o}{v_i} = -\frac{g_m R_C}{1 + g_m R_1} = -5.$$

But $g_m = 40 I_C = 200$ mS,

$$\therefore 5 = \frac{200 R_C}{1 + 200 R_1},$$

$\therefore R_1 = 95\ \Omega$ and $R_2 = 880\ \Omega - 95\ \Omega$,

$\therefore R_2 = 785\ \Omega.$

$$R_{in} = \frac{h_{fe}}{g_m}(1 + g_m R_1)$$

$$= \frac{200}{200 \times 10^{-3}}(1 + 200 \times 10^{-3} \times 95)$$

$$= \underline{20\ k\Omega}.$$

Example 2.5

Explain the Miller effect by reference to the c–b capacitance of a transistor connected in CE, and calculate the equivalent effective capacitance at the input.

A variable-frequency sinusoidal oscillator with an output resistance of 1 kΩ and open-circuit output voltage of 10 mV is applied to a common emitter amplifier which operates at a d.c. collector current of 1 mA, with a collector load resistor of 5 kΩ. The output of the amplifier at low frequency is 1 V. Estimate the low-frequency transistor current gain h_{fe}.

If the significant transistor capacitances are C_{be} = 200 pF and C_{bc} = 10 pF, determine the 3 dB frequency for the amplifier.

Solution 2.5

The Miller effect is the effect of the inter-electrode capacitance feedback between collector and base of a common-emitter amplifier. Since the collector and base voltages are in antiphase, the Miller effect greatly reduces the voltage gain at high frequencies where the impedance of the capacitor becomes much lower. The effect may be seen by reference to diagram (a) where the collector–base capacitance C_{bc} is shown added externally. We have

$$A_v = \frac{v_o}{v_i} = -g_m R_C.$$

The magnitude of the current in the capacitor is given by

$$i = \frac{v_i - v_o}{Z} = \frac{v_i + g_m R_C v_i}{Z};$$

$$\therefore \frac{v_i}{i} = \frac{Z}{1 + g_m R_C} = \frac{1}{j\omega C_{bc}(1 + g_m R_C)}.$$

Thus the equivalent effective capacitance at the input is

$$\underline{C_{bc}(1 + g_m R_C)},$$

together with the b–e capacitance C_{be} already there.

The equivalent circuit for the amplifier is shown in diagram (b).

Since I_C = 1 mA, then g_m = 40 mS.

Now

$$v_o = -g_m v_{be} R_C = -\frac{g_m R_C r_i}{R_S + r_i} v_s.$$

$$\therefore 1\ V = \frac{40 \times 5 \times r_i}{1 + r_i} \times 0.01\ V,$$

where resistances are in kΩ,

$$\therefore r_i = 1\ k\Omega,$$

$$\therefore h_{fe} = g_m r_i = \underline{40}.$$

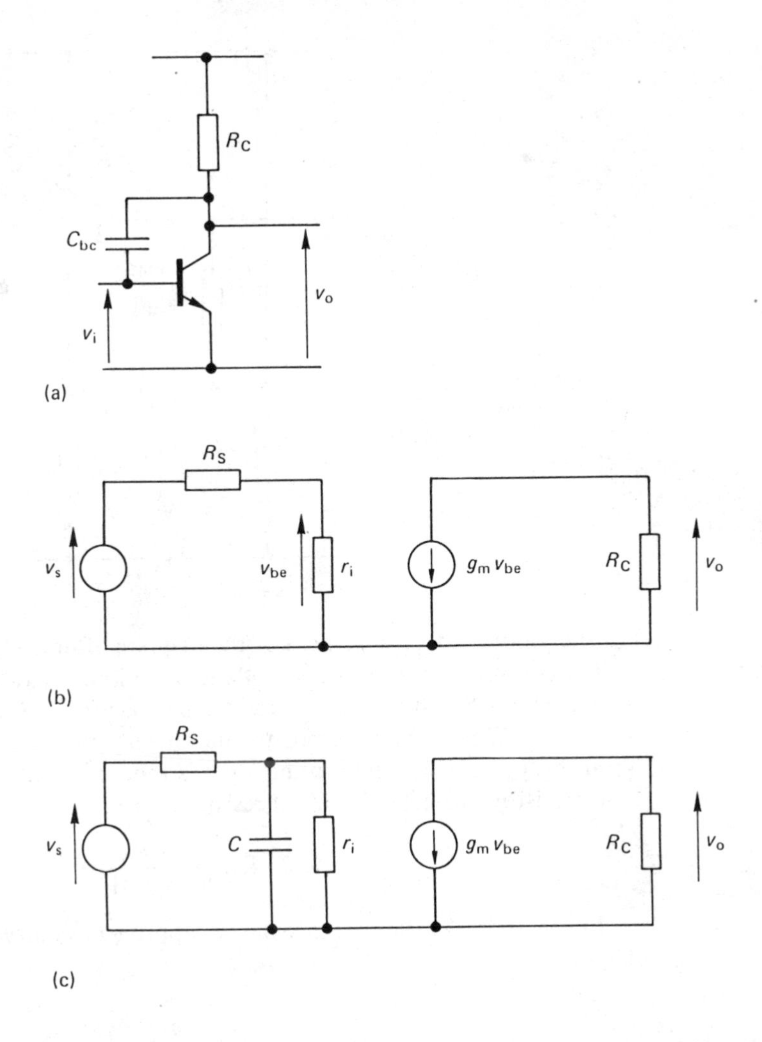

At high frequency the transistor amplifier equivalent circuit becomes that shown in diagram (c). By Miller's theorem,

$$C = C_{bc}\,(1 + g_m R_C) - C_{be}$$
$$= 10\,(1 + 40 \times 5) + 200 = 2210 \text{ pF.}$$

The 3 dB frequency is at $f = \dfrac{1}{2\pi CR}$, where $R = R_S \parallel r_i$.

$$\therefore 3 \text{ dB frequency} = \underline{144 \text{ kHz}}.$$

The effect of C_{bc} across R_C is considered negligible.

Example 2.6

Explain the 'bootstrap' principle and show how it may be used to increase the input resistance of an emitter follower circuit.

Calculate the effective input resistance of an emitter follower with bootstrapping applied.

Solution 2.6

The input resistance of an emitter follower circuit is the input resistance seen at the input to the base of the transistor in parallel with the resistance of the bias circuit. This bias circuit may significantly reduce the overall input resistance.

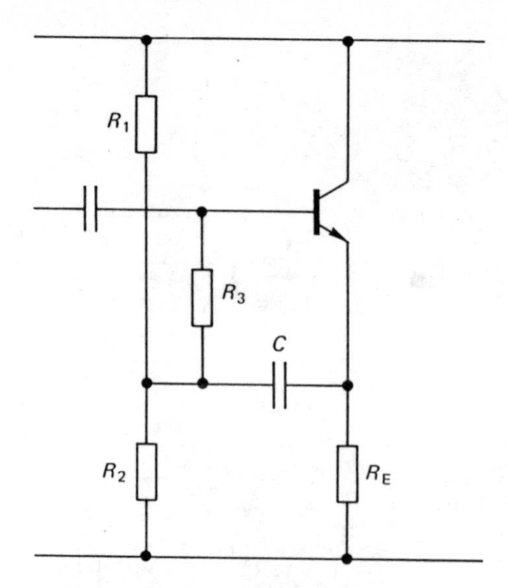

The 'bootstrap' arrangement of the diagram effectively overcomes this problem. If we consider the resistance of the bias circuit with C removed then it is equal to $R_3 + R_1 \parallel R_2$. By the addition of the capacitor C, then for a.c. signals, one end of R_3 is effectively connected to the output (the emitter). If the voltage gain is given by $v_o/v_i = A_v$, then, using Miller's theorem, and considering R_3 as the feedback resistor, the effective input resistance presented by R_3 is given by

$$R_{eff} = \frac{R_3}{1 - A_v} .$$

In the emitter follower, the gain A_v approaches unity and therefore R_{eff} is very large.

2.12 Unworked Problems

Problem 2.1

Draw the equivalent circuit of the amplifier shown in the diagram. Under its small-signal operating conditions, the transistor has the following values of common-emitter hybrid parameters: $h_{ie} = 1$ kΩ, $h_{fe} = 100$, h_{re} and h_{oe} negligible.

When the output is loaded by the 2 kΩ resistor R_L the amplifier has a voltage gain $G_v = v_o/v_i = -10$ at frequencies where the reactances of C_1 and C_2 are negligible. Determine the required value of R_E and calculate the current gain, $G_i = i_o/i_i$. (EC Part 1)

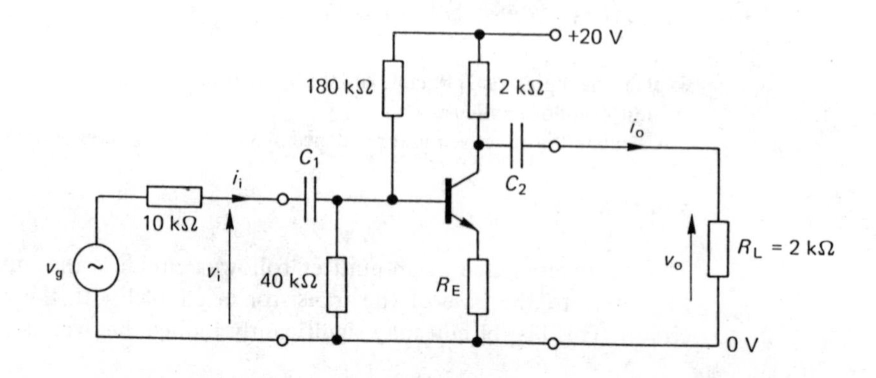

Problem 2.2

The small-signal properties of an *npn* transistor can be represented by the simplified equivalent circuit shown in the diagram. The short-circuit current gain is β and $r_e = 26/I_E$ is the dynamic emitter resistance (in Ω), where I_E is the quiescent current (in mA).

Show that the small-signal voltage gain of a common emitter stage is given by the expression $G_v = -38.5V_L$, where V_L is the quiescent voltage across the collector load resistor. Derive an expression for the input resistance of the stage.

Derive circuit component values for an amplifier configuration to give a voltage gain of -50 and an input resistance of 1 kΩ. Use a 10 V d.c. supply and assume that $\beta = 100$.

(EC Part 2)

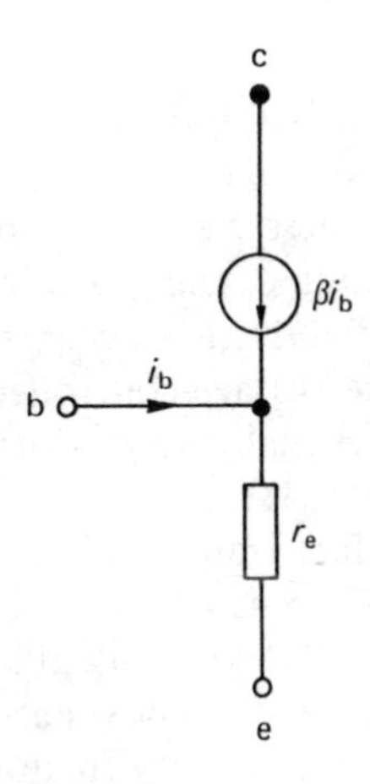

Problem 2.3

Design a single-transistor, common-emitter amplifier given the following design parameters:

(a) quiescent collector current = 2 mA;
(b) supply voltage = 15 V;
(c) base voltage with respect to $+V_{CC} = -12$ V;
(d) d.c. voltage gain = -4;
(e) a.c. voltage gain = 40;
(f) ambient operating temperature = 27 °C;
(g) $h_{fe} = 300$.

You may assume that $V_{BE} = 0.7$ V and that $KT/qn = 25$ mV at 27 °C.

Determine the minimum load impedance that could be a.c. coupled to the amplifier to keep the load to no-load voltage ratio > 0.9.

Problem 2.4

Describe why a common-collector (or emitter follower) amplifier stage is used as a buffer amplifier.

Show that when the source resistance is small, the input and output resistances are given by the expressions

$$R_{in} \approx R_E h_{fe}, \qquad R_{out} \approx h_{ie}/h_{fe}$$

where R_E is the emitter resistor, and the effect of h_{re} may be neglected. Justify any further assumptions made in deriving the expressions, using typical parameter values.

(CEI Part 2)

3 Field Effect Transistors

3.1 Introduction

The field effect transistor (FET) is a semiconductor device that depends for its operation on the control of current by an electric field. The FET has three terminals, these being the source, the drain and the gate. The gate acts as the control terminal, and the voltage applied between the gate and the source controls the flow of current between source and drain.

FETs are voltage-controlled rather than current-controlled devices and the gate current is virtually zero, thus giving the FET a very high input impedance (of the order of 10^{12} Ω).

Field effect transistors can replace bipolar transistors in many applications, can be designed to handle large currents, and are almost exclusively the devices used in large-scale integration circuits such as microprocessors and memory devices.

The FET is also known as a 'unipolar' transistor since current is transported by carriers of one polarity (majority carriers). In the bipolar transistor, both majority and minority carriers are involved.

The FET may often produce a performance superior to that of the bipolar transistor.

3.2 Types of FET

There are two basic types of FET, namely the junction FET (JFET), and the metal-oxide-semiconductor FET (MOSFET).

Each of these is available as either *n*-channel or *p*-channel, with symbols as shown in Fig. 3.1.

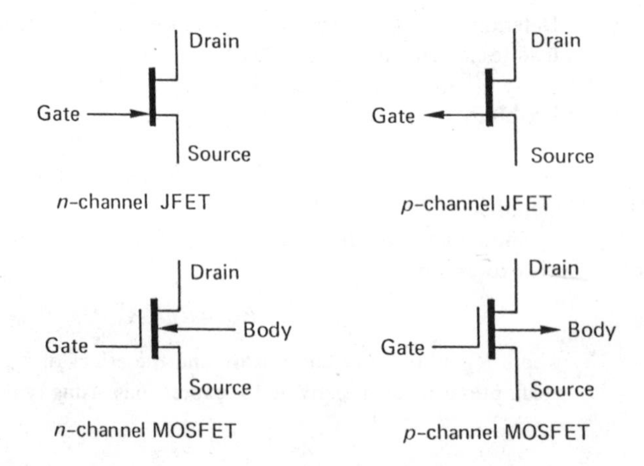

Figure 3.1

(a) The JFET

In its simplest form the JFET consists of a piece of high-resistivity silicon that constitutes a channel for the flow of majority carriers. The magnitude of this current is controlled by a voltage applied to the gate, which is a *pn* junction formed along the channel. The FET gate-source junction is operated in the reverse bias region, and thus the gate current is practically zero.

(b) The MOSFET

The MOSFET depends for its operation on the fact that it is not necessary to form a semiconductor junction on the channel of an FET in order to achieve gate control of the channel current. Instead, the gate electrode is a metallic region separated from the conducting channel by a thin silicon dioxide layer (hence MOS).

Since the gate is electrically insulated from the source–drain circuit, its voltage polarity may go either positive or negative without gate current flowing. This allows for the construction of two types of MOSFET, these being as follows:

(a) Enhancement type: this conducts only when forward biased and is cut off at zero volts between gate and source.
(b) Depletion type: this conducts in both forward and reverse directions and is cut off only when the gate–source is reverse biased by several volts.

JFETs are always depletion mode devices. The gate–source forward bias must not exceed a positive voltage of about 0.5 V or else the junction will conduct and the input impedance will thus be drastically reduced.

3.3 JFET Characteristics

Figure 3.2(a) shows an FET with a short circuit between the gate and the source. The current I_D is controlled by variation of V_{DS}. Figure 3.2(b) shows an idealised characteristic thus obtained. It is referred to as the output characteristic.

In the region below pinch-off, as V_{DS} is increased from zero, the channel acts as a low resistance and the drain current rises until it reaches the level of the limiting current, I_{DSS}. At this point, I_D begins to be pinched off and the characteristic moves into the saturation region.

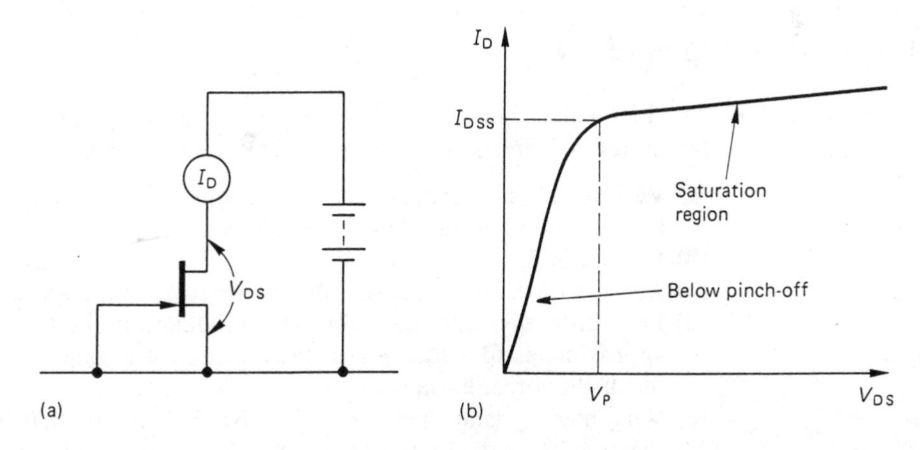

Figure 3.2

I_{DSS} is the drain saturation current with the gate voltage V_{GS} at zero.

The pinch-off voltage V_P may be defined as the value of V_{DS} at which the maximum drain current I_{DSS} flows.

The drain current I_D is also a function of the gate voltage V_{GS}.

In Fig. 3.3(a) a fixed value of V_{DS} is applied across the device and V_{GS} is varied. If a negative voltage V_{GS} is applied to the gate then the drain current I_D

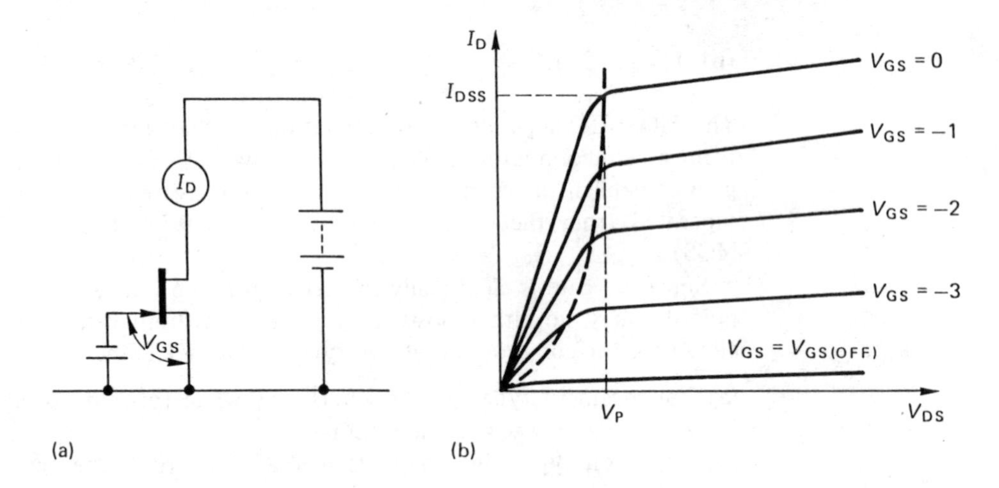

Figure 3.3

reduces. At a value of $|V_{GS}| \geqslant |V_P|$, the channel current is almost entirely cut off. This cut-off voltage is referred to as the gate cut-off voltage, $V_{GS(off)}$. (V_P has been widely used in the past, but $V_{GS(off)}$ is now more commonly accepted since it eliminates the ambiguity with the drain pinch-off voltage V_P. $V_{GS(off)}$ and V_P are equal in magnitude but opposite in polarity.)

The dependency of the drain current I_D on both V_{DS} and V_{GS} is shown by the family of curves in Fig. 3.3(b). This is referred to as the output characteristic. Note that above saturation V_{DS} has little effect and V_{GS} essentially controls I_D. This is the normal operating region for FET amplifiers.

3.4 Advantages of FETs

The FET enjoys certain inherent advantages over bipolar transistors. The particular features of interest to the designer are as follows:

(a) Very high input impedance at low frequencies (in practice the input impedance is limited by the shunt gate bias resistor).
(b) Low noise.
(c) Absence of thermal runaway, due to negative temperature coefficient.
(d) Low distortion and negligible intermodulation products (transfer function approximates to a square law, thus producing strong second-order and negligible higher-order harmonics).
(e) Very high dynamic range (⊳ 100 dB). This means that it can amplify small signals because the FET produces very little noise, and can amplify large signals because FETs have negligible intermodulation distortion.

(f) Zero temperature coefficient quiescent point (i.e. changes in temperature do not alter the quiescent operating point).
(g) Junction capacitance is independent of device current (i.e. inherent stability allowing the design of stable high frequency oscillators).

The disadvantages are low transconductance as compared with bipolar transistors.

3.5 FET Parameters

The major parameters include the following:

I_{DSS}	Drain current with gate shorted to source;
$V_{GS(off)}$	Gate-source cut-off voltage;
I_{GSS}	Gate-source current with drain shorted to source;
BV_{GSS}	Gate-source breakdown voltage with drain shorted to source;
g_{fs}	Common-source forward transconductance;
C_{gs}	Gate-source capacitance;
C_{gd}	Gate-drain capacitance;
r_{ds}	Drain-source resistance.

3.6 FET Equations

A typical transfer characteristic for an FET showing I_D versus V_{GS} is given in Fig. 3.4. It represents the case where the device is in the saturation region (i.e. above pinch off).

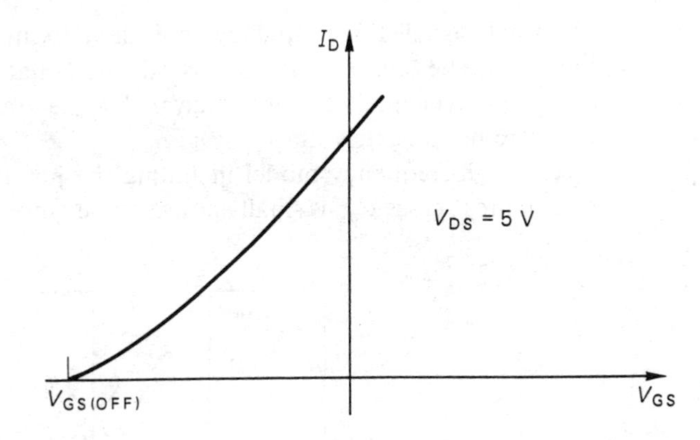

Figure 3.4

The curve approximates very well to a square law and may be described by the following equation:

$$I_D = k(V_{GS} - V_{GS(off)})^2.$$

Now, when $V_{GS} = 0, I_D = I_{DSS}$,

$$\therefore k = \frac{I_{DSS}}{V^2_{GS(off)}},$$

$$\therefore I_D = I_{DSS}\left(1 - \frac{V_{GS}}{V_{GS(off)}}\right)^2.$$

The drain current of a JFET operating in the region below pinch-off is given by the equation

$$I_{\mathrm{D}} = I_{\mathrm{DSS}}\left(\frac{V_{\mathrm{DS}}}{V_{\mathrm{GS(off)}}}\right)^{1/2}.$$

The transconductance g_{fs} of an FET is a measure of the effect of gate voltage on drain current

$$g_{\mathrm{fs}} = \frac{\Delta I_{\mathrm{D}}}{\Delta V_{\mathrm{GS}}} \qquad (V_{\mathrm{DS}} = \text{constant}).$$

From the equation for I_{D} we have

$$g_{\mathrm{fs}} = \frac{\mathrm{d}I_{\mathrm{D}}}{\mathrm{d}V_{\mathrm{GS}}} = -\frac{2I_{\mathrm{DSS}}}{V_{\mathrm{GS(off)}}}\left(1 - \frac{V_{\mathrm{GS}}}{V_{\mathrm{GS(off)}}}\right).$$

If we call g_{fs0} the value of the transconductance at $V_{\mathrm{GS}} = 0$ then

$$g_{\mathrm{fs0}} = -\frac{2I_{\mathrm{DSS}}}{V_{\mathrm{GS(off)}}}$$

and

$$g_{\mathrm{fs}} = g_{\mathrm{fs0}}\left(1 - \frac{V_{\mathrm{GS}}}{V_{\mathrm{GS(off)}}}\right).$$

The drain source resistance r_{ds} of an FET is the incremental channel resistance (i.e. small-signal equivalent). It is capable of a wide range of values, depending on the bias, but in the saturation region is generally of the order of several megohms.

3.7 The FET Model

The small-signal, low-frequency equivalent circuit may be represented as shown in Fig. 3.5. The input resistance is so large that its effect may be ignored. The incremental channel current is given by the transconductance g_{fs} multiplied by the incremental gate voltage.

The high-frequency model including the junction capacitances is shown in Fig. 3.6. In most cases C_{ds} is small enough to be ignored.

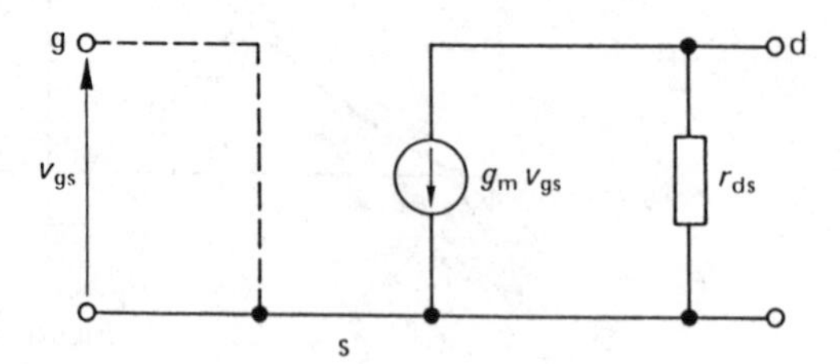

Figure 3.5

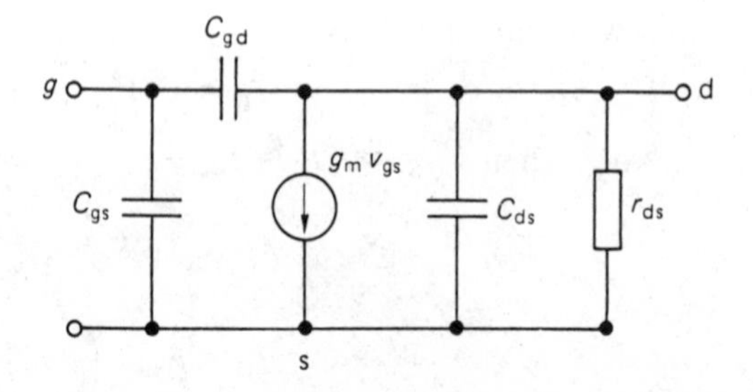

Figure 3.6

3.8 MOSFET Characteristics

The transfer characteristic is a graph of drain current against gate-source voltage for the MOSFET as shown in Fig. 3.7 for both the enhancement and depletion types.

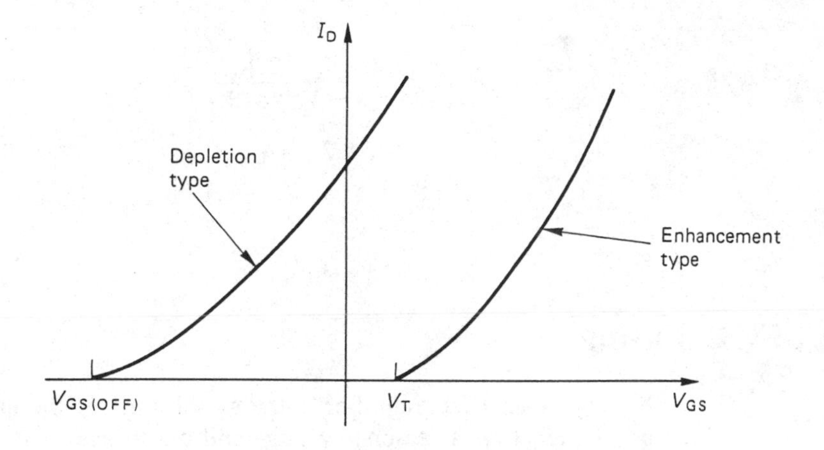

Figure 3.7

$V_{GS(off)}$ is the 'gate-source cut-off voltage' and is the voltage at which the drain current approaches zero (it is only specified for depletion mode devices),

V_T is the 'threshold voltage' and is the gate-source voltage at which the drain current approaches zero in the enhancement mode device.

The output characteristic for a MOSFET is a graph of drain current against drain-source voltage. Curves for both the depletion and enhancement mode types are shown in Fig. 3.8.

Because there is no junction involved, V_{GS} can be made positive or negative without causing gate current to flow.

Note that in the enhancement mode MOSFET the current I_D is zero when V_{GS} = 0V.

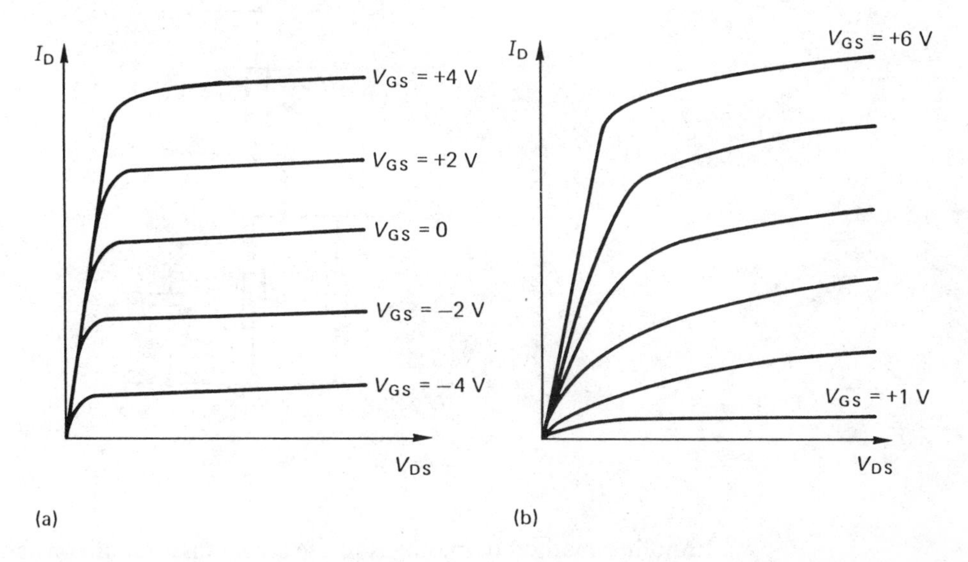

Figure 3.8 MOSFET output characteristics: (a) *n*-channel *depletion* mode, (b) *n*-channel *enhancement* mode

The equations relating I_D and g_{fs} with V_{GS} and I_{DSS} are the same for the depletion-mode MOSFET as for the JFET.

For an enhancement-type MOSFET,

$$I_D = k(V_{GS} - V_T)^2,$$

$$g_{fs} = \frac{dI_D}{dV_{GS}} = 2k(V_{GS} - V_T)$$

$$= \frac{2I_D}{V_{GS} - V_T},$$

or

$$g_{fs} = 2\sqrt{kI_D}.$$

3.9 FET Biasing

Biasing of an FET amplifier is the establishing of the quiescent (d.c.) operating point such that a reasonably large and undistorted output voltage swing can be achieved and the quiescent point does not change significantly when one considers the normal production spread of device characteristics.

(a) Common-source Amplifier

One method of biasing a common-source amplifier is to use the self-bias circuit of Fig. 3.9. The resistor R_G connects the gate to the common rail and serves mainly to isolate the input signal from the common rail. No voltage is dropped across R_G because the gate current is essentially zero.

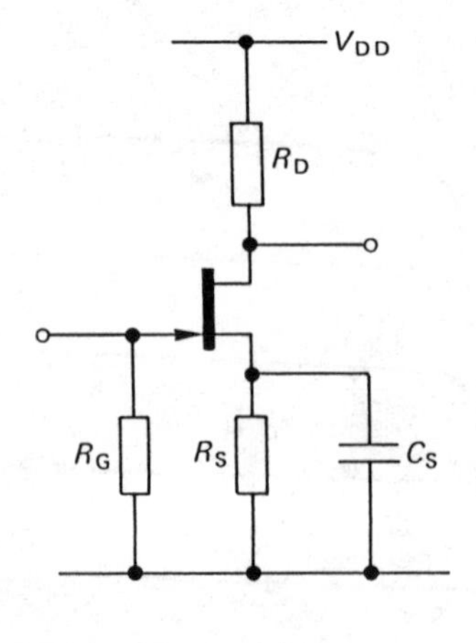

Figure 3.9

Another method of biasing is to use a combination of constant voltage and self-biasing, as shown in Fig. 3.10(a) and (b). This method reduces the sensitivity to variations in FET characteristics with device production spread.

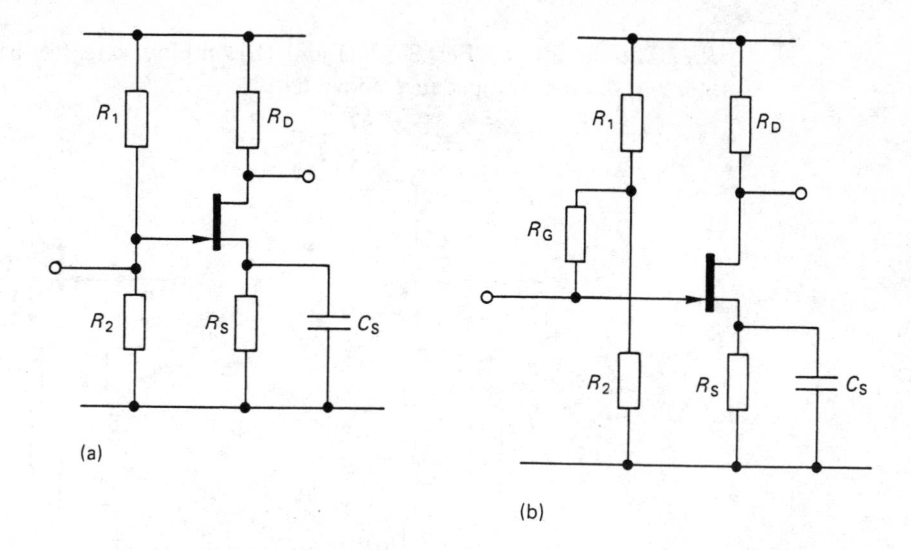

Figure 3.10

(b) Common-drain Amplifier Source Follower

This circuit is particularly useful owing to its high input impedance and low output impedance.

There are two basic connections for source followers: with and without gate feedback.

Figure 3.11(a) and (b) show two of the several possible arrangements for a source follower without gate feedback. Their input impedance is thus equal to

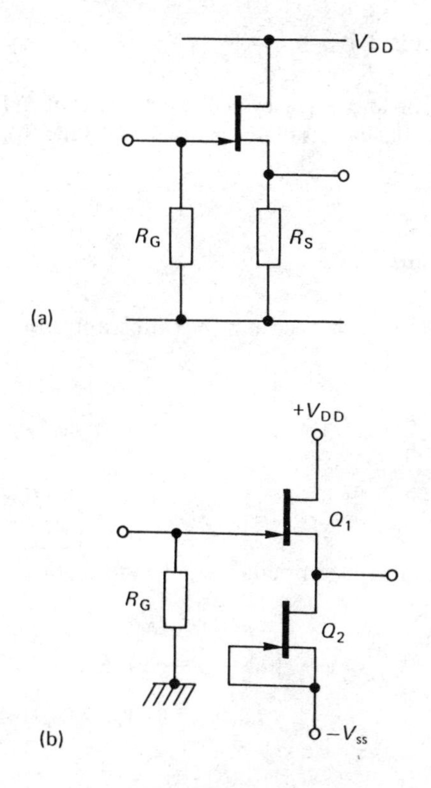

Figure 3.11

R_G. The circuits of Fig. 3.12(a) and (b) employ gate feedback to the gate to increase the input impedance above R_G.

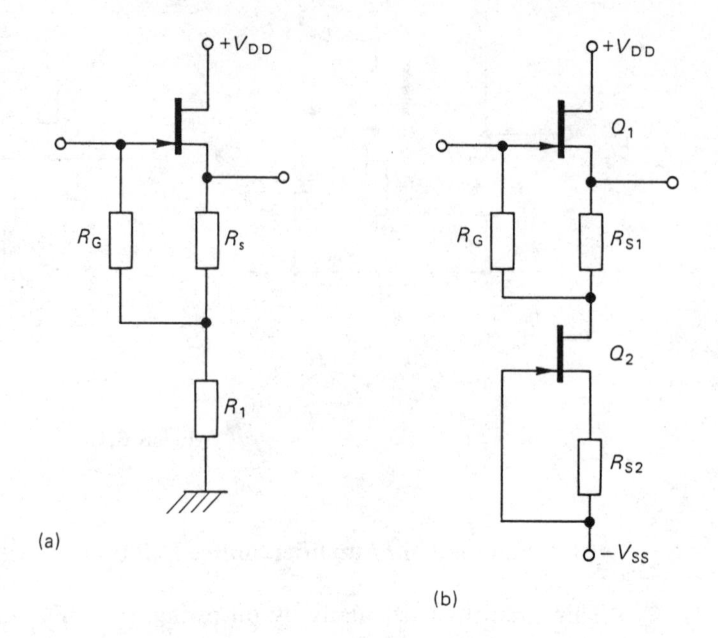

Figure 3.12

3.10 Worked Examples

Example 3.1

Draw the low-frequency equivalent circuit of an FET and hence calculate the mid-band voltage gain of the circuits shown in diagrams (a) and (b). Calculate the output impedance in each case.

Solution 3.1

Diagram (a) is a common-source amplifier with an equivalent circuit as shown in diagram (c).

$$v_o = -g_m v_{gs} \frac{r_{ds} R_D}{r_{ds} + R_D},$$

$$\therefore A_v = -\frac{g_m r_{ds} R_D}{r_{ds} + R_D}.$$

The output impedance is found from

$$v_{o(oc)} = g_m v_{gs} R, \qquad \text{where } R = R_D \parallel r_{ds}$$

$$= g_m v_i R,$$

$$i_{o(sc)} = g_m v_{gs} = g_m v_i.$$

$$\therefore R_o = \frac{v_{o(oc)}}{i_{o(sc)}} = R = \frac{R_D r_{ds}}{R_D + r_{ds}}.$$

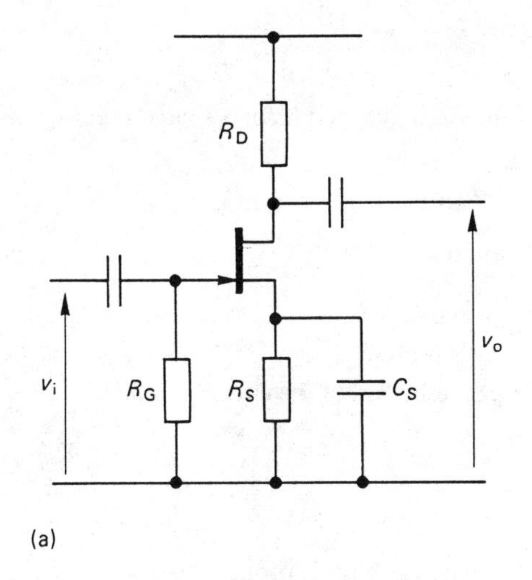

(a)

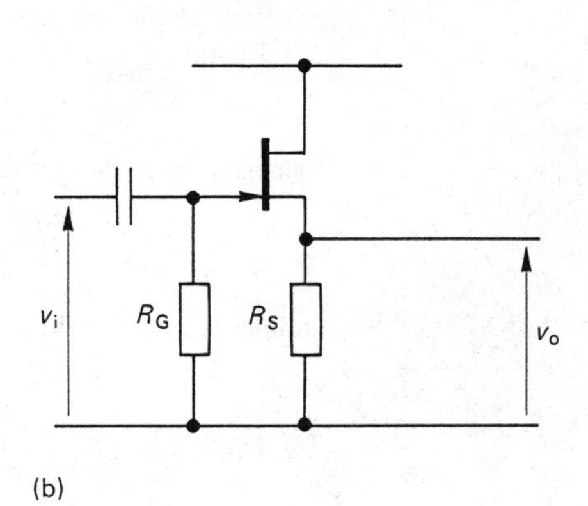

(b)

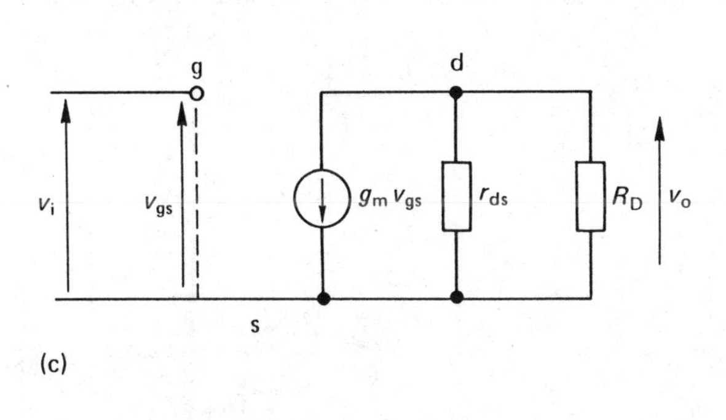

(c)

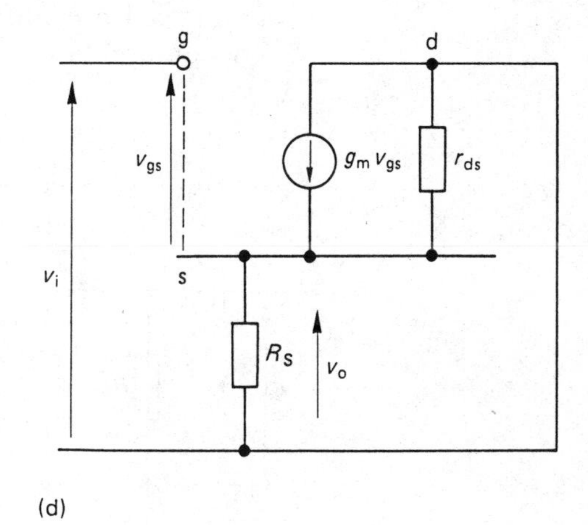

(d)

Diagram (b) is a common-drain amplifier (source follower) with an equivalent circuit as shown in diagram (d).

$$v_o \approx g_m v_{gs} R, \qquad \text{where } R = \frac{R_S r_{ds}}{R_S + r_{ds}}.$$

Also,

$$v_{gs} = v_i - v_o,$$

$$\therefore \frac{v_o}{g_m R} = v_i - v_o$$

$$\therefore A_v = \frac{v_o}{v_i} = \frac{g_m R}{1 + g_m R}.$$

The output impedance is found from

$$v_{o(oc)} = g_m v_{gs} R$$

$$= g_m (v_i - v_{o(oc)}) R,$$

$$\therefore v_{o(oc)} = \frac{g_m R}{1 + g_m R} v_i,$$

$$i_{o(sc)} = g_m v_{gs} = g_m v_i,$$

$$\therefore R_o = \frac{v_{o(oc)}}{i_{o(sc)}} = \frac{R}{1 + g_m R}.$$

Example 3.2

Explain the purpose of a source follower, and draw the circuit for a source follower using a JFET with

$$g_m = 5 \text{ mS} \quad \text{and} \quad r_{ds} = 8 \text{ k}\Omega.$$

Calculate the output impedance using the equation

$$R_o = \frac{v_{o(oc)}}{i_{o(sc)}}$$

and hence calculate the voltage gain if the output resistance of the circuit is 100 Ω.

Solution 3.2

The purpose of a source follower is to provide a high input impedance and a low output impedance buffer such as may be used between two amplifier stages.

The source follower and its equivalent circuit are shown in diagrams (a) and (b).

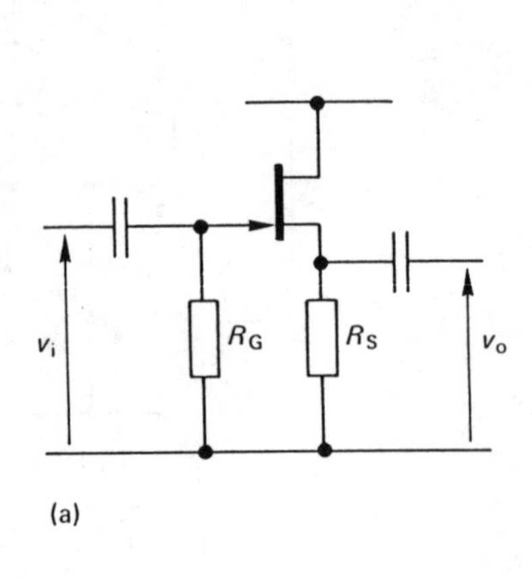

(a)

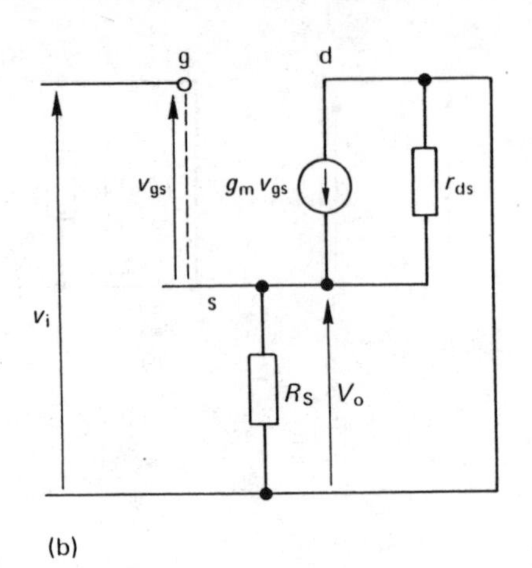

(b)

Now $\quad i_{o(sc)} = g_m v_i$

and $\quad v_{o(oc)} = g_m v_{gs} r_{ds} = g_m (v_i - v_o) r_{ds}$ (removing R_S).

$$\therefore \quad v_{o(oc)} = \frac{g_m r_{ds} v_i}{1 + g_m r_{ds}},$$

$$\therefore \quad R_o = \frac{v_{o(oc)}}{i_{o(sc)}} = \frac{r_{ds}}{1 + g_m r_{ds}}$$

$$= \frac{8 \times 10^3}{1 + 5 \times 10^{-3} \times 8 \times 10^3} = \underline{195\ \Omega}.$$

$$\therefore \quad R_o' = R_o \parallel R_S = 100\ \Omega.$$

$$\therefore \quad R_S = 205\ \Omega.$$

Now $\quad \dfrac{v_o}{v_i} = \dfrac{g_m R}{1 + g_m R} \quad$ where $R = R_S \parallel r_{ds} = 200\ \Omega$.

$$= \frac{5 \times 10^{-3} \times 200}{1 + 5 \times 10^{-3} \times 200} = \underline{0.5}.$$

Example 3.3

The circuit diagram shows a 'bootstrapped' source follower using an *n*-channel JFET.

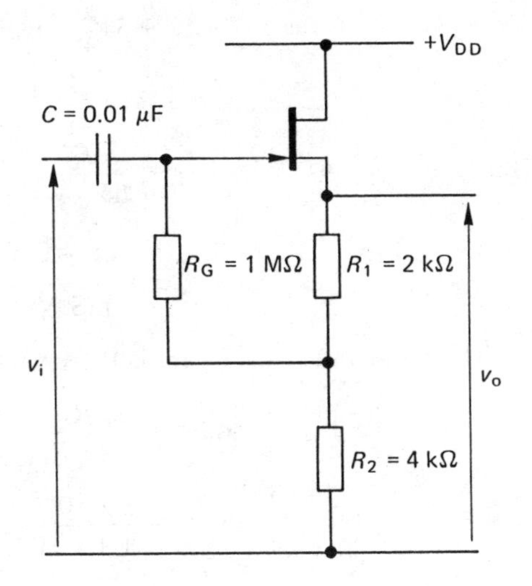

(a) Explain the purpose of the bootstrapping.
(b) Calculate the mid-band voltage gain.
(c) Calculate the low-frequency 3 dB breakpoint.

Assume that the FET operation is defined by the equation

$$I_D = I_{DSS}\left(1 - \frac{V_{GS}}{V_P}\right)^2,$$

where $I_{DSS} = 1$ mA and $V_P = -1$ V.

Solution 3.3

(a) The bootstrapping provides positive feedback to the gate to increase the circuit input resistance.

(b) To calculate the mid-band voltage gain, use the equation

$$A_v = \frac{g_m R_S}{1 + g_m R_S},$$

since r_{ds} is not mentioned and we may therefore assume $r_{ds} \gg R_S$.

Also, $R_G \gg R_1$ and R_2 and does not therefore effect the voltage gain.

Now, $$R_S = R_1 + R_2 = 6\ \text{k}\Omega.$$

We may find g_m as follows:

$$I_D = I_{DSS}\left(1 - \frac{V_{GS}}{V_P}\right)^2$$

$$= 1 \times 10^{-3}\,(1 + V_{GS})^2. \qquad \ldots(1)$$

Also, since the gate potential is approximately equal to the potential at the junction of R_1 and R_2, then:

$$V_{GS} = -I_D R_1 = -2 \times 10^3\, I_D. \qquad \ldots(2)$$

Eliminating I_D between 1 and 2 gives

$$(1 + V_{GS})^2 = -\frac{V_{GS}}{2},$$

$$\therefore\ 2V_{GS}^2 + 5V_{GS} + 2 = 0,$$

$$\therefore\ V_{GS} = -0.5\ \text{V}.$$

But
$$g_m = \frac{dI_D}{dV_{GS}} = \frac{-2I_{DSS}}{V_P}\left(1 - \frac{V_{GS}}{V_P}\right)$$
$$= 2(1 - 0.5)$$
$$= 1\ \text{mS}.$$

The mid-band voltage gain is then given by

$$A_v = \frac{g_m R_s}{1 + g_m R_s}$$
$$= \frac{1 \times 10^{-3} \times 6 \times 10^3}{1 + 1 \times 10^{-3} \times 6 \times 10^3} = \underline{0.86}.$$

(c) The low-frequency 3 dB point is determined by the capacitor C and the input resistance.

Now,
$$R_{in} = \frac{v_i}{i_i}$$

and
$$i_i = \frac{v_i - \frac{4}{6}v_o}{R_G} = \frac{v_i - \frac{4}{6}A_v v_i}{R_G},$$

$$\therefore R_{in} = \frac{R_G}{1 - \frac{4}{6}A_v} = 2.33\ \text{M}\Omega,$$

$$\therefore f_1 = \frac{1}{2\pi C R_{in}} = \underline{6.8\ \text{Hz}}.$$

Example 3.4

The constant-current circuit shown in diagram (a) uses a JFET whose operation is described by the equation

$$I_D = I_{DSS}\left(1 - \frac{V_{GS}}{V_P}\right)^2,$$

with $I_{DSS} = 8$ mA and $V_P = -4$ V.

(a) Explain its operation.
(b) Calculate the required value of R to give a current of 0.5 mA.
(c) If the FET drain-source resistance r_{ds} is equal to 50 kΩ at $I_D = 0.5$ mA, determine the incremental resistance of the circuit for the value of R calculated in (b).
(d) Describe briefly the advantages of using this type of circuit in preference to a resistor and give two applications.

Solution 3.4

(a) The JFET is operated at a point on the output characteristic such that it is beyond the 'knee'. The current I is thus relatively constant independent of V_{DS}, and determined by the resistor R.

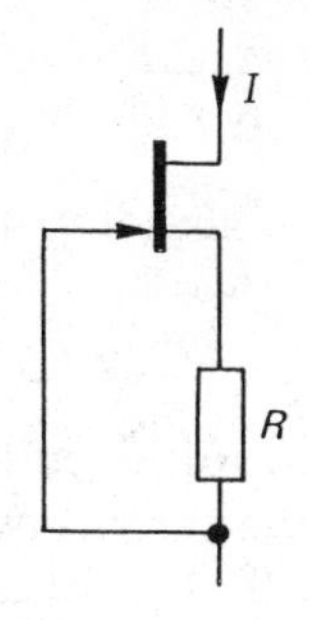

(a)

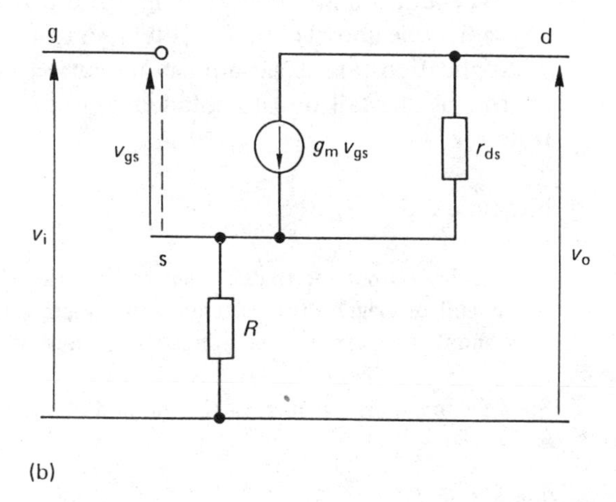

(b)

(b) From the equation for I_D,

$$0.5 \times 10^{-3} = 8 \times 10^{-3} \left(1 + \frac{V_{GS}}{4}\right)^2$$

$$\therefore \; V_{GS} = -3 \text{ V}.$$

But $$R = \frac{-V_{GS}}{I_D} = \frac{3 \text{ V}}{0.5 \text{ mA}} = \underline{6 \text{ k}\Omega}.$$

(c) The output resistance may be determined by considering the circuit as an amplifier with equivalent circuit as shown in diagram (b).

The output resistance is found from

$$R_o = \frac{v_{o(oc)}}{i_{o(sc)}}.$$

With output open circuit,

$$v_{o(oc)} = -g_m v_i r_{ds}.$$

With output short circuit,

$$v_{gs} = v_i + Ri_{o(sc)}.$$

But $$i_{o(sc)} = -\frac{g_m v_{gs} r_{ds}}{R + r_{ds}},$$

$$\therefore \; i_{o(sc)} = -\frac{g_m r_{ds}}{R + r_{ds}} (v_i + Ri_{o(sc)})$$

$$\therefore\ i_{o(sc)} = \frac{g_m r_{ds} v_i}{R + (1 + g_m R)\, r_{ds}}$$

$$\therefore\ R_o = \frac{v_{o(oc)}}{i_{o(sc)}} = R + (1 + g_m R)\, r_{ds}$$

$$= \underline{356\ \text{k}\Omega}.$$

(d) Advantages of this type of circuit are simplicity, reasonably constant current, the determination of the current by choice of resistance, the provision by the resistor of negative feedback to increase the effective output resistance, and the I_D vs. V_{DS} curve is flatter when the gate is more reverse biased.

Disadvantages are that, owing to the I_{DSS} spread, the current is somewhat unpredictable and also that it is temperature sensitive (typically 5 per cent).

Applications are as an adjustable current source, as part of one type of current mirror, as the tail of a long-tailed pair, or as part of the source load in a source follower.

Example 3.5

An FET has output and transfer characteristics as shown in diagrams (a) and (b). Using the self-bias circuit of diagram (c), calculate appropriate values for R_G, R_S and R_D, assuming V_{DD} = 30 V and that a quiescent operating drain current of 0.4 mA is required.

State the major disadvantage of this method of biasing, and suggest a better alternative.

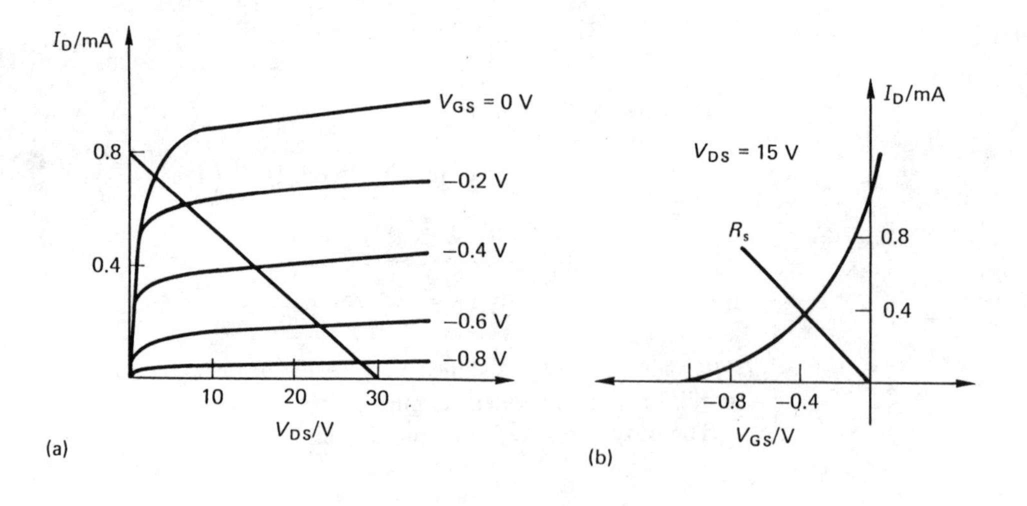

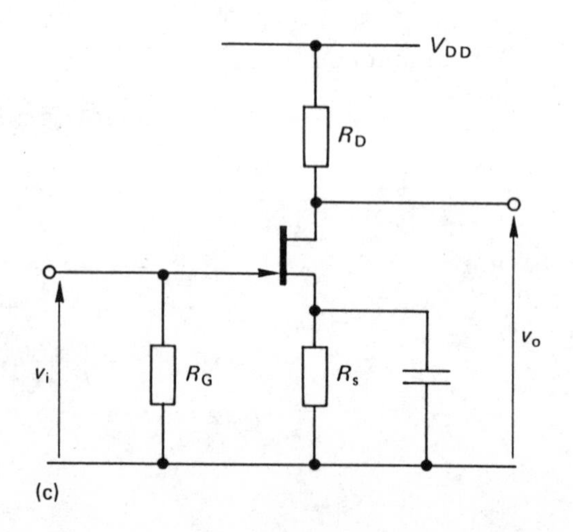

Solution 3.5

For $I_{DQ0} = 0.4$ mA, from the output characteristic choose

$$R_D = \frac{15 \text{ V}}{0.4 \text{ mA}} = 37.5 \text{ k}\Omega$$

$$\simeq \underline{39 \text{ k}\Omega}.$$

From the transfer characteristic to give

$$I_{DQ0} = 0.4 \text{ mA choose}$$

$$R_S \simeq \frac{0.4 \text{ V}}{0.4 \text{ mA}} = \underline{1 \text{ k}\Omega}.$$

Choose R_G to suit the loading acceptable to the signal source driving the FET input (say, 1 MΩ).

The disadvantage with this method of biasing is due to the variation in the output and transfer characteristics over the normal production spread of a given FET. A better method of biasing is to use a combination of constant voltage and self-biasing as shown in diagrams (d) and (e). The principal advantage of these circuits is that they reduce sensitivity to variation in FET characteristics. The resistors R_1 and R_2 act as a voltage divider giving a gate voltage V_{GG}. If the source resistance is now superimposed on the transfer characteristic, it crosses the V_{GS} axis at some positive voltage V_{GG} as shown in diagram (f). If R_1 and R_2 are chosen such that

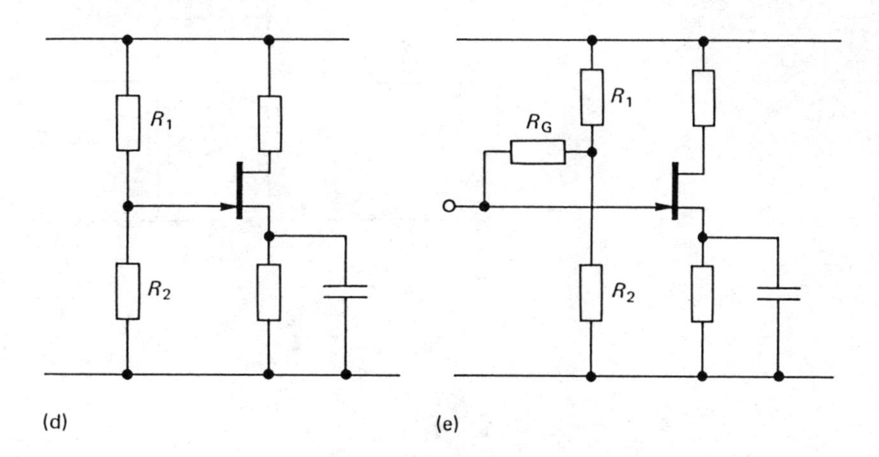

(d)

(e)

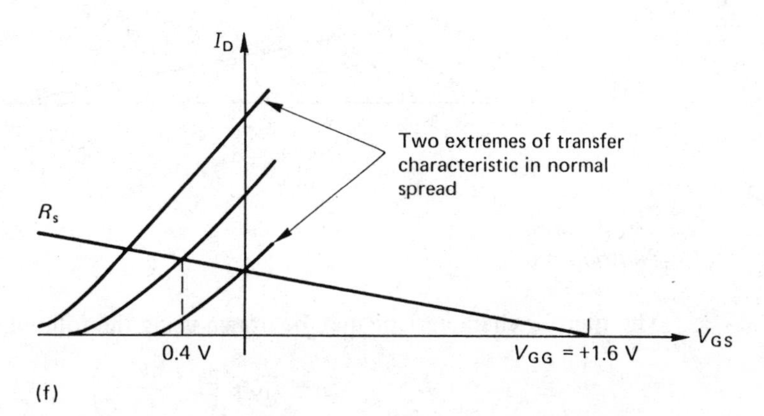

(f)

V_{GG} = + 1.6 V say, then a suitable value of R_S to cross the transfer characteristic at I_D = 0.4 mA may be found from the equation

$$V_{GS} = V_{GG} - I_D R_S,$$

$$\therefore R_S = \frac{1.6 + 0.4}{0.4} = 5 \text{ k}\Omega.$$

Since the slope of this load line is now *much more horizontal*, the drain current is less sensitive to variations in the transfer characteristic.

In the circuit of diagram (e) the resistor R_G serves merely to increase the circuit input impedance, while allowing R_1 and R_2 to be reasonably small values.

Example 3.6

The circuit of diagram (a) has V_P = –4 V and I_{DSS} = 10 mA. Sketch an appropriate transfer characteristic for the device, and from it calculate values for I_D, V_{GS} and V_{DS}.

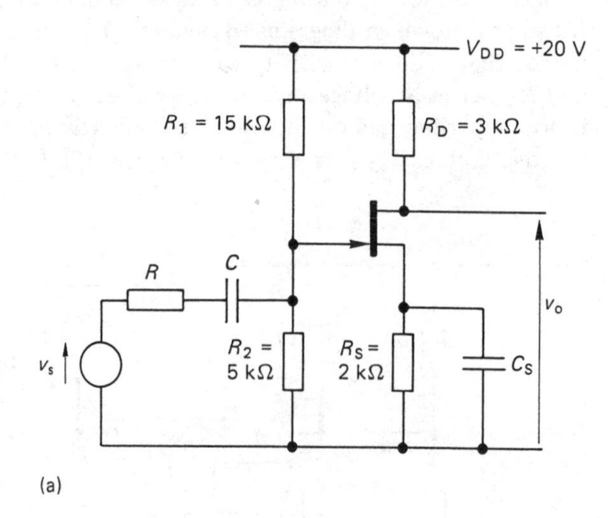

(a)

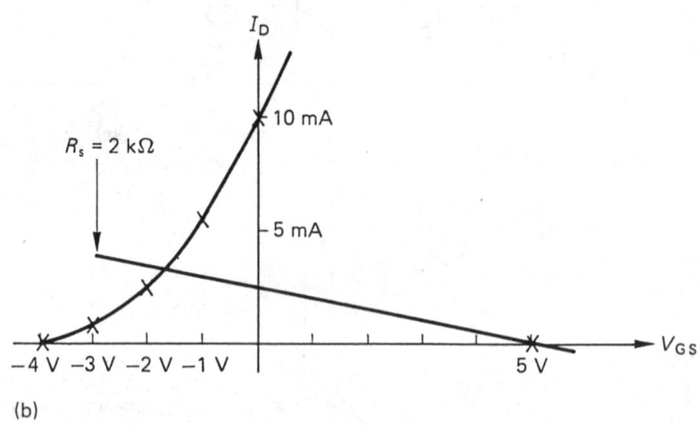

(b)

Solution 3.6

The transfer characteristic may be drawn using the following equation:

$$I_D = I_{DSS}\left(1 - \frac{V_{GS}}{V_P}\right)^2,$$

where I_{DSS} = 10 mA and V_P = −4 V. The values in the table are calculated using this equation.

V_{GS} (V)	I_D (mA)
0	10
−1	5.62
−2	2.5
−3	0.63
−4	0

The corresponding transfer characteristic is shown in diagram (b).

$$V_{GG} = \frac{R_2}{R_1 + R_2} V_{DD} = 5 \text{ V}.$$

Drawing the bias line for R_S = 2 kΩ on the graph, the required values for I_D and V_{GS} may be obtained graphically as the point at which the line crosses the transfer characteristic.

$$\therefore I_D \approx \underline{3.3 \text{ mA}}, \qquad V_{GS} \approx \underline{-1.6 \text{ V}}$$

and

$$V_{DS} = V_{DD} - I_D R_D - I_D R_S$$
$$= \underline{3.4 \text{ V}}.$$

Example 3.7

Describe fully the construction and operating principle of an MOS field-effect transistor (MOST). Sketch the cross-section of the device and give typical characteristics. Distinguish between *p*- and *n*-channel devices, and between depletion and enhancement modes.

Design an audio-frequency common-source amplifier, using a MOST with a mutual conductance of 4 mS and a drain slope resistance of 10 kΩ, to have a voltage gain of 25 dB. Specify all critical components. (CEI Part 2)

Solution 3.7

The construction of the MOSFET is shown in diagram (a). For, say, an *n*-channel MOSFET the drain and source regions are *n*-type diffusions into a *p*-type silicon layer. The metal gate is separated from the silicon by an oxide insulation. This means very low gate current.

The conduction is via an induced *n*-channel between the *n*-type diffusions. The channel induced is a function of the positive gate-to-source voltage. If a positive charge is placed on the gate then a negative charge consisting of electrons must appear in the *p*-type silicon, producing an inversion layer which forms a conduct-

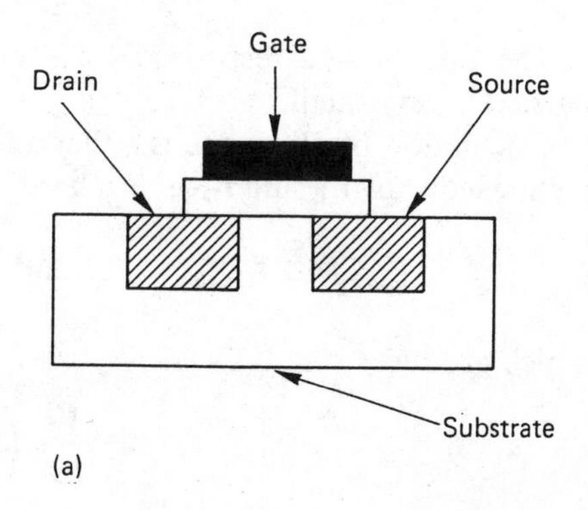

(a)

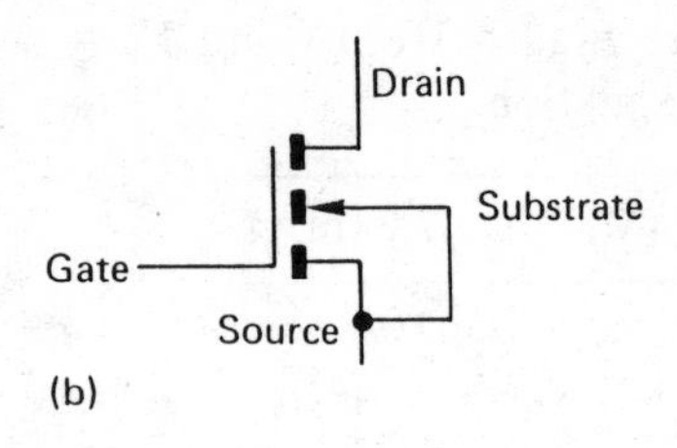

(b)

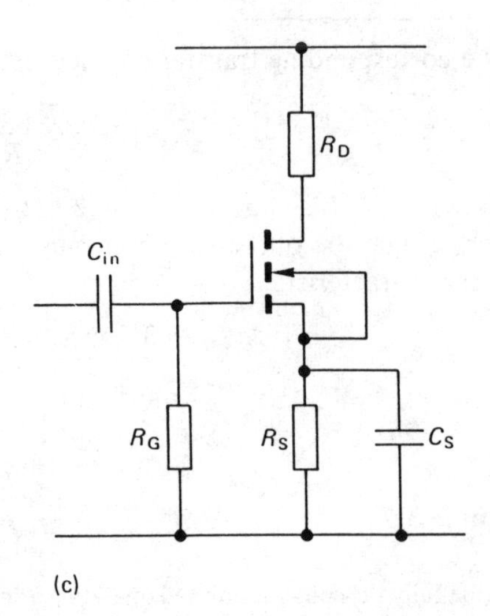

(c)

ing channel between drain and source. The size of this channel is a function of the gate charge.

In a *p*-channel device, *p*-regions are diffused into an *n*-type silicon substrate.

The characteristics and the definition of the terms 'depletion and enhancement modes' are given in the text.

The amplifier design is shown in diagram (c) using a depletion mode MOSFET.

$$|A_v| = \frac{g_m R_D r_{ds}}{R_D + r_{ds}}.$$

Now, $20 \log A_v = 25$ dB.

$$\therefore A_v = 17.78,$$

$$\therefore 17.78 = \frac{4 \times 10^{-3} \times R_D \times 10 \times 10^3}{R_D + 10 \times 10^3},$$

$$\therefore R_D = 8 \text{ k}\Omega.$$

The gate biasing resistor R_G can be large (say, 1 MΩ), since the gate leakage current is very small.

Calculation of R_S requires either a set of characteristic curves for the device, or a knowledge of V_P and I_{DSS}. We have

$$V_{GS} = -I_D R_S \quad \text{and} \quad I_D = I_{DSS}\left(1 - \frac{V_{GS}}{V_P}\right)^2,$$

which produces

$$R_S = \frac{V_P}{I_D}\left(1 - \left(\frac{I_D}{I_{DSS}}\right)^{1/2}\right).$$

For, say, $I_D = 1$ mA, $V_P = -2$ V and $I_{DSS} = 2$ mA, then $R_S \approx 600\ \Omega$.
Typical capacitor values are:

$$C_{in} = 0.1\ \mu F, \qquad C_S = 100\ \mu F.$$

3.11 Unworked Problems

Problem 3.1

Compare briefly the main features of bipolar and m.o.s. field effect transistors, and sketch graphs showing a family of output characteristics for each type.

The small-signal amplifier shown is to operate with $V_{DS} = -9$ V, $V_{GS} = +2$ V, a quiescent drain current $I_D = 1.5$ mA, and to have an open-circuit voltage gain $v_o/v_i = -20$. For $I_D = 1.5$ mA the transistor has transconductance $g_m = 5$ mS and a dynamic drain resistance $r_d = 12$ kΩ.

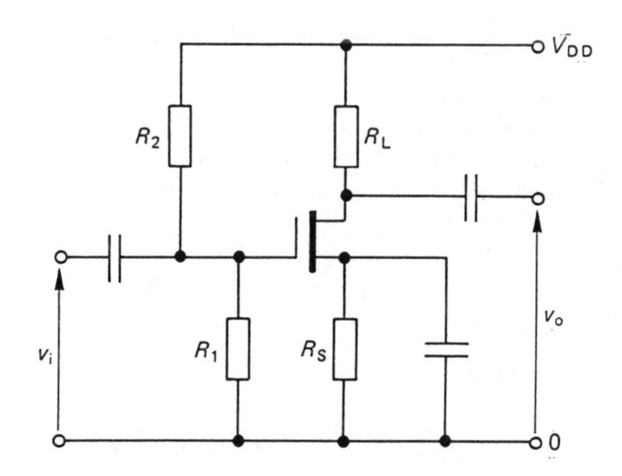

Draw the equivalent circuit of the stage. Calculate the required value of R_L and select suitable values for V_{DD}, R_1, R_2 and R_S. Assume that the capacitors have negligible reactance.

(EC Part 1)

Problem 3.2

The operation of the JFET in the diagram is defined by the equation

$$I_D = I_{DSS}\left(1 - \frac{V_{GS}}{V_P}\right)^2,$$

where $\qquad I_{DSS} = 8$ mA $\qquad$ and $\qquad V_P = -4$ V.

Determine values for R_D and R_S so that the transistor operates at

$$I_D = 0.5\ \text{mA}, \qquad V_{DS} = 7\ \text{V}.$$

Determine the voltage gain, v_o/v_i, if the FET drain resistance $r_{ds} = 20$ kΩ.

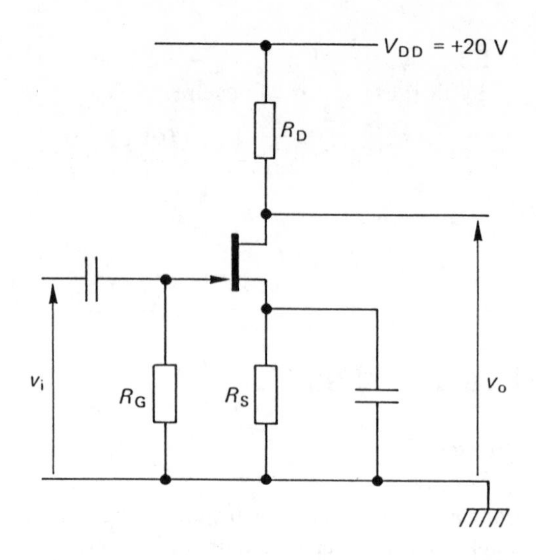

Explain briefly, without mathematical derivation, what would be the effect on the circuit performance of removing the capacitor across R_S.

Problem 3.3

Describe briefly how the low-frequency small-signal equivalent circuit for a field effect transistor (FET) can be modified in order to represent its high-frequency performance. Show that the input capacitance of a common-source amplifier is given by $C_{in} = C_{gs} + (1 - A)\,C_{gd}$ where the symbols have their usual meanings.

Calculate the drain resistance of an FET amplifier having a mid-band gain of −50 when the external load is 50 kΩ. If the capacitance associated with this load is 10 pF, calculate the upper cut-off frequency of the amplifier when it is supplied from a source of resistance (a) 10 kΩ, (b) 50 Ω. The transistor has parameters g_m = 5 mS, r_d = 40 kΩ, C_{gs} = 10 pF, C_{gd} = 1.5 pF, C_{ds} = 2 pF. (CEI Part 2)

Problem 3.4

The bootstrapped source follower shown uses a JFET whose drain current is given by

$$I_D = 5\,(1 + V_{GS}/4)^2,$$

with I_D in mA, V_{GS} in volts over the range $-4\text{ V} \leqslant V_{GS} \leqslant 0$.

Sketch the d.c. characteristics for v_o against v_i over the full active region of the transistor and find the value of the capacitor that, when placed in series with the input, will produce a low-frequency 3 dB point of 20 Hz. Assume the source resistance to be negligible.

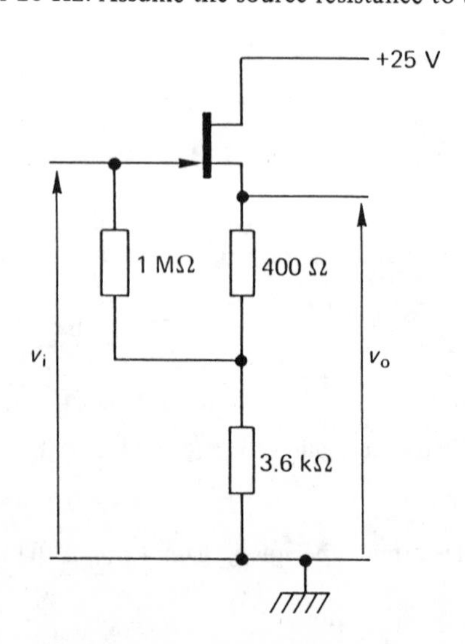

4 Frequency Response of Amplifiers

Amplifiers may be classified in a number of ways according to frequency range, method of operation, ultimate use, type of load, or method of interstage coupling.

Types of frequency classification include d.c. (from zero frequency), audio (20 Hz to 20 kHz), video or pulse (up to a few MHz), r.f. (a few kHz to hundreds of MHz) and v.h.f. (hundreds or thousands of MHz).

The types of amplifier considered in this section are mainly untuned audio and video voltage amplifiers with a resistive load operating in class A.

4.1 Amplifier Frequency Response

We will consider the frequency characteristics in three regions: mid-band frequencies (over which the amplification is reasonably constant), the low-frequency region and the high-frequency region.

Consider the case for a mid-band gain of unity.

(a) Low-frequency Response

If the amplifier is capacitive-coupled then below mid-band the amplifier will behave like a simple high-pass circuit as shown in Fig. 4.1.

$$A_L(jf) = \frac{v_o}{v_i} = \frac{R_1}{R_1 + 1/(j2\pi f C_1)}$$

$$= \frac{1}{1 - j/(2\pi f\, C_1 R_1)} = \frac{1}{1 - j\left(\frac{f_L}{f}\right)},$$

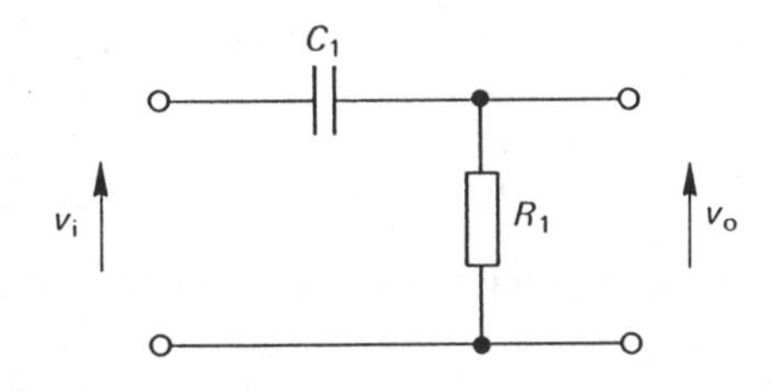

Figure 4.1 High-pass circuit

where $f_L = \dfrac{1}{2\pi C_1 R_1}$.

$$|A_L| = \frac{1 +}{\sqrt{1 + \left(\frac{f_L}{f}\right)^2}}$$

$$\phi_L = \arctan\left(\frac{f_L}{f}\right)$$

The lower 3 dB frequency or break point is where $f = f_L$, at which

$$|A_L| = \frac{1}{\sqrt{2}}.$$

(b) High-frequency Response

In the high-frequency region above the mid-band the amplifier stage may be approximated by a simple low-pass circuit as shown in Fig. 4.2.

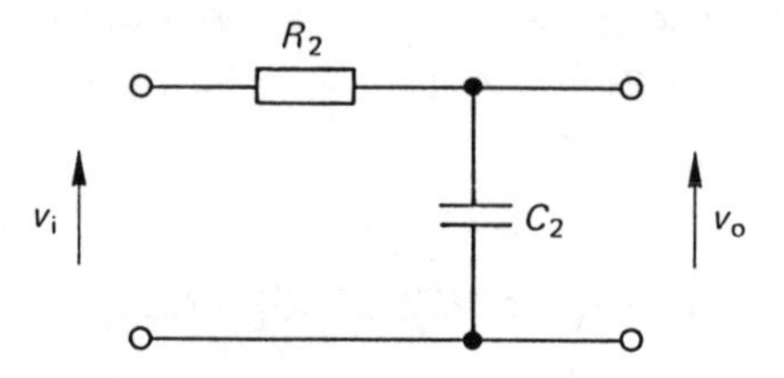

Figure 4.2 Low-pass circuit

$$A_H(jf) = \frac{v_o}{v_i} = \frac{1}{1 + j2\pi f C_2 R_2}$$

$$= \frac{1}{1 + j\left(\frac{f}{f_H}\right)}$$

where $f_H = \dfrac{1}{2\pi C_2 R_2}$.

$$|A_H| = \frac{1}{\sqrt{1 + \left(\frac{f}{f_H}\right)^2}}$$

$$\phi_H = -\arctan\left(\frac{f}{f_H}\right).$$

The upper 3 dB frequency or break point is where $f = f_H$, at which

$$|A_H| = \frac{1}{\sqrt{2}}.$$

A typical gain expression for a capacitive-coupled amplifier is thus

$$A(jf) = \frac{A_0}{\left(1 + j\frac{f}{f_H}\right)\left(1 - j\frac{f_L}{f}\right)}.$$

The corresponding Bode plot shows the dependency of the gain on frequency. The Bode plot is a graph of dB gain against the log of frequency. The fine-line shows the idealised straight-line approximation to the Bode plot (see Fig. 4.3).

The bandwidth is $f_H - f_L$.

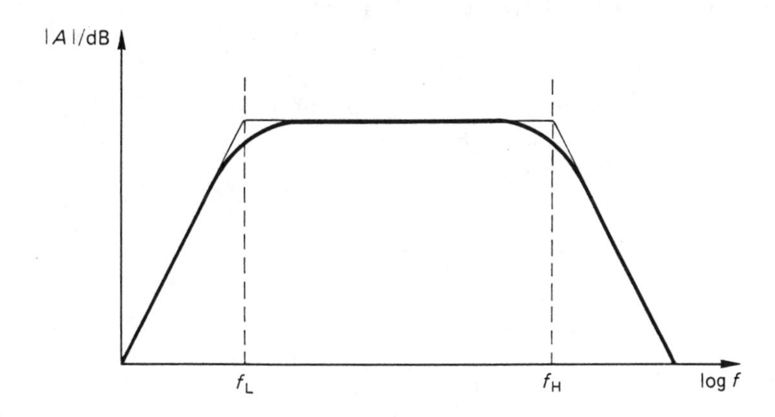

Figure 4.3 Bode plot for typical amplifier

4.2 Bode Plots

For a typical single-pole transfer function,

$$A(jf) = \frac{A_0}{1 + j\left(\frac{f}{f_H}\right)},$$

the gain–frequency curve is idealised by two straight lines as shown in Fig. 4.4. The break point is at $f = f_H$. At frequencies below f_H the gain is assumed constant. At frequencies above f_H the gain falls at a constant rate of 20 dB/decade. The

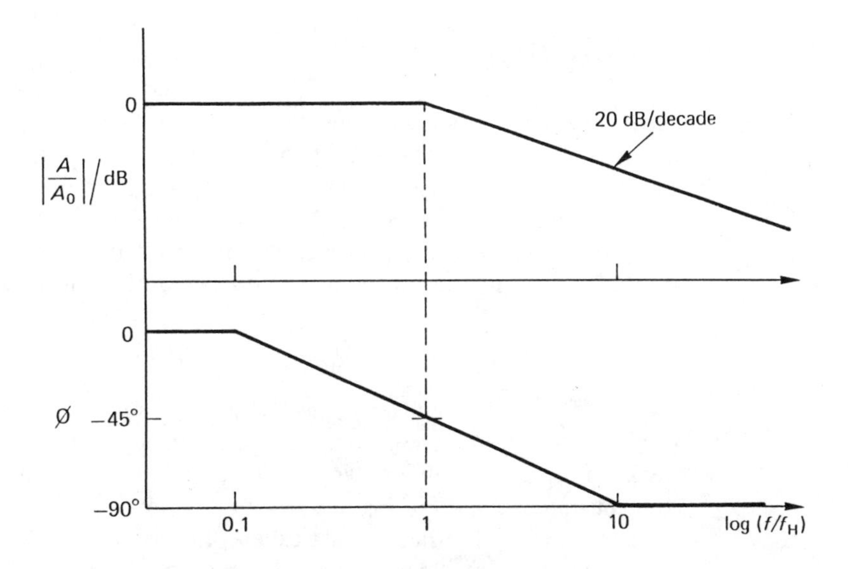

Figure 4.4 Bode plot

phase-frequency curve is also idealised by straight lines. For frequencies below $0.1f_H$ the phase is assumed to be zero. For frequencies above $10f_H$ the phase is assumed to be $-90°$. The phase-log-frequency is assumed to be a straight line between these two points, passing through $-45°$ at $f = f_H$.

4.3 Step Response of Amplifiers

The step response of an amplifier is related to the frequency response. If we consider the step response of the CR network of Fig. 4.5(a) the response is shown in Fig. 4.5(b), where

$$v_o = v_i (1 - e^{-t/\tau}),$$

where τ = CR.

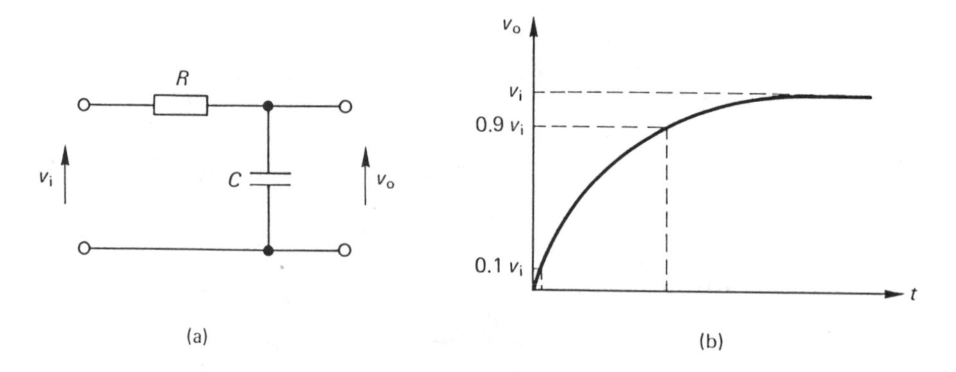

Figure 4.5 Step response

If the rise time t_r is defined as the time to rise from $0.1v_i$ to $0.9v_i$ then

$$t_r = 2.2CR = \frac{2.2}{2\pi f_H} = \frac{0.35}{f_H}.$$

4.4 Cascaded Amplifiers

Capacitor coupling between the stages of a multistage amplifier as shown in Fig. 4.6 affects the overall low-frequency response. The following analysis assumes that the emitter bypass capacitors are arbitrarily large so that they act as an effective short circuit to a.c. signals. The equivalent circuit is shown in Fig. 4.7. Using Thévenin's theorem, the equivalent circuit of the coupled stage is as shown in Fig. 4.8. For this circuit the low-frequency 3 dB point is given by

$$f_L = \frac{1}{2\pi (R_{C1} + R_i') C}.$$

If the effect of the emitter bypass capacitor is taken into account in the analysis of the low-frequency response, then the gain-log frequency plot has a further break point (a zero) due to $R_E C_E$ which causes the gain to level off; a typical Bode diagram is shown in Fig. 4.9.

When several amplifiers are cascaded, the overall gain expression is not necessarily the simple product of several gain expressions $A(f)$. Interaction between stages can cause significant changes in the individual responses. One amplifier

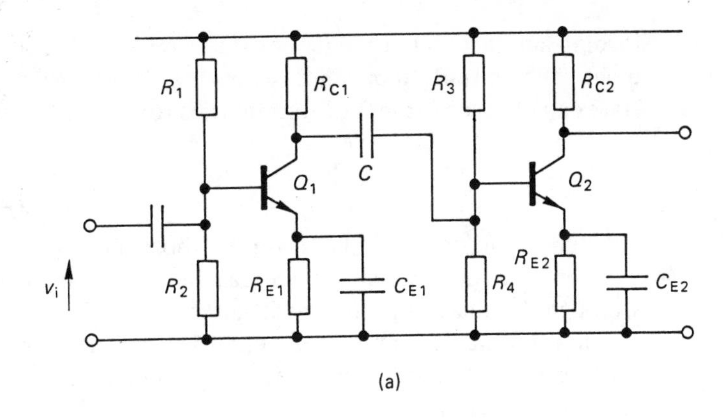

Figure 4.6 RC coupled amplifier

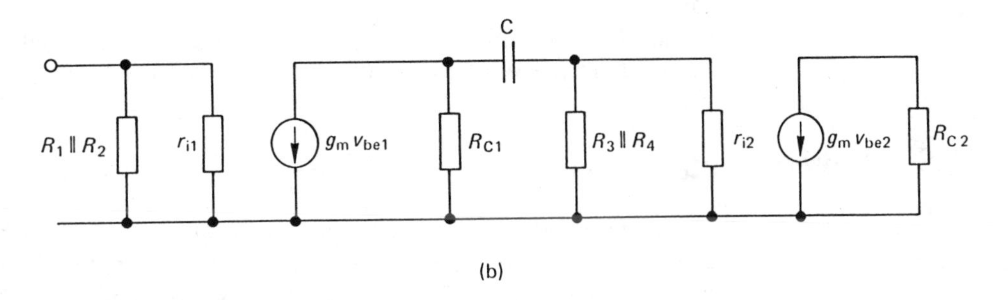

Figure 4.7 Equivalent circuit of RC coupled amplifier

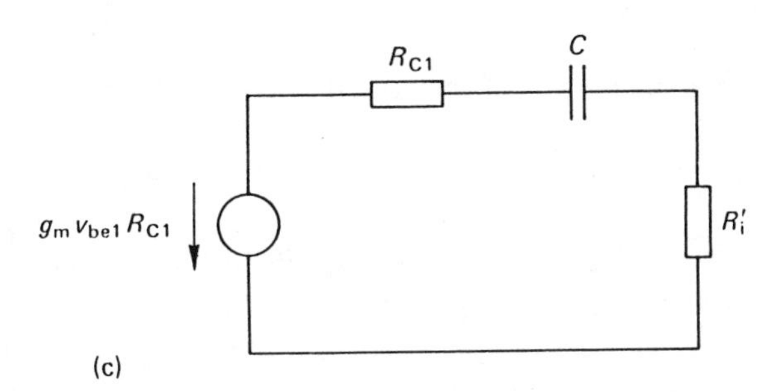

Figure 4.8 Thevenin's equivalent

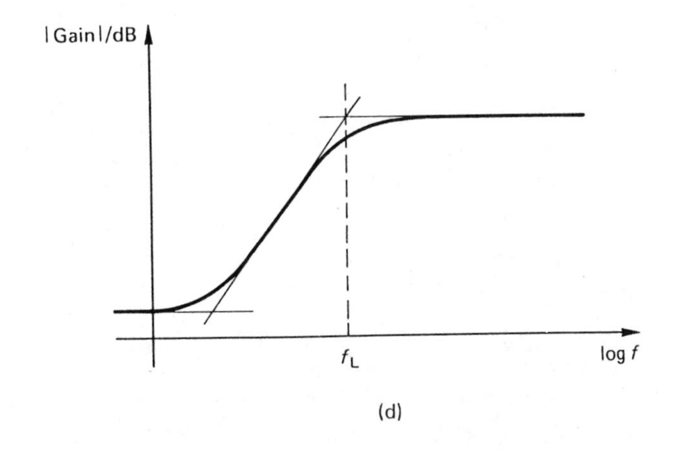

Figure 4.9 Effect of emitter bypass capacitor on low-frequency response

loading another can change both the mid-band gain and the 3 dB frequency unless the output impedance of each stage is low. If identical non-interacting stages are cascaded, the 3 dB frequencies for an n-stage amplifier become

$$f_{Hn} = (\sqrt{2^{1/n} - 1})\, f_H, \qquad f_{Ln} = \frac{f_L}{\sqrt{2^{1/n} - 1}};$$

i.e. there is a shrinkage in bandwidth at both ends of the frequency response.

When non-identical stages are cascaded, similar bandwidth reduction occurs, although the analysis is more awkward.

The response of a multistage amplifier can often be approximated to that of a single stage if the frequency response is dominated by the cut-off frequency of one particular amplifier.

If the amplitude response of an n-stage amplifier is plotted on log-linear graph paper then the resulting graph will approach a straight line with a slope of $20n$ dB/decade.

It may be shown that the optimum bandwidth of cascaded amplifiers occurs when the individual amplifier bandwidths are equal.

4.5 Tuned Amplifiers

A narrow-bandwidth tuned amplifier can be obtained using a parallel resonant circuit as the load of an amplifier stage, as shown in Fig. 4.10(a). If the coil resistance losses and transistor output resistance are represented by the combined resistance R shown in diagram (b) then the parallel impedance is given by

$$Z(j\omega) = \frac{R}{1 + j\,(\omega CR - R/\omega L)};$$

or, in terms of the circuit Q-factor, where

$$Q = \omega_0 CR = \frac{R}{\omega_0 L}$$

and ω_0 is the amplifier resonant frequency,

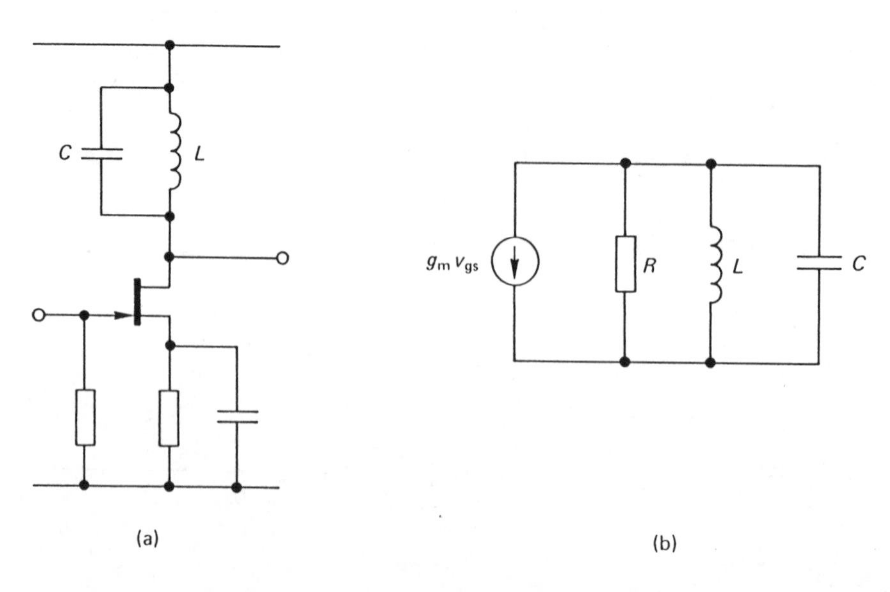

Figure 4.10 Tuned amplifier and equivalent circuit

then $$Z(\mathrm{j}f) = \frac{R}{1 + \mathrm{j}Q\left(\frac{f}{f_0} - \frac{f_0}{f}\right)}.$$

Using this equation, it may be shown that the bandwidth of the amplifier (the frequency difference between the 3 dB points) is given by

$$B = \frac{f_0}{Q} = \frac{1}{2\pi CR}.$$

The effect of the circuit Q-factor on frequency response is shown in Fig. 4.11. The product of gain x bandwidth is constant and is given by $\frac{g_m}{2\pi C}$ so that high gain is possible only at the expense of reduced bandwidth and vice versa.

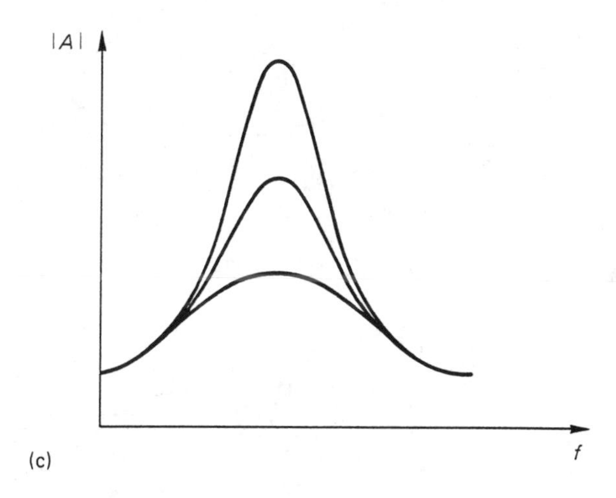

Figure 4.11 Frequency response of tuned amplifier

4.6 Worked Examples

Example 4.1

Discuss briefly the factors that could influence the choice between a bipolar and a field effect transistor as the active device in a small-signal a.c. amplifier. What properties of a dual-gate FET make it suitable for use in a v.h.f. amplifier stage?

Estimate the operating point of the transistor in the circuit shown in diagram (a) and the mid-band voltage gain of the stage. Determine the values of the capacitors that would give a low-frequency 3 dB cut-off at 50 Hz. The transistor has parameters $h_{fe} = 150$, $h_{ie} = 4000\ \Omega$ and $h_{oe} = 20\ \mu S$ and the source resistance is $600\ \Omega$. (CEI Part 2)

Solution 4.1

See the text (sections 2.9 and 3.4) for a comparison of bipolar and field effect transistors.

In circuit diagram (a), the operating point is found as follows:

$$V_B = \frac{10}{10 + 27} \times 15\ \mathrm{V} = 4.05\ \mathrm{V},$$

$$\therefore\ V_E = V_B - 0.6\ \mathrm{V} = 3.45\ \mathrm{V},$$

$$\therefore\ I_E = \frac{V_E}{R_E} \approx 1\ \mathrm{mA}.$$

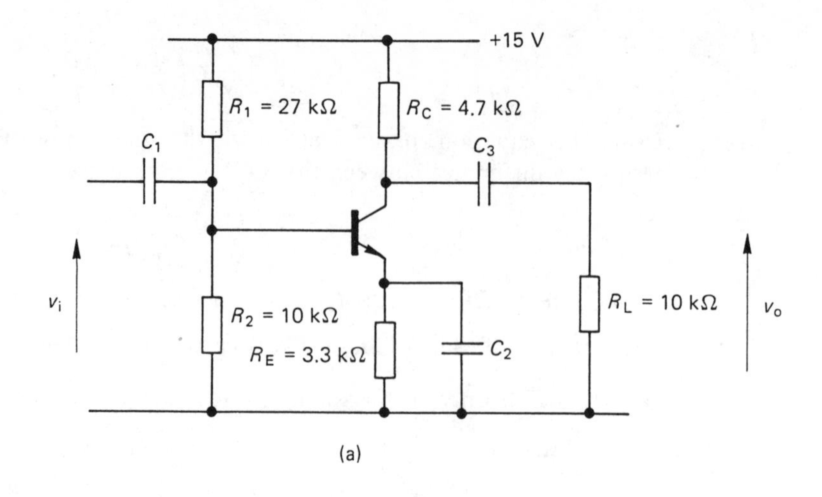

(a)

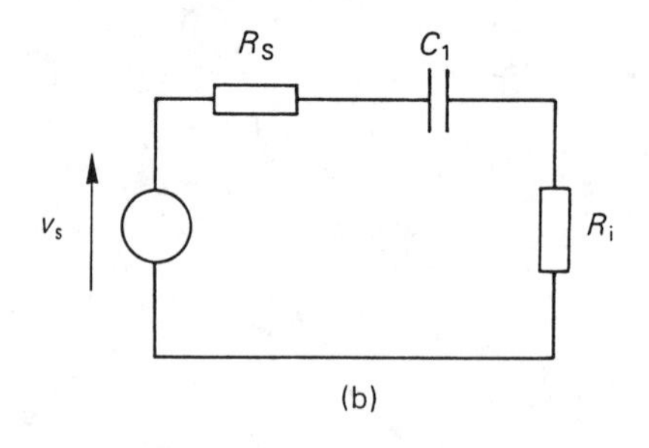

(b)

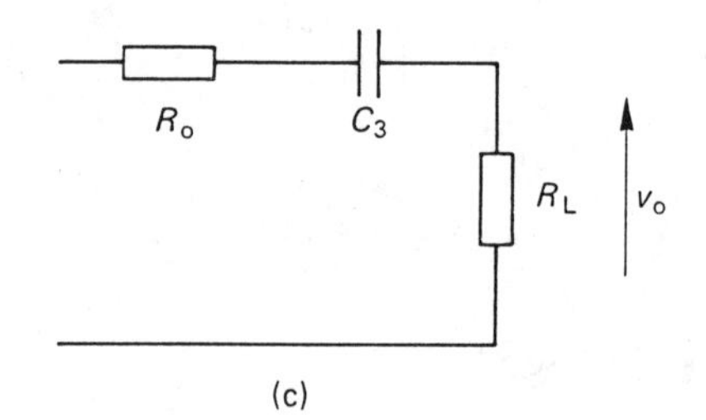

(c)

Now $I_C \approx I_E$,

$$\therefore V_C = 15 \text{ V} - 4.7 \text{ k}\Omega \times 1 \text{ mA},$$
$$= \underline{10.3 \text{ V}}.$$

$$A_v = -\frac{h_{fe}R}{h_{ie}}, \qquad \text{where } R = \frac{1}{h_{oe}} \parallel R_C \parallel R_L,$$

$$\therefore R = \frac{10^6}{20}\ \Omega \parallel 4.7 \text{ k}\Omega \parallel 10 \text{ k}\Omega$$

$$= 3.2 \text{ k}\Omega,$$

$$\therefore A_v = -\frac{150 \times 3.2}{4} = \underline{-120}.$$

For the purposes of the low-frequency analysis, we assume that we are to consider the effects of C_1 and C_3 independently, and to ignore the 'zero' due to C_2.

The effect of C_1 may be considered using the equivalent circuit of diagram (b), where

$$R_i = h_{ie} \parallel R_1 \parallel R_2 = 2.6 \text{ k}\Omega.$$

∴ The low-frequency cut-off due to the input stage is when

$$f_L = \frac{1}{2\pi C_1 (R_S + R_i)}.$$

$$\therefore\ C_1 = \frac{1}{2\pi \times 50\,(600 + 2.6 \times 10^3)}$$

$$\approx \underline{1\ \mu F.}$$

The effect of C_3 may be considered using the equivalent circuit of diagram (c), where

$$R_o = \frac{1}{h_{oe}} \parallel R_C = 50\ k\Omega \parallel 4.7\ k\Omega = 4.3\ k\Omega.$$

The low-frequency cut-off due to the output stage is when

$$f_L = \frac{1}{2\pi C_3 (R_o + R_L)},$$

$$\therefore\ C_3 = \frac{1}{2\pi \times 50\,(4.3 \times 10^3 + 10 \times 10^3)}$$

$$\approx \underline{0.22\ \mu F.}$$

Example 4.2

A variable-frequency sinusoidal oscillator with an output resistance of 2.2 kΩ and open-circuit output voltage of 10 mV is applied to the input of a common-emitter amplifier. For the frequencies below 10 kHz the amplifier output voltage remains virtually constant at 1 V. As the oscillator frequency is increased, the output voltage falls, because of the Miller effect, reaching an amplitude of 0.5 V at 100 kHz.

Estimate the low-frequency current gain, h_{fe}, and the collector-base capacitance of the transistor used, given that it is operated with a collector load resistance of 4.7 kΩ and a d.c. collector current of 1 mA.

Solution 4.2

The equivalent circuit is shown in the diagram, where, owing to the Miller effect, the capacitance C is given by

$$C = C_{be} + (1 + g_m R_C)\, C_{bc} \approx (1 + g_m R_C)\, C_{bc}.$$

Now $$g_m = 40 I_C = 40 \times 1\ mA$$

$$= 40\ mS,$$

and $$A_v = \frac{v_o}{v_s} = \frac{1\ V}{10\ mV} = 100.$$

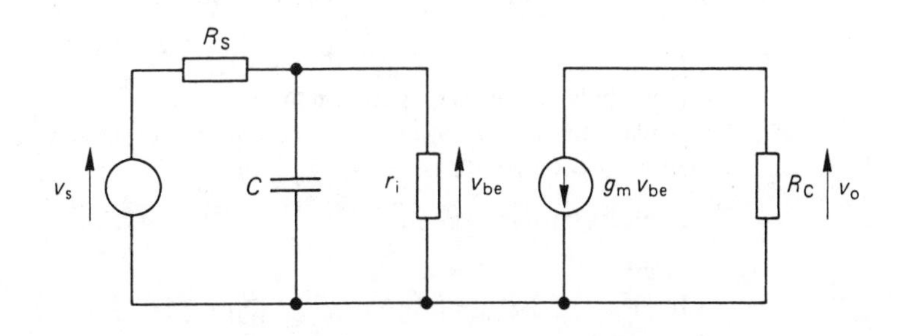

But $$\frac{v_o}{v_{be}} = -g_m R_C,$$

$$\therefore \frac{v_o}{v_s} = -g_m R_C \left(\frac{r_i}{r_i + R_S}\right) = -100,$$

$$\therefore 40 \times 10^{-3} \times 4.7 \times 10^3 \left(\frac{r_i}{r_i + 2.2 \times 10^3}\right) = 100,$$

$$\therefore r_i = 2.5 \text{ k}\Omega,$$

$$\therefore h_{fe} = g_m r_i = 40 \times 10^{-3} \times 2.5 \times 10^3$$

$$= \underline{100}.$$

At high frequencies,

$$v_{be} = \left(\frac{r_i}{r_i + R_S}\right)\left(\frac{1}{1 + j\omega C R_i}\right) v_s,$$

where $R_i = r_i \parallel R_S = 2.5 \text{ k}\Omega \parallel 2.2 \text{ k}\Omega = 1.17 \text{ k}\Omega$,

$$\therefore A_v(f) = \frac{A_{v0}}{1 + j\omega C R_i},$$

$$\therefore \quad 50 = \frac{100}{\sqrt{1 + (\omega/\omega_H)^2} = 2}, \quad \text{at } f = 100 \text{ kHz},$$

where $\omega_H = \dfrac{1}{CR_i}$,

$$\therefore \sqrt{1 + (\omega/\omega_H)^2} = 2,$$

$$\frac{\omega}{\omega_H} = \sqrt{3}.$$

$$\therefore CR_i = \frac{\sqrt{3}}{2\pi \times 100 \times 10^3},$$

$$\therefore C = 2.36 \text{ nF},$$

$$\therefore C_{bc} = \frac{C}{1 + g_m R_C} = \underline{12.5 \text{ pF}}.$$

Example 4.3

Show how the low-frequency small-signal equivalent circuit for the FET can be modified to represent its high-frequency performance.

(a) Calculate the drain resistance for an FET amplifier having a mid-band gain of −50.
(b) Show that the input capacitance of a common-source amplifier is given by

$$C_{in} = C_{gs} + (1 - A)\, C_{gd},$$

the symbols having their usual meanings.
(c) Calculate the upper cut-off frequency of the amplifier when it is supplied from a source resistance of (i) 10 kΩ, (ii) 50 Ω.

The transistor parameters are:

$g_m = 5$ mS, $r_{ds} = 40$ kΩ,
$C_{gs} = 10$ pF, $C_{gd} = 1.5$ pF, $C_{ds} = 2$ pF.

Solution 4.3

The high-frequency equivalent circuit for the FET is shown in the diagram.

(a) $$A_{v0} = -g_m R, \qquad \text{where } R = R_D \parallel r_{ds}$$

$$= -50.$$

$$\therefore R = \frac{50}{5 \times 10^{-3}} = 10\ \text{k}\Omega,$$

$$\therefore R_D = \underline{13.3\ \text{k}\Omega.}$$

(b) Using Miller's theorem, the capacitance C_{gd} between the input and output of the amplifier of voltage gain A is equivalent to a capacitance $C_{gd}\ (1 - A)$ connected across the input terminals where $A = -g_m R$. This is in parallel with C_{gs}.

The equivalent capacitance at the output is $\frac{A - 1}{A}\ C_{gd}$ in parallel with C_{ds}.

(c) With C_{gd} referred to the input and output terminals, the equivalent circuit is that of the diagram, where

$$C_1 = C_{gs} + (1 + g_m R)\ C_{gd} = 86.5\ \text{pF},$$

$$C_2 = C_{ds} + \left(\frac{1 + g_m R}{g_m R}\right) C_{gd} = 3.5\ \text{pF}.$$

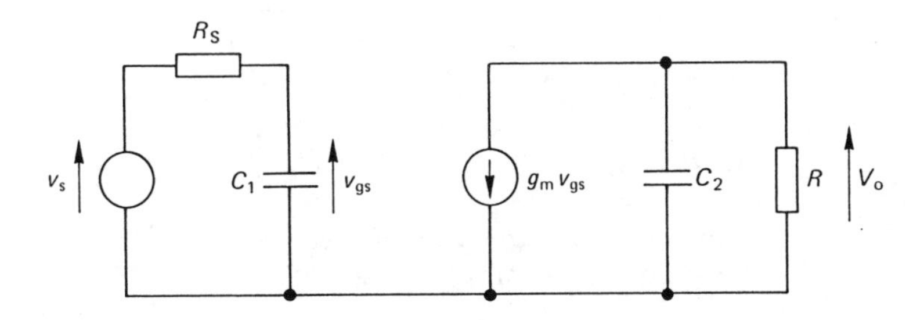

(i) With $R_S = 10\ \text{k}\Omega$ the $C_1 R_S$ time constant effectively determines the frequency response of the amplifier since $C_2 R$ is small in comparison. The upper cut-off frequency is thus

$$\frac{1}{2\pi C_1 R_S} = \underline{184\ \text{kHz}.}$$

(ii) With $R_S = 50\ \Omega$, the $C_2 R$ time constant dominates, since $C_1 R_S$ is now small in comparison. The upper cut-off frequency is thus

$$\frac{1}{2\pi C_2 R} = 4.5\ \text{MHz};$$

Example 4.4

For the circuit of diagram (a) it is desired to have a low 3 dB cut-off frequency not greater than 10 Hz. Calculate the minimum required value for the coupling capacitor C_C. Assume $h_{ie} = 1000\ \Omega$ for each transistor.

If the bandwidth is to be limited to 10 kHz by inclusion of a capacitor across R_{C2}, calculate a suitable capacitor value.

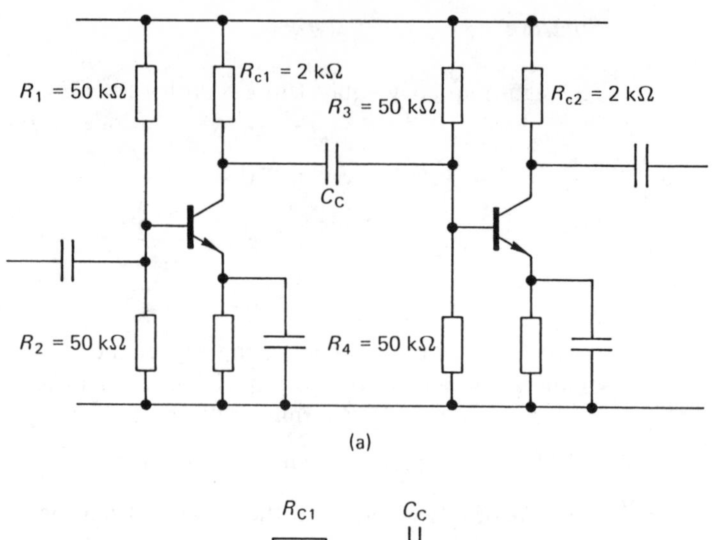

(a)

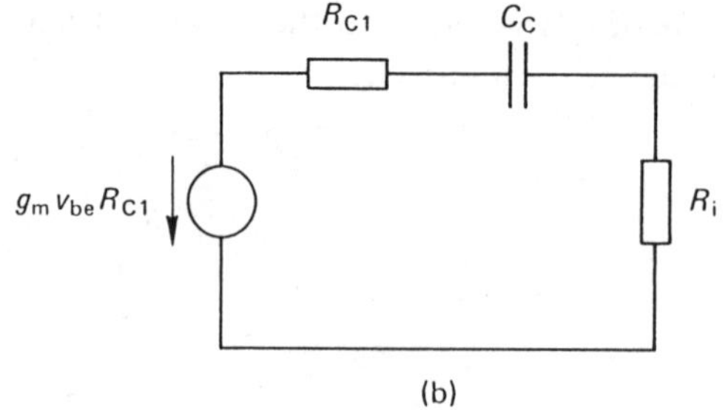

(b)

Solution 4.4

The equivalent circuit of the section that determines the low-frequency cut-off is shown in diagram (b), where

$$R_i = R_3 \parallel R_4 \parallel h_{ie} = 50\ k\Omega \parallel 50\ k\Omega \parallel 1\ k\Omega = 0.96\ k\Omega.$$

$$\therefore f_L = \frac{1}{2\pi (R_{C1} + R_i) C_C},$$

$$\therefore C_C \geqslant \frac{1}{2\pi \times 10 (2 \times 10^3 + 0.96 \times 10^3)}$$

$$\geqslant \underline{5.3\ \mu F}.$$

To limit the bandwidth to 10 kHz requires the addition of a capacitor C_x in parallel with R_{C2}.

$$\therefore f_H = \frac{1}{2\pi R_{C2} C_x},$$

$$\therefore C_x = \frac{1}{2\pi \times 10 \times 10^3 \times 2 \times 10^3}$$

$$= \underline{8\ nF}.$$

The effects of interelectrode capacitance may generally be ignored at this frequency.

Example 4.5

Show that the 3 dB frequency for three identical, non-interacting amplifier stages connected in cascade is given by 0.51 f_H, where f_H is the 3 dB frequency of each amplifier individually. Establish a general equation for n identical non-interacting stages.

Three identical cascaded amplifier stages have an overall upper 3 dB frequency of 30 kHz and a lower 3 dB frequency of 30 Hz. Calculate f_L and f_H for each stage. (Assume that the stages are non-interacting.)

Solution 4.5

Let the high 3 dB frequencies of three amplifiers separately be f_{H1}, f_{H2} and f_{H3}. Let the high 3 dB frequency for the three stages cascaded together be f_{HC}, this being the frequency at which the overall voltage gain is $1/\sqrt{2}$ of its mid-band value. To obtain the overall transfer function of the three non-interacting stages, the transfer gains of the individual stages are multiplied together.

$$\therefore \frac{1}{\sqrt{1+\left(\frac{f_{HC}}{f_{H1}}\right)^2}} \times \frac{1}{\sqrt{1+\left(\frac{f_{HC}}{f_{H2}}\right)^2}} \times \frac{1}{\sqrt{1+\left(\frac{f_{HC}}{f_{H3}}\right)^2}} = \frac{1}{\sqrt{2}}.$$

If we choose identical amplifiers with

$$f_{H1} = f_{H2} = f_{H3} = f_H,$$

then

$$\frac{1}{\left(\sqrt{1+\left(\frac{f_{HC}}{f_H}\right)^2}\right)^3} = \frac{1}{\sqrt{2}},$$

$$\therefore f_{HC} = f_H\sqrt{2^{1/3} - 1} = 0.51 f_H.$$

In general, for n identical non-interacting stages,

$$\underline{f_{HC} = f_H \sqrt{2^{1/n} - 1}.}$$

Now

$$f_{HC} = 0.51 f_H$$

$$\therefore f_H = \frac{30 \text{ kHz}}{0.51} \approx \underline{58.8 \text{ kHz}.}$$

$$f_{LC} = \frac{f_L}{0.51} = 30 \text{ Hz},$$

$$\therefore f_L = \underline{15.3 \text{ Hz}.}$$

Example 4.6

Give a general definition for the quality factor (Q) of a reactive circuit in terms of energies. Hence derive an expression for the Q of a parallel tuned circuit in terms of the circuit parameters.

Determine (i) the voltage gain v_o/v_i at resonance of the tuned amplifier shown in diagram (a), (ii) the resonant frequency, and (iii) the 3 dB bandwidth. The transistor has parameters $h_{fe} = 50$, $h_{oe} = 50\ \mu S$, $h_{re} = 0$, $h_{ie} = 1\ k\Omega$. Assume that the coil is (a) lossless, (b) has a series resistance of 4 Ω.

(CEI Part 2)

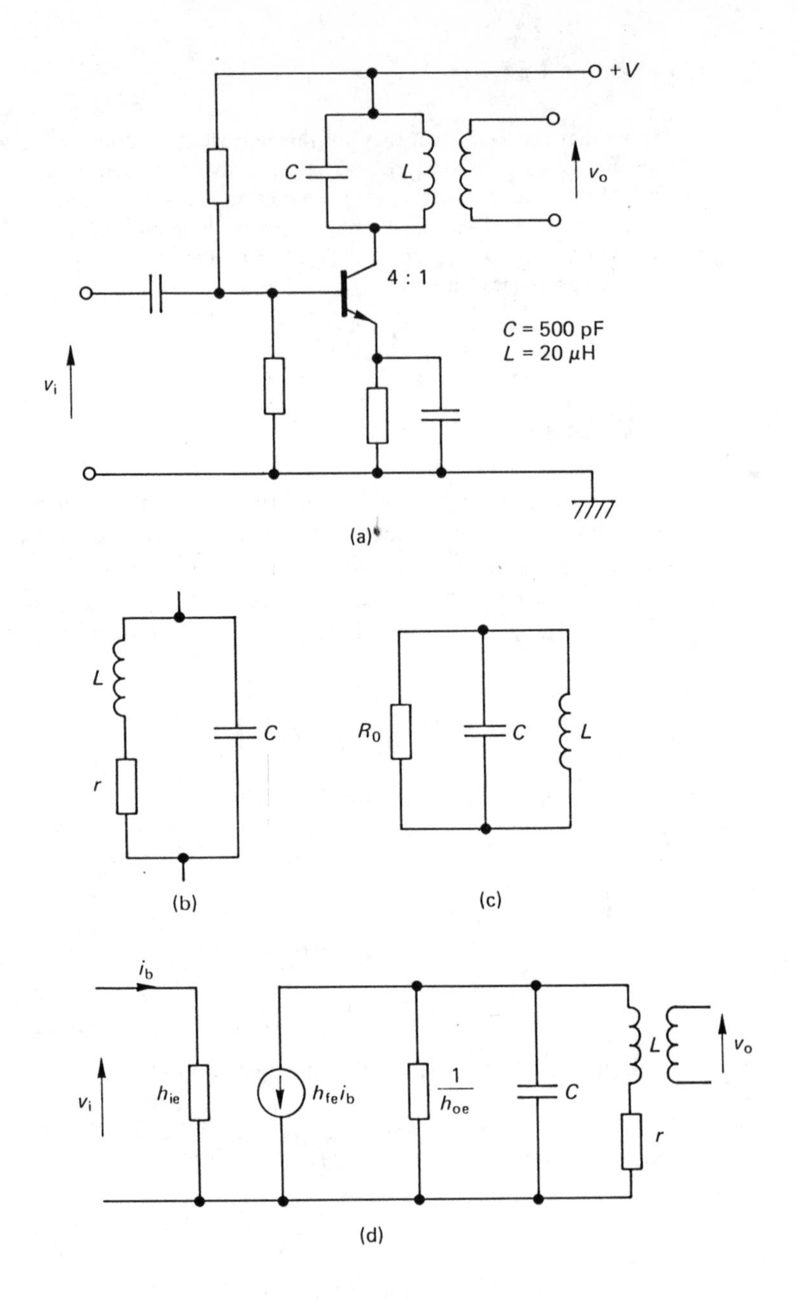

Solution 4.6

One definition of quality factor is

$$Q = 2\pi \times \frac{\text{maximum energy stored}}{\text{energy dissipated per cycle}} .$$

Now maximum energy stored $= \frac{1}{2} L I_m^2$,
where I_m = peak current,

and energy dissipated per cycle = power × period of cycle

$$= \frac{I_{rms}^2 r}{f} = \frac{I_m^2 r \times 2\pi}{2\omega} ,$$

$$\therefore Q = \frac{2\pi \times \frac{1}{2} L I_m^2 \times 2\omega}{I_m^2 r \times 2\pi}$$

$$= \frac{\omega L}{r},$$

where the component values are as shown in diagram (b). Replacing the series resistor by the equivalent parallel resistor R_0 as shown in diagram (c),

$$Q = \omega C R_0 = \frac{R_0}{\omega L}.$$

The equivalent circuit for the tuned amplifier is shown in diagram (d).

(a) Lossless circuit with $r = 0$:

(i) The impedance of the tuned circuit at resonance is given by

$$R_0 = \frac{L}{Cr}.$$

If $r = 0$, then R_0 is theoretically infinite at resonance.

$$\therefore \frac{v_o}{v_i} = \frac{1}{4} \times \frac{h_{fe}\, i_b \times (1/h_{oe})}{h_{ie} i_b}$$

$$= \frac{1}{4} \times \frac{50}{50 \times 10^{-6} \times 10^3}.$$

$$\therefore A_v = \underline{250}.$$

(ii) $f_0 = \dfrac{1}{2\pi\sqrt{LC}} = \underline{1.592 \text{ MHz}}.$

(iii) Bandwidth $= \dfrac{f_0}{Q}$

where $\quad Q = \dfrac{R}{\omega_0 L} \quad$ and $\quad R = \dfrac{1}{h_{oe}}.$

$$\therefore Q = \frac{1}{50 \times 10^{-6} \times 2\pi \times 1.592 \times 10^6 \times 20 \times 10^{-6}} = 100.$$

$$\therefore \text{Bandwidth} = \frac{1.592 \times 10^6}{100} = \underline{15.92 \text{ kHz}}.$$

(b) Series resistance $r = 4\ \Omega$.

(i) Impedance of tuned circuit at resonance is $R_0 = \dfrac{L}{Cr} = 10\ \text{k}\Omega$.

$\therefore$ Load impedance is $10\ \text{k}\Omega \parallel \dfrac{1}{h_{oe}}$

$= 6.67\ \text{k}\Omega.$

$$\therefore A_v = \frac{1}{4} \times \frac{50 \times 6.67 \times 10^3}{10^3}$$

$$= 83.3.$$

(ii) $$f_0 = \frac{1}{2\pi}\sqrt{\frac{1}{LC} - \frac{r^2}{L^2}} = \underline{1.586 \text{ MHz}}.$$

(iii) $$Q = \frac{R}{\omega_0 L} \text{ with } R = 6.67\ \text{k}\Omega$$

$$\therefore Q = 33.3,$$

$$\therefore \text{bandwidth} = \frac{f_0}{Q} = \underline{47.6 \text{ kHz}}.$$

Example 4.7

Diagram (a) shows a hybrid π equivalent circuit for a transistor used in the common-emitter configuration. Explain the physical significance of each of the components and show under what circumstances the circuit may be approximated to that shown in diagram (b). Derive an expression for C.

Derive expressions for (i) f_β, the frequency at which the short-circuit current gain has fallen to $1/\sqrt{2}$ of its low-frequency value, and (ii) f_T, the current gain-bandwidth product.

Estimate the current and voltage gains at 5.5 MHz of a video amplifier having a load resistance $R_L = 1\ \text{k}\Omega$ and a source resistance $R_S = 2\ \text{k}\Omega$. The transistor has $h_{fe} = 50$, $h_{ie} = 1.7\ \text{k}\Omega$, $f_T = 90$ MHz, $C_{b'c} = 1.3$ pF. (CEI Part 2)

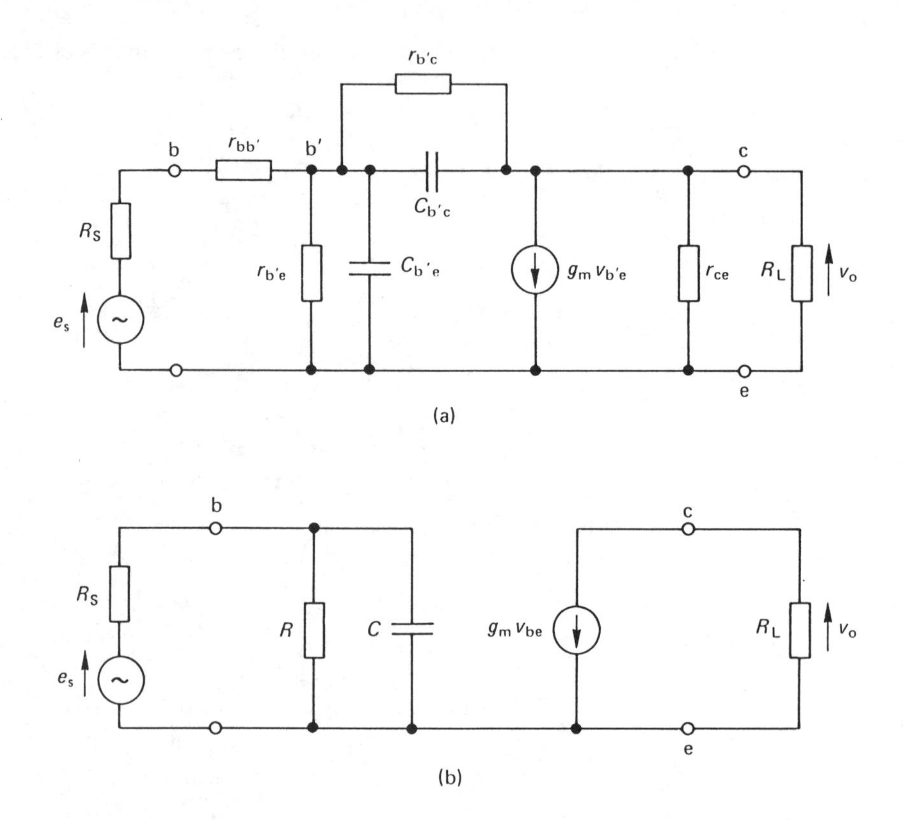

Solution 4.7

The hybrid-π model shown is that which is commonly used for the transistor at high frequencies. It is an approximation that gives a reasonable compromise between accuracy and simplicity. The internal node b′ is not physically accessible.

$r_{bb'}$ represents the 'ohmic base-spreading resistance' between the external base and b′.

$g_m\ v_{b'e}$ represents the current generator to account for the collector current change due to the small change in $v_{b'e}$.

$r_{b'e}$ represents the resistance from b′ to the emitter terminal.

$r_{b'c}$ takes into account the Early effect due to collector voltage variation affecting the base width.

$C_{b'c}$ and $C_{b'e}$ represent the junction capacitances.

The circuit may be approximated to the more simplified version if we ignore the Early effect, assume that the output resistance is sufficiently high to have negligible effect, replace the base resistances by a single resistance and apply Miller's theorem to the junction capacitances.

Using Miller's theorem,

$$C \approx C_{b'e} + (1 + g_m R_L)\, C_{b'c}.$$

(i) The short-circuit current i_{sc} is given by

$$i_{sc} = -g_m v_{be},$$

where

$$v_{be} = \frac{i_i R}{1 + j\omega CR}$$

$$\therefore\ A_i = \frac{i_{sc}}{i_i} = -\frac{g_m R}{1 + j\omega CR} = -\frac{h_{fe}}{1 + j\omega CR}$$

$$= -\frac{h_{fe}}{1 + j(f/f_\beta)}.$$

$$\therefore\ \text{Bandwidth } f_\beta = \underline{\frac{1}{2\pi CR}}.$$

(ii)

$$f_T = h_{fe} f_\beta = \frac{g_m R}{2\pi CR}$$

$$= \underline{\frac{g_m}{2\pi C}}$$

For the specific values given,

$$f_\beta \approx \frac{f_T}{h_{fe}} = \frac{90}{50} = 1.8 \text{ MHz}.$$

The current gain is given by

$$|A_i| = \frac{h_{fe}}{(1 + (f/f_\beta)^2)^{1/2}} = \frac{50}{\left(1 + \left(\frac{5.5}{1.8}\right)^2\right)^{1/2}}$$

$$= \frac{50}{3.2} = \underline{15.6}.$$

The voltage gain is found from

$$v_o = -g_m v_{be} R_L$$

$$= -g_m R_L e_s \left(\frac{R}{R + R_S}\right) \frac{1}{(1 + (\omega C R_P)^2)^{1/2}},$$

where $g_m \approx \dfrac{h_{fe}}{h_{ie}} = \dfrac{50}{1.7} \times 10^{-3} = 29.4$ mS

and

$$C = C_{b'e} + C_{b'c}\,(1 + g_m R_L)$$

where

$$C_{b'e} = \frac{g_m}{2\pi f_T} - C_{b'c} = 52 - 1.3 = 50.7 \text{ pF}$$

$\therefore$

$$C = 50.7 + 1.3\,(1 + 29.4 \times 1)$$

$$= \underline{90.2 \text{ pF}}$$

$$R \approx h_{ie} = 1.7 \text{ k}\Omega,$$

and $R_P = \dfrac{R R_S}{R + R_S} = \dfrac{1.7 \times 2}{1.7 + 2} = 0.92 \text{ k}\Omega.$

$$\therefore \; \omega C R_P = 2\pi \times 5.5 \times 10^6 \times 90.2 \times 10^{-12} \times 0.92 \times 10^3$$

$$= 2.87$$

$$\therefore \; |A_v| = 29.4 \times 1 \times \left(\frac{1.7}{1.7 + 2}\right)\left(\frac{1}{(1 + 2.87^2)^{1/2}}\right)$$

$$= \underline{4.4}$$

4.7 Unworked Problems

Problem 4.1

Define the Q of a tuned circuit in terms of energy stored and dissipated per cycle. Hence derive an expression for the Q of a tuned circuit consisting of elements L, C and R in parallel.

In the small-signal tuned amplifier shown, the field effect transistor has a transconductance g_m = 6 mS and C = 1 nF. All other capacitors have negligible reactance. The peak response occurs at an angular frequency $\omega_0 = 5 \times 10^6$ rad/s and the 3 dB bandwidth is 1.25×10^5 rad/s. Draw the equivalent circuit of the amplifier. Calculate the dynamic drain resistance of the transistor and the maximum gain of the amplifier. The inductor L and capacitor C may be assumed to be loss-free. (EC Part 1)

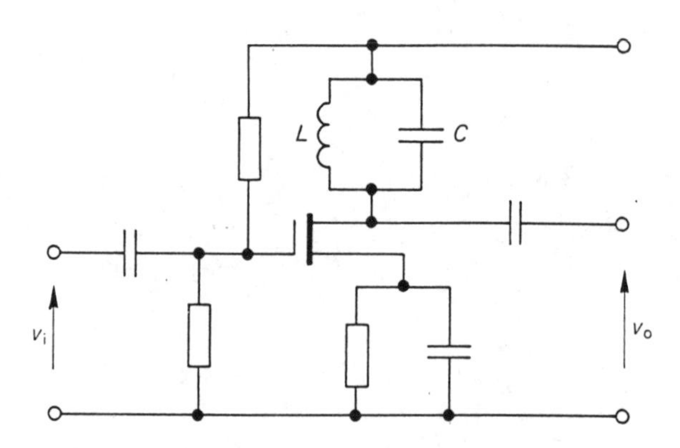

Problem 4.2

(a) Derive from first principles an expression for the Quality Factor (Q) for a circuit consisting of R, L and C all in parallel.

(b) The diagram opposite shows a circuit for a tuned amplifier which is to have a centre frequency of 477 kHz. The inductor L has a Q of 50. Estimate the Q of the amplifier.
(Assume that the identical transistors have $h_{ie} = 26\,(1 + h_{fe})/I_E$ where I_E is the emitter current in mA, h_{fe} = 100, and h_{re} and h_{oe} are both zero. Unlabelled capacitors have negligible reactance at the signal frequency.

(c) Suggest how the design can be modified to increase the Q of the circuit without altering the Q of the inductor. (CEI Part 2)

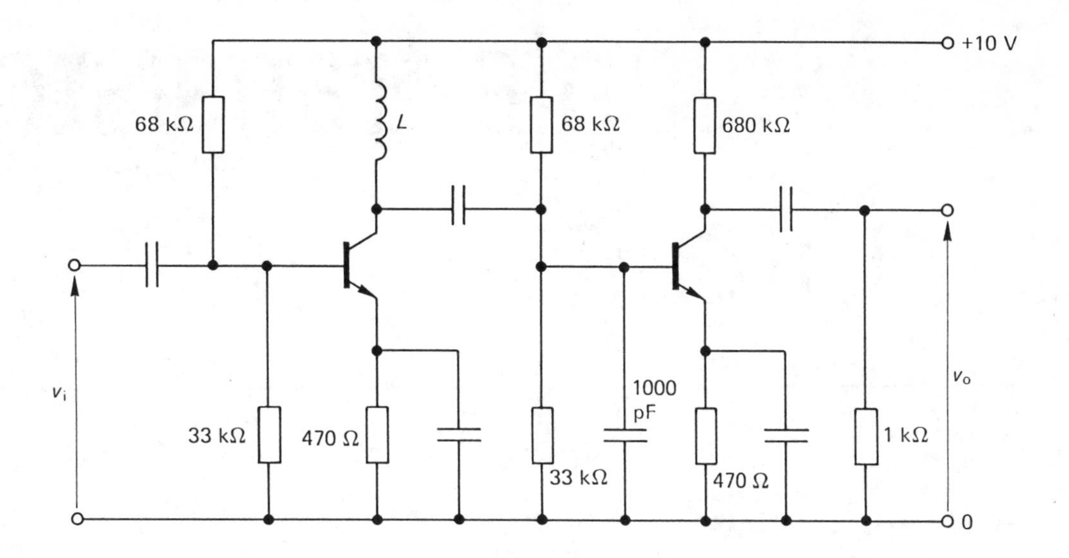

Problem 4.3

A three-stage amplifier is made up using op-amps having gain-bandwidth products of G_1 = 10^8 Hz, G_2 = 4×10^8 Hz, G_3 = 8×10^8 Hz. A d.c. gain of 1000 is required, with maximum bandwidth. Assuming that for maximum bandwidth of the three-stage amplifier, each stage bandwidth should be the same, determine the d.c. gains of each of the three stages and the bandwidth of the three-stage amplifier. (EC Part 2)

5 Multiple-transistor Circuits

5.1 Introduction

Transistors when used singly have a number of limitations which include poor bandwidth, low input impedance, high output impedance, relatively low gain and inability to amplify low-frequency or d.c. signals due to the negative-feedback effect of the stabilisation circuit.

Circuits using two or more transistors overcome many of these limitations and enable the design of wide-band, d.c. and high-gain amplifiers.

Transistor configurations using two transistors include the differential amplifier (emitter-coupled amplifier), used to amplify d.c. and a.c. signals; the cascode and differential amplifier, used to reduce the Miller effect and improve the amplifier bandwidth; and the Darlington pair and multistage amplifier, used to improve amplifier gain and input and output impedances.

5.2 The Differential Amplifier

The differential amplifier is shown in Fig. 5.1. It amplifies the difference between the two input signals. The ideal difference amplifier would respond only to the difference between the two input signals and be entirely independent of the individual signal levels (i.e. reject common-mode signals).

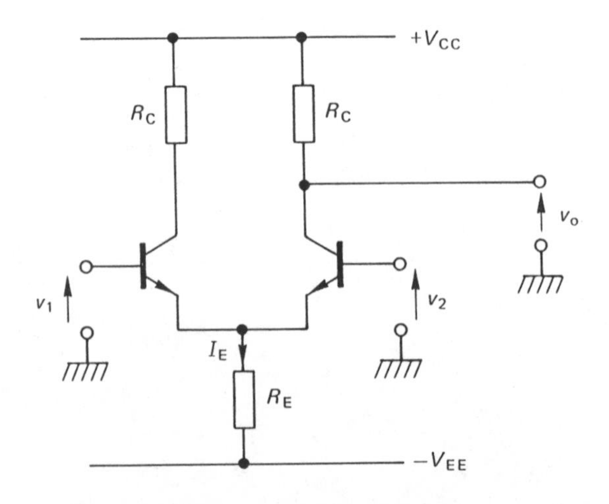

Figure 5.1 Differential amplifier

A practical amplifier cannot fully achieve this, and the output will depend on both the input difference voltage,

$$v_d = v_1 - v_2,$$

and the mean signal-level (or common-mode input voltage),

$$v_c = 0.5\,(v_1 + v_2).$$

The relationship between these voltages is shown in Fig. 5.2.

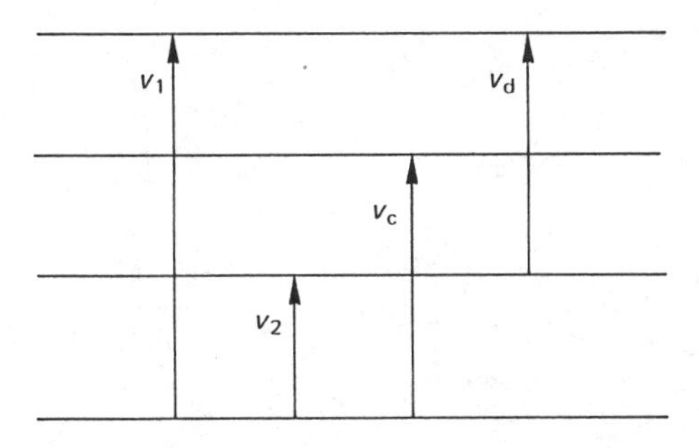

Figure 5.2 Relationship between differential-amplifier voltages

A practical difference amplifier thus has both a difference gain A_d and a common-mode gain A_c, and so

$$v_o = A_d v_d + A_c v_c.$$

A figure of merit used to indicate the effectiveness of a difference amplifier is the common-mode rejection ratio,

$$\text{CMRR} = \left|\frac{A_d}{A_c}\right|,$$

which should be as large as possible. It may be shown that

$$\text{difference gain } A_d = 0.5 g_m R_C = 20\, I_E R_C.$$

If the output is taken from one collector only then the difference gain is given by $10 I_E R_C$.

$$\text{Common-mode gain } A_c = -\frac{R_C}{2R_E},$$

$$\text{Common-mode rejection ratio} = \left|\frac{A_d}{A_c}\right| = 20 I_E R_E.$$

This expression demonstrates the importance of a high impedance, R_E, in the emitter circuit. One method of achieving this is to use an emitter resistance of several megohms. This however requires a very large negative supply voltage to give a reasonable emitter current. A better method is to use a constant current sink.

It is difficult to design differential amplifiers using discrete transistors because of drift with temperature (h_{fe}, V_{BE} and I_{CO} all vary with temperature). A shift in any of these quantities causes changes of the output voltage and cannot be distinguished from a change in the input signal voltage. This problem is almost overcome by having the two transistors on the same piece of silicon and under this condition any parameter changes due to temperature variation tend to cancel out.

5.3 The Cascode Circuit

The cascode circuit of Fig. 5.3 may be thought of as a common-emitter stage, Q_1, feeding a common-base stage, Q_2. The voltage gain of the CE stage is approximately unity since its collector load is small, being the input resistance of the CB stage. Very little Miller-effect multiplication therefore takes place. The CB stage has voltage gain, but no phase inversion and therefore no Miller-effect multiplication. The arrangement provides a significant improvement in amplifier bandwidth.

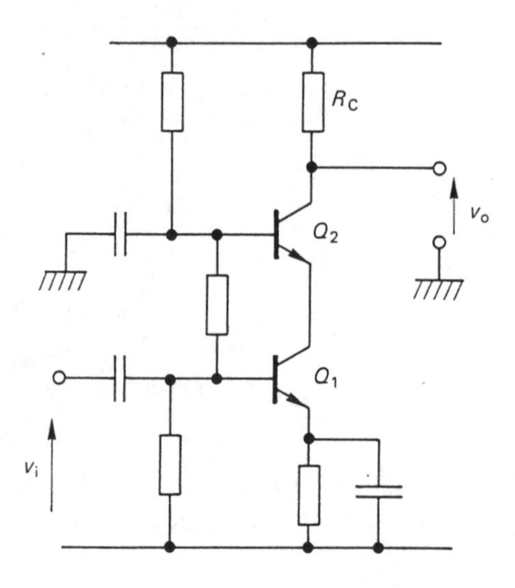

Figure 5.3 Cascode circuit

5.4 The Current Mirror

The circuit of Fig. 5.4 shows the transistor used as a constant current source (or sink). It is biased such that it operates beyond the 'knee' of the output characteristic of Fig. 2.7, and therefore variations of collector voltage have little effect on the collector current. It thus acts like a high-impedance current source. This makes it ideal as a replacement for the emitter resistor in the differential amplifier of Fig. 5.1, since it provides a means of defining the current I_0 while maintaining a high resistance, thus giving a high CMRR.

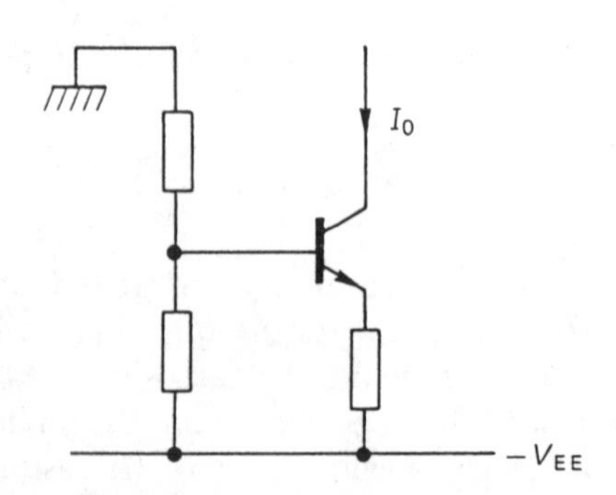

Figure 5.4 Constant-current source

An improvement on the constant current source is the current mirror which uses two transistors, as shown in Fig. 5.5. Usually the two transistors are on the same piece of silicon, so that they are well matched. The short circuit between collector and base of Q_1 biases Q_1 on as well as Q_2. Since the devices are matched, the current I_0 in Q_2 is almost exactly equal to I in Q_1. The user thus 'programs' the current I by the choice of the resistor R, and the mirror 'sources' the same current ($I = I_0 \approx V_{EE}/R$). It provides a means of accurately defining a high-impedance current source. It thus forms an ideal alternative to the emitter resistor in the differential amplifier of Fig. 5.1.

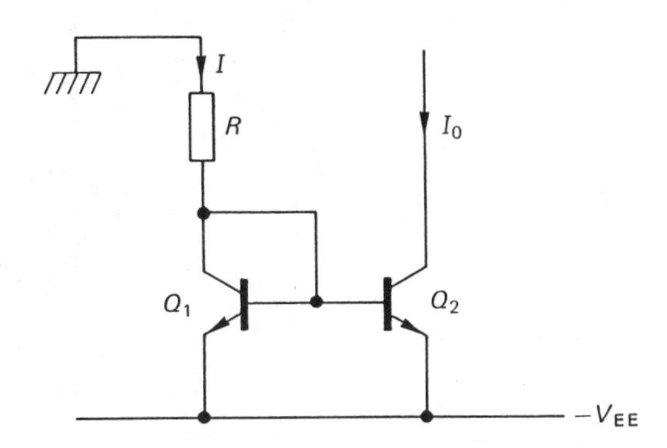

Figure 5.5 Current mirror

5.5 The Darlington Pair

When two transistors are connected as shown in Fig. 5.6, they are referred to as a Darlington pair. The current gain of the resultant circuit is the product of the individual current gains. The combination greatly increases the voltage gain of the common-emitter stage, or alternatively provides a very high input resistance when used in an emitter follower stage. In an emitter follower, the resultant input resistance is given approximately by $h_{fe}^2 R_E$. Similarly, the output resistance of the emitter follower is reduced to

$$\frac{R_S}{h_{fe}^2} + \frac{2h_{ie}}{h_{fe}}.$$

Darlington integrated-circuit transistor pairs are available with a resultant h_{fe} as high as 30 000.

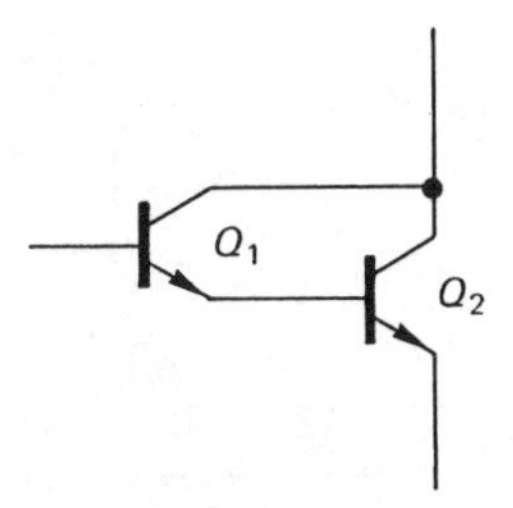

Figure 5.6 Darlington pair

5.6 The Phase Splitter and Complementary Pair

The phase splitter circuit is shown in Fig. 5.7. It provides two outputs, one in phase with the input signal and the other out of phase. It is a useful circuit from which to drive complementary transistor pairs as discussed in Chapter 10 on power amplifiers.

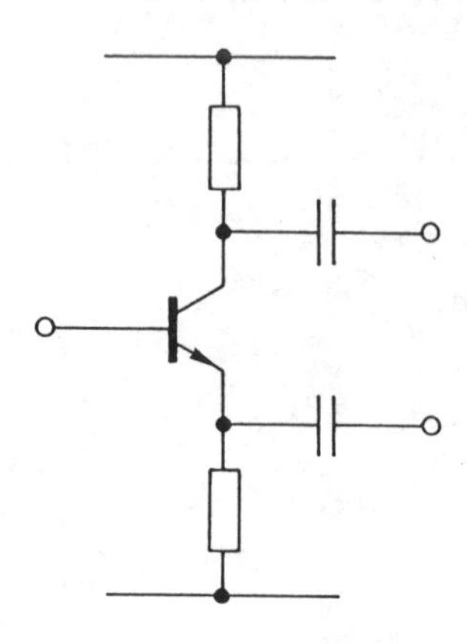

Figure 5.7 Phase splitter

5.7 Cascading Transistor Amplifiers

When the amplification of a single transistor stage is not sufficient for a particular application, then two or more stages may be connected together in cascade. Alternatively, it may be necessary to add other stages to alter the input or output impedance of the circuit.

The circuit of Fig. 5.8 shows a common-emitter with a common-collector output stage. This provides an amplifier with a good voltage gain and a low output impedance. The CC stage (emitter follower) does not require any additional bias circuit in this case, since suitable bias is provided by the collector resistor of the previous stage.

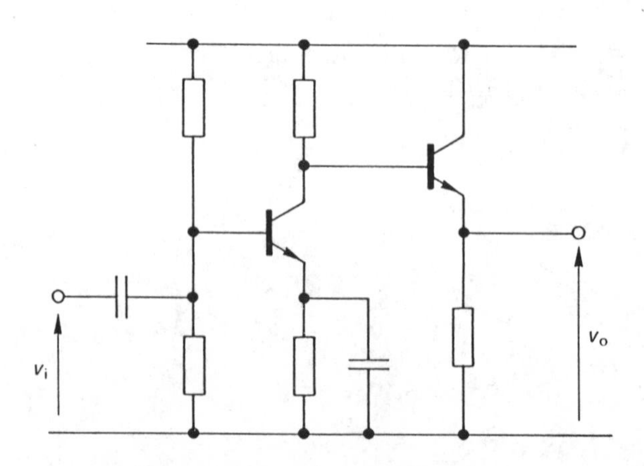

Figure 5.8 Cascaded amplifier, CE-CC

5.8 Worked Examples

Example 5.1

'Transistors perform best when used in pairs.' Discuss this statement with regard to the use of double-transistor circuits used in reducing the Miller-effect limitation on transistor frequency response.

Describe why such circuits would be useful in r.f. tuned amplifiers.

Solution 5.1

Transistors used singly are limited particularly in respect of their gain–bandwidth product. The common-emitter amplifier has a limited bandwidth due to Miller feedback via the collector–base capacitance.

Two methods of reducing the Miller feedback effect are the emitter-coupled amplifier and the cascode circuit, both of which use two transistors.

In the emitter-coupled amplifier of diagram (a), no Miller effect multiplication takes place since the collector of Q_1 and base of Q_2 are both grounded (as far as a.c. signals are concerned). The circuit may be thought of as a common-collector stage driving a common-base stage, giving good gain with no phase inversion, and a reasonably high input resistance.

In the cascode circuit of diagram (b), Q_1 may be thought of as a common-emitter stage feeding Q_2 connected as a common-base stage. The voltage gain of the *CE* stage is approximately unity, since its collector load is the low input resistance of a common-base stage. Very little Miller-effect multiplication therefore takes place. The CB stage using Q_2 has voltage gain but no phase inversion and therefore no Miller-effect multiplication.

Both of these circuits give good voltage gain, usable over a significantly wider frequency range. For this reason they are found useful in radio-frequency tuned amplifiers.

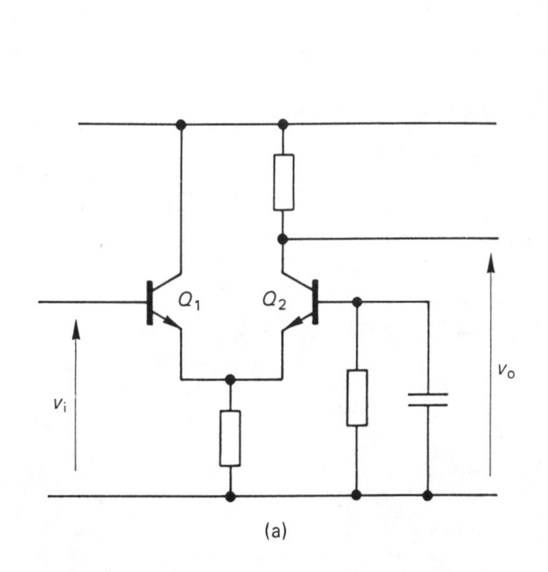

(a)

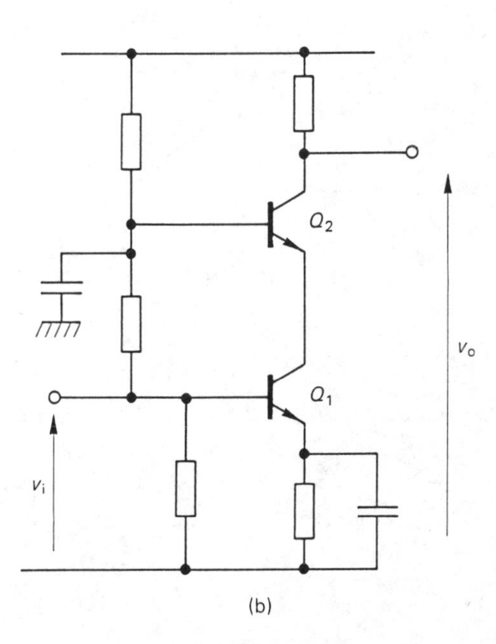

(b)

Example 5.2

For the two-stage amplifier of the diagram using a CE-CC configuration, calculate the input and output impedance, and the individual and overall voltage gains. Assume $h_{fe} = 50$.

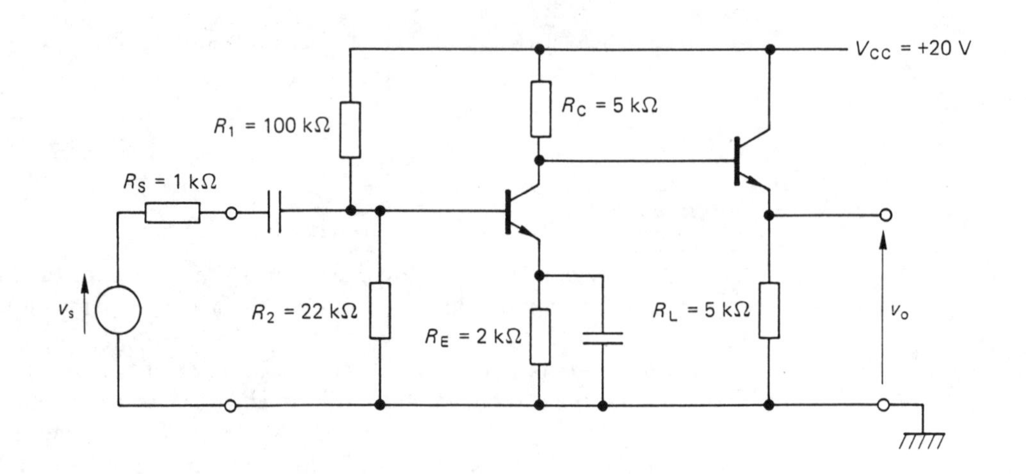

Solution 5.2

Consider first the static conditions:

$$V_{B1} = \frac{22}{122} \times 20 \text{ V} = 3.6 \text{ V},$$

$$\therefore\ V_{E1} \approx 3 \text{ V}, \qquad I_E \approx 1.5 \text{ mA}$$

$$g_{m1} = 40 \times 1.5 = 60 \text{ mS}$$

Similarly, $\qquad g_{m2} \approx 96 \text{ mS}.$

The second stage loads the first stage.

$$\therefore\ R_{i2} = \frac{h_{fe}}{g_{m2}} + h_{fe} R_L$$

$$= \frac{50}{96} + 50 \times 5 \text{ k}\Omega \approx 250 \text{ k}\Omega.$$

$$A_{v2} = \frac{g_{m2} R_L}{1 + g_{m2} R_L} = \frac{96 \times 5}{1 + 96 \times 5} \approx \underline{0.98}.$$

Consider the first stage:

$$R_{i1} = \frac{h_{fe}}{g_{m1}} = \frac{50}{60 \times 10^{-3}} \approx 830\ \Omega,$$

$$A_{v1} = g_{m1} R, \qquad \text{where } R = R_C \parallel R_{i2}$$

$$= 5 \text{ k}\Omega \parallel 250 \text{ k}\Omega.$$

$$= 60 \times 4.9 \approx \underline{294}$$

$$\therefore \text{ Overall gain } A_v = A_{v1} A_{v2}$$

$$= \underline{288}.$$

$$R_{o2} = \frac{R_{o1}}{h_{fe}} + \frac{1}{g_{m2}} = \frac{5 \times 10^3}{50} + \frac{1}{60 \times 10^{-3}}$$

$$\approx 117\ \Omega,$$

$$R_{o2}' = R_{o2} \parallel R_L \approx \underline{114\ \Omega}.$$

$$R_{in} = R_{i1} \parallel R_1 \parallel R_2 \approx \underline{790\ \Omega}.$$

Example 5.3

An emitter-coupled amplifier has collector load resistors $R_C = 10\ k\Omega$ and an emitter resistor $R_E = 10\ k\Omega$.

Assuming that the transistor operation is defined by the equation

$$I_C = I_S \exp(40V_{BE}),$$

analyse the circuit to obtain the difference gain, the common-mode gain and the CMRR.

Assume $h_{fe} \gg 1$ and that the supply voltages available are ±10 V.

Draw a circuit to replace R_E so that the CMRR is increased while maintaining the same d.c. transistor currents.

Solution 5.3

The circuit is the emitter-coupled amplifier of Fig. 5.1, where $g_m = 40I_C$, $V_{CC} = +10$ V, and $V_{EE} = -10$ V.

With $\quad R_E = 10\ k\Omega, \qquad I_E \approx \dfrac{10\ V}{10\ k\Omega} = 1$ mA,

$$\therefore\ I_C \approx 0.5 \text{ mA},$$

$$\therefore\ g_m \approx 20 \text{ mS}.$$

The difference gain is found as follows:

$$v_d = v_1 - v_2,$$

where v_1 and v_2 are the voltages at the base of Q_1 and Q_2 respectively. Let $v_1 = \frac{1}{2} v_d$ and $v_2 = -\frac{1}{2} v_d$. Making use of the circuit symmetry, we can bisect the circuit as shown in diagram (a). (R_E is not involved, since $i_{C1} = -i_{C2}$ and there is therefore no voltage change across R_E. The emitter is thus effectively grounded for small signals.) Using the equivalent circuit of diagram (b),

$$v_o = \frac{R_C\, g_m\, v_d}{2}.$$

$$A_d = \frac{v_o}{v_d} = \frac{g_m R_C}{2}$$

$$= \frac{20 \times 10^{-3} \times 10 \times 10^3}{2} = \underline{100}.$$

The common-mode gain is found as follows:

$$v_c = v_1 = v_2.$$

We can bisect the circuit as shown in diagram (c), since there is *twice* the change of current in R_E. The equivalent circuit is shown in diagram (d).

Analysing the currents at the node,

$$\frac{v_c - v_e}{r_i} + g_m v_{be} - \frac{v_e}{2R_E} = 0.$$

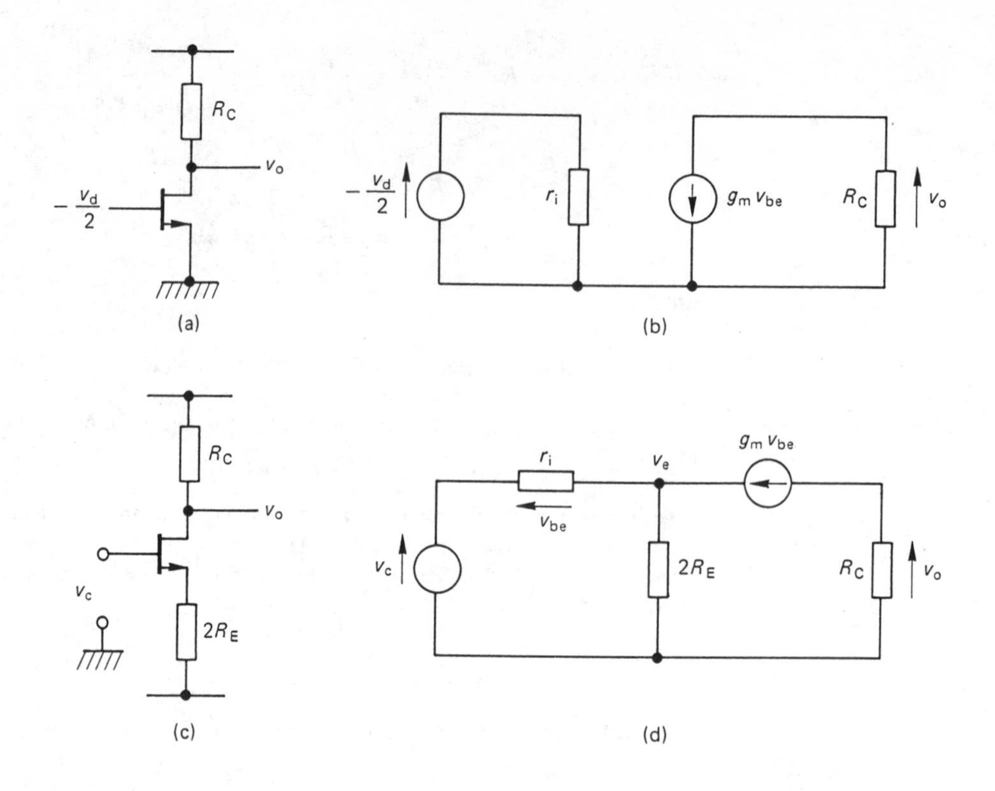

Also,

$$v_c = v_e + v_{be}.$$

$$\therefore\ v_{be} = \frac{v_c}{2R_E\left(\dfrac{1}{r_i} + g_m + \dfrac{1}{2R_E}\right)}$$

$$= \frac{v_c}{1 + 2g_m R_E\left(1 + \dfrac{1}{g_m r_i}\right)}$$

$$\approx \frac{v_c}{1 + 2g_m R_E} \qquad \text{(since } g_m r_i = h_{fe} \gg 1\text{)}.$$

$$\therefore\ v_o = -g_m v_{be} R_C$$

$$= -\frac{g_m R_C}{1 + 2g_m R_E}\, v_c.$$

$$\therefore\ A_c = \frac{v_o}{v_c} = -\frac{g_m R_C}{1 + 2g_m R_E}.$$

Since

$$g_m = \frac{1}{r_e},$$

where r_e is the intrinsic emitter resistance,

$$A_c = -\frac{R_C}{r_e + 2R_E} \approx -\frac{R_C}{2R_E}$$

$$= -\frac{10 \times 10^3}{2 \times 10 \times 10^3} = \underline{-0.5}.$$

$$\text{CMRR} = \left|\frac{A_d}{A_c}\right| = \frac{100}{0.5} = \underline{200}.$$

A circuit that could replace R_E and give an effective resistance of a much higher value is shown in diagram (e). It uses a current mirror in which the 1 mA current in the 10 kΩ resistor 'programs' a 1 mA constant current in the 'tail' of the amplifier.

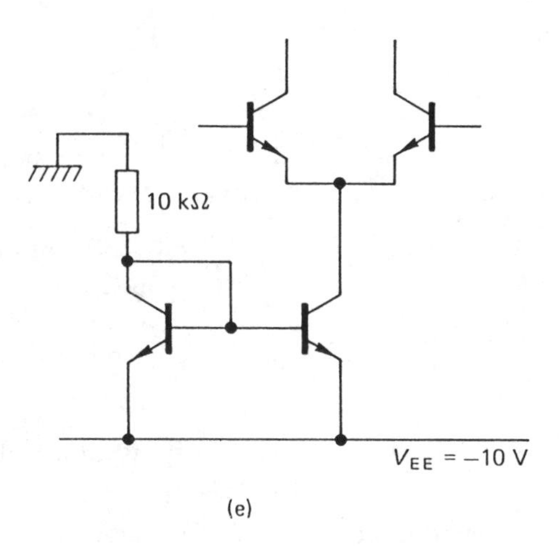

(e)

Example 5.4

The combination of a common-emitter stage and a common-base stage shown in diagram (a) is known as a cascode circuit. By assuming typical parameter values, and making suitable approximations, show that the circuit can be represented by the following overall parameters, and hence determine typical values for these parameters:

$$h_{11} \approx h_{ie}, \qquad h_{12} \approx h_{re}h_{rb}, \qquad h_{21} \approx h_{fe}, \qquad h_{22} \approx h_{ob}.$$

Describe why this circuit is useful as the active device in an r.f. tuned amplifier. (CEI Part 2)

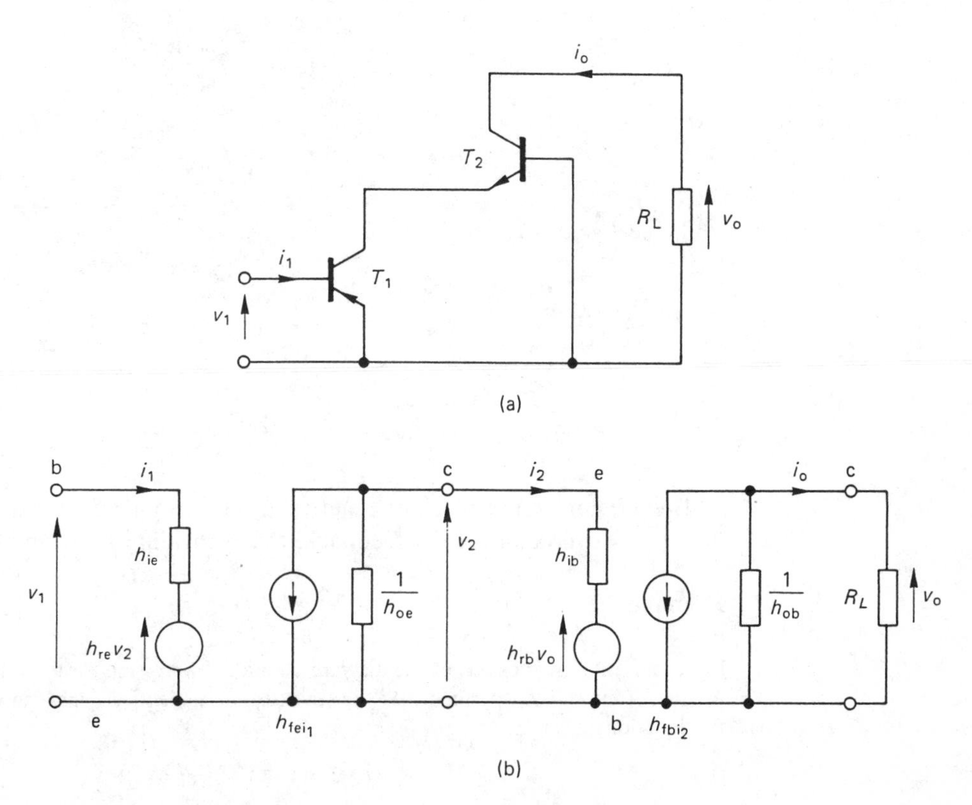

(b)

Solution 5.4

The parameters are those defined for a two-port device, where

$$v_1 = h_{11} i_1 + h_{12} v_o$$

and

$$i_o = h_{21} i_1 + h_{22} v_o,$$

where

$$h_{11} = \left.\frac{v_1}{i_1}\right|_{v_o=0} = \text{input resistance with output short-circuit,}$$

$$h_{12} = \left.\frac{v_1}{v_o}\right|_{i_1=0} = \text{fraction of output voltage at input with input open-circuit,}$$

$$h_{21} = \left.\frac{i_o}{i_1}\right|_{v_o=0} = \text{current gain with output short-circuit, and}$$

$$h_{22} = \left.\frac{i_o}{v_o}\right|_{i_1=0} = \text{output conductance with input open-circuit.}$$

The equivalent circuit of the cascode connection may be represented as shown in diagram (b), where the h-parameters are with reference to CE for T_1 and CB for T_2.

By analysis of the circuit,

$$h_{11} = \left.\frac{v_1}{i_1}\right|_{v_2=0} \approx h_{ie}.$$

Also, with $i_1 = 0$, $\quad v_2 = h_{rb} v_o \quad$ and $\quad v_1 = h_{re} v_2$

$$\therefore \; h_{12} = \left.\frac{v_1}{v_o}\right|_{i_1=0} \approx h_{re} h_{rb},$$

$$h_{21} = \left.\frac{i_o}{i_1}\right|_{v_o=0} \approx h_{fe} h_{fb} \approx h_{fe},$$

(since $h_{fb} \approx 1$)

$$h_{22} = \left.\frac{i_o}{v_o}\right|_{i_1=0} \approx h_{ob}.$$

Typical values are:

$$h_{11} \approx \underline{1000\ \Omega},$$

$$h_{12} \approx 2.5 \times 10^{-4} \times 3.0 \times 10^{-4} \approx \underline{7.5 \times 10^{-8}},$$

$$h_{21} \approx \underline{50},$$

$$\underline{h_{22} \approx 0.5\ \mu\text{A/V giving } \frac{1}{h_{22}} \approx 2\ \text{M}\Omega.}$$

The circuit is useful as the active device in an r.f. tuned amplifier since it effectively overcomes Miller feedback, thus providing good frequency response.

Example 5.5

In the current mirror circuit of the diagram obtain an expression for the value of R_2 required to produce a current I_0. Assume that the transistors are identical high-gain devices, specified by the equation

$$I_C = I_S \exp(40\ V_{BE}).$$

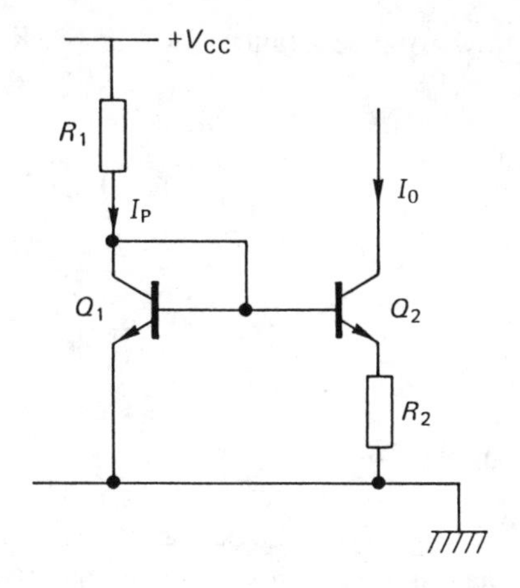

What are the major advantages of this circuit as compared with the basic current mirror?

Design the circuit to give $I_0 = 5\ \mu A$ with $V_{CC} = 30$ V, using the minimum total resistance.

Solution 5.5

In the conventional current mirror, with R_2 replaced by a short circuit, the load current I_0 is equal to the 'programming' current I_P.

Adding resistor R_2 allows an output current I_0 to be generated that is a fraction of the programming current I_P.

The transistors are then operating at different current densities and the currents are given by the Ebers–Moll equation.

$$\therefore \quad V_{BE1} = \frac{1}{40} \log_e \frac{I_P}{I_S}$$

and

$$V_{BE2} = \frac{1}{40} \log_e \frac{I_0}{I_S}$$

$$\therefore V_{BE1} - V_{BE2} = \frac{1}{40} \log_e \frac{I_P}{I_0}$$

Also

$$V_{BE1} - V_{BE2} \approx I_0 R_2$$

$$\therefore \quad I_0 R_2 = \frac{1}{40} \log_e \frac{I_P}{I_0}$$

$$\therefore \quad R_2 = \frac{1}{40 I_0} \log_e \frac{I_P}{I_0}$$

But

$$I_P = \frac{V_{CC} - V_{BE1}}{R_1} \approx \frac{V_{CC}}{R_1}$$

$$\therefore R_2 \approx \frac{1}{40 I_0} \log_e \frac{V_{CC}}{I_0 R_1}$$

With $V_{CC} = +30$ V and $I_0 = 5\ \mu A$,

$$R_2 = \frac{1}{40 \times 5 \times 10^{-6}} \log_e \frac{30}{5 \times 10^{-6} R_1},$$

$$\therefore R_1 = 6 \times 10^6 \exp(-0.2 \times 10^{-3} R_2).$$

Now total resistance $R_T = R_1 + R_2$. This is minimum when $\frac{dR_T}{dR_2} = 0$, leading to

$$R_2 = \frac{1}{0.2 \times 10^{-3}} \log_e (1.2 \times 10^3)$$

$$= \underline{35.5 \text{ k}\Omega}$$

and $$R_1 = \underline{5 \text{ k}\Omega}.$$

Example 5.6

For what purpose is the Darlington emitter follower connection used?

Draw an equivalent circuit and hence calculate the input resistance and the current and voltage gains of the circuit shown in diagram (a). Each transistor has

$h_{fe} = 70$, $h_{oe} = 20\ \mu S$, $h_{ie} = 1.5\ k\Omega$; h_{re} is negligible.

Reasonable approximations may be made in the calculation.

Give a circuit diagram showing a method of biasing the base of the input transistor without causing a large reduction of the input resistance of the amplifier. (CEI Part 2)

Solution 5.6

The purpose of the Darlington emitter follower is to provide a large input impedance. The circuit effectively consists of two cascaded emitter followers with the input impedance of the second transistor constituting the emitter load of the first.

The equivalent circuit is shown in diagram (b).

For the second stage, assuming that $h_{oe}R_E \leqslant 0.1$ and $h_{fe}R_E \gg h_{ie}$, then the current gain is given by

$$A_{i2} = 1 + h_{fe}$$

The input impedance R_{i2} is given by:

$$R_{i2} = (1 + h_{fe})\, R_E .$$

The effective load for transistor Q_1 is R_{i2}.

For the first transistor, we must take into account the effect of h_{oe}, since it is not likely that $h_{oe}R_{i2} \leqslant 0.1$.

Using the current divider rule,

$$i_{b2} = (1 + h_{fe})\, i_{b1} \frac{1/h_{oe}}{R_{i2} + 1/h_{oe}}$$

$$\therefore A_{i1} = \frac{1 + h_{fe}}{1 + h_{oe}R_{i2}}$$

$$= \frac{1 + h_{fe}}{1 + h_{oe}(1 + h_{fe})R_E}$$

$$\approx \frac{1 + h_{fe}}{1 + h_{oe}h_{fe}R_E},$$

since $h_{oe}R_E \leqslant 0.1$.

The overall current gain is given by

$$A_i = A_{i1}A_{i2} \approx \frac{(1 + h_{fe})^2}{1 + h_{oe}h_{fe}R_E}.$$

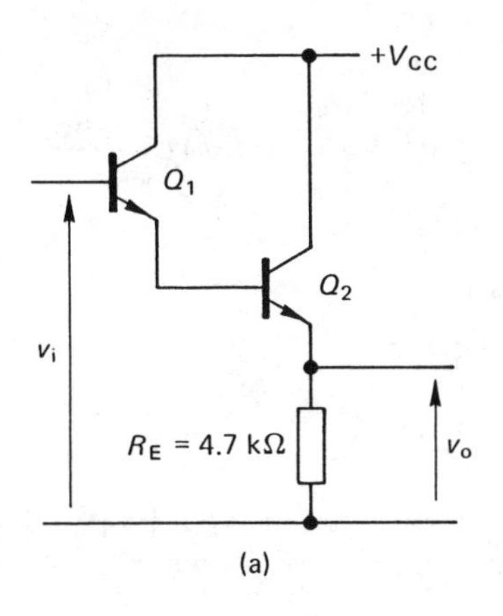

(a)

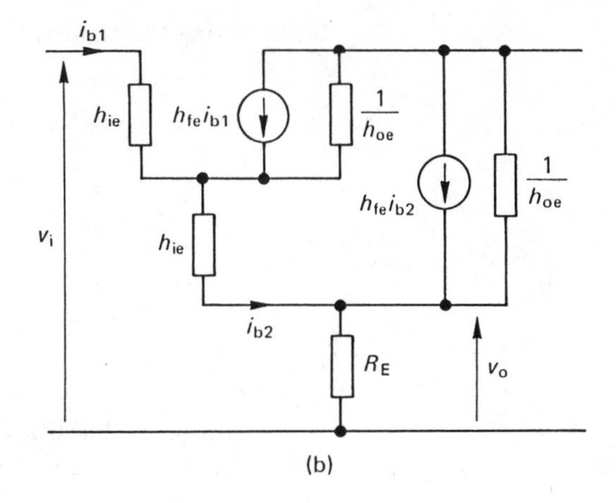

(b)

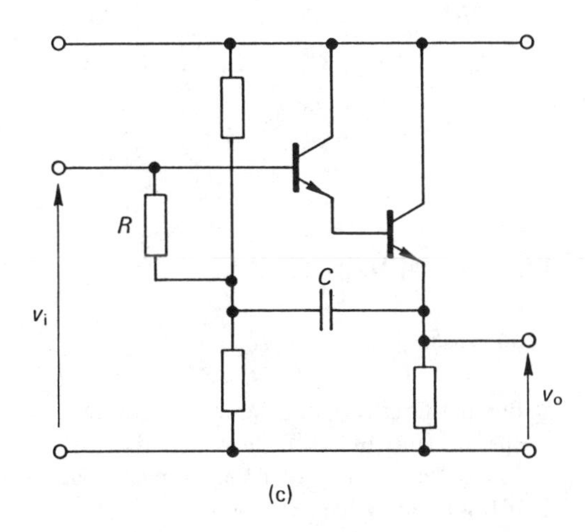

(c)

The circuit input resistance is given by

$$R_{i1} = h_{ie} + A_{i1} R_{i2}$$

$$\approx \frac{(1 + h_{fe})^2 R_E}{1 + h_{oe} h_{fe} R_E} .$$

To calculate the voltage gain we have

$$A_{v2} = 1 - \frac{h_{ie}}{R_{i2}} ,$$

$$A_{v1} = 1 - \frac{h_{ie}}{R_{i1}} \approx 1 - \frac{h_{ie}}{A_{i1} R_{i2}} .$$

$$\therefore A_v = A_{v1} A_{v2}$$

$$\approx 1 - \frac{h_{ie}}{R_{i2}} .$$

Substituting the figures given,

$$A_i = \frac{(1 + 70)^2}{1 + 20 \times 10^{-6} \times 70 \times 4.7 \times 10^3} = \underline{665,}$$

$$A_{i1} = 9.4,$$

$$R_{i2} = 329 \text{ k}\Omega,$$

$$R_{i1} = A_{i1} R_{i2} = \underline{3.1 \text{ M}\Omega},$$

$$A_v = 1 - \frac{1.5}{329} = \underline{0.995}.$$

A method of biasing the base, without causing a large reduction in the input resistance, is to use a bootstrap arrangement as shown in diagram (c), in which

$$R_{eff} = \frac{R}{1 - A_v}.$$

5.9 Unworked Problems

Problem 5.1

Define the term common-mode rejection ratio (CMRR) and explain its significance in assessing the performance of a difference amplifier.

The difference amplifier shown in the diagram uses two identical transistors. Show that the CMRR is given by the expression

$$0.5\ [1 + 2R_E\ (1 + h_f)/h_i]$$

if h_o and h_r are neglected.

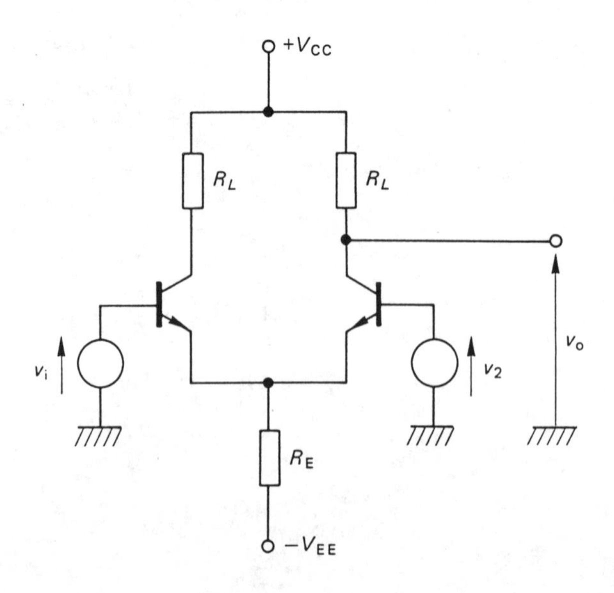

Hence determine the value of R_E necessary to attain a CMRR of 100 dB for $h_f = 100$ and $h_i = 1.5$ kΩ. How could this value of R_E be realised in a practical circuit? (CEI Part 2)

Problem 5.2

Determine the small-signal difference and common-mode gains of the source-coupled amplifier shown. Assume the transistors to be identical, their operation being defined by:

$$I_{DS} = I_{DSS}\,(1 - V_{GS}/V_P)^2,$$

with $V_P = -2$ V and $I_{DSS} = 2$ mA.

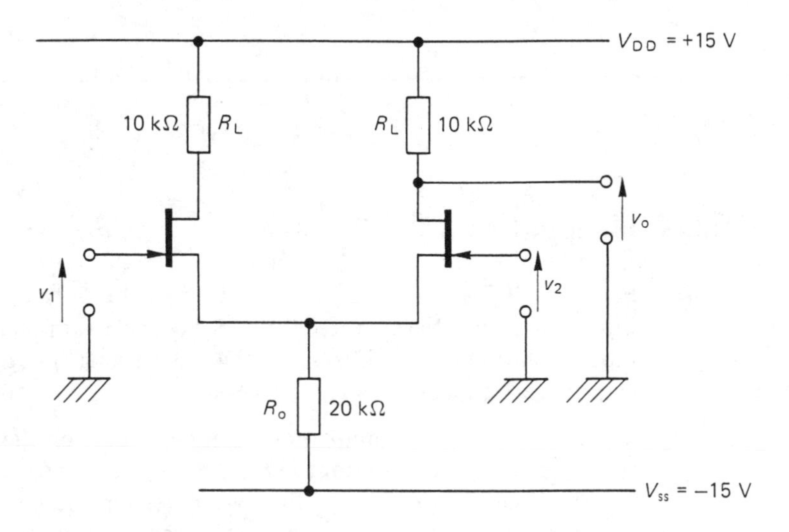

Problem 5.3

The diagram shows a two-stage amplifier circuit using a CE–CC configuration. The transistor parameters at the quiescent point are

$$h_{ie} = 2\ \text{k}\Omega, \qquad h_{fe} = 50;$$

other h paramters may be neglected for both devices. Find the input and output impedances and individual as well as overall voltage and current gains.

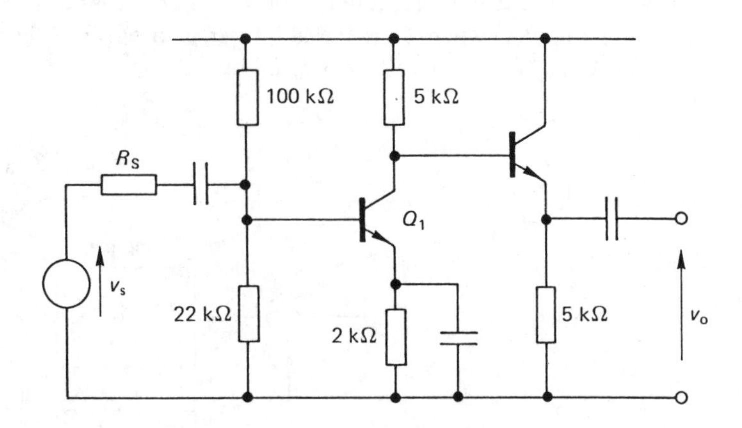

6 Feedback Amplifiers and Stability

6.1 Effects of Negative Feedback

Feedback is the combining of a portion of the output signal of an amplifier with the input signal. Negative feedback is the process of coupling the output with the input such that it cancels some of the input signal. The advantages to be gained by using negative feedback are as follows:

(a) An increase in the input resistance, depending on the feedback type.
(b) A reduction in the output resistance, depending on the feedback type.
(c) Stabilisation against amplifier parameter variations.
(d) Increased bandwidth (i.e. improved frequency response).
(e) Improved linearity and reduced distortion.
(f) Reduction in noise.

These advantages are obtained at the expense of a reduction in overall amplifier gain.

Amplifiers are often designed with a gain considerably higher than that required. Negative feedback is then applied to reduce the gain to the chosen value with the subsequent improvement in the performance outlined above. With negative feedback, the overall amplifier characteristics become determined more by the feedback components and less by the amplifier itself.

The schematic representation of an amplifier with negative feedback is shown in Fig. 6.1, where A is the amplifier gain without feedback and β is the feedback ratio. The output voltage is given by

$$v_o = A\, v_i$$

where $$v_i = v_s - v_f = v_s - \beta\, v_o .$$

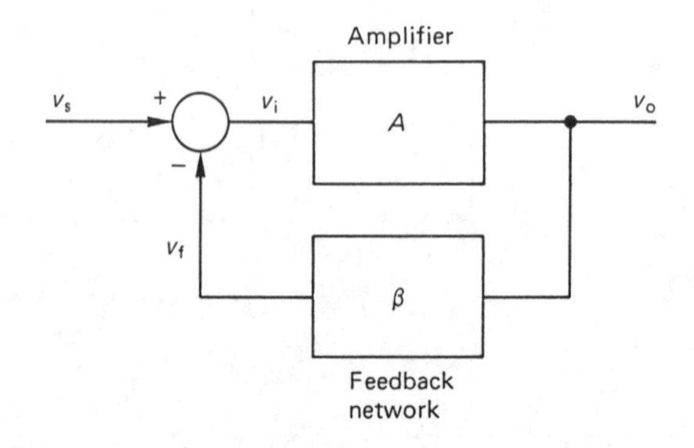

Figure 6.1 Schematic feedback amplifier

The gain A_F with feedback is then given by

$$A_F = \frac{v_o}{v_s} = \frac{A}{1 + A\beta},$$

where $A\beta$ is defined as the loop gain.

6.2 Types of Feedback

From a consideration of the input and output variables of an amplifier, there are four types:

(a) A voltage amplifier where the output voltage V_o is proportional to the input voltage V_s

$$V_o = A_v V_s$$

The requirement is generally for a high input resistance and a low output resistance.

(b) A current amplifier where the output current I_L is proportional to the input current I_s:

$$I_L = A_i I_s.$$

The requirement is generally for a low input resistance and a high output resistance.

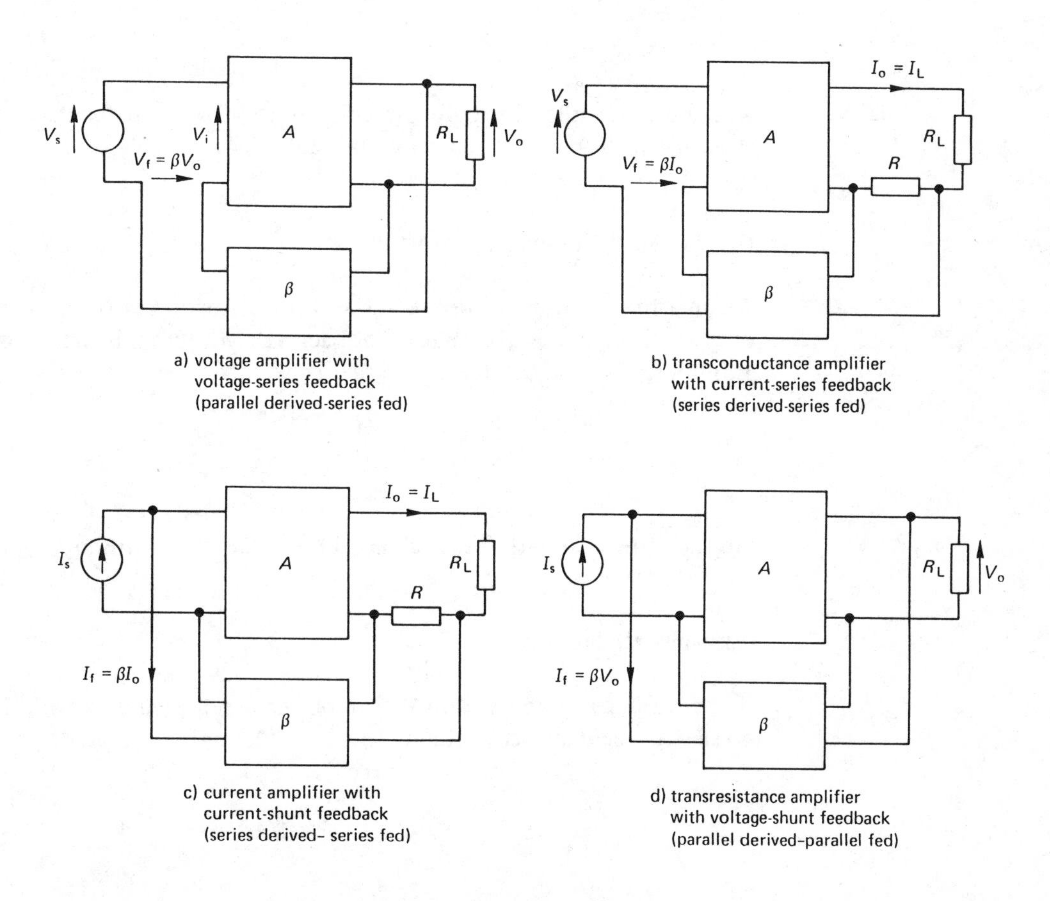

Figure 6.2 Feedback amplifier topologies

(c) A transconductance amplifier where the output current I_L is proportional to the input voltage V_s:

$$I_L = G_m V_s$$

The requirement is generally for a high input resistance and a high output resistance.

(d) A transresistance amplifier where the output voltage V_o is proportional to the input current I_s:

$$V_o = R_m I_s.$$

The requirement is generally for a low input resistance and a low output resistance.

In a feedback amplifier the feedback signal may be derived from either the output voltage or the output current. Similarly, the input may be connected either in series or in shunt with the input signal. There are thus four feedback amplifier topologies, as shown in Fig. 6.2.

6.3 Characteristics of Negative-feedback Amplifiers

The characteristics of negative-feedback amplifiers are as follows:

(a) Reduced Sensitivity to Amplifier Parameter Variation

$$\frac{dA_F}{A_F} = \frac{1}{1 + A\beta} \frac{dA}{A}$$

where dA/A is the sensitivity to gain variation without feedback and dA_F/A_F is the sensitivity to gain variation with feedback.

(b) Reduced Non-linear Distortion

Assume that without feedback the output distortion is D and with feedback the output distortion is D_F. With feedback, the output distortion is D due to the amplifier, and $-A\beta D_F$ due to the feedback.

$$\therefore D_F = D - A\beta D_F$$

$$= \frac{D}{1 + A\beta}.$$

i.e. the distortion with feedback is $1/(1 + A\beta)$ of that without feedback.

(c) Noise Reduction

If the noise generated within the amplifier is represented as N in Fig. 6.3 and the noise with feedback as N_F then

$$N_F = N - A\beta N_F,$$

$$= \frac{N}{1 + A\beta};$$

i.e. the noise with feedback is $1/(1 + A\beta)$ of that without feedback.

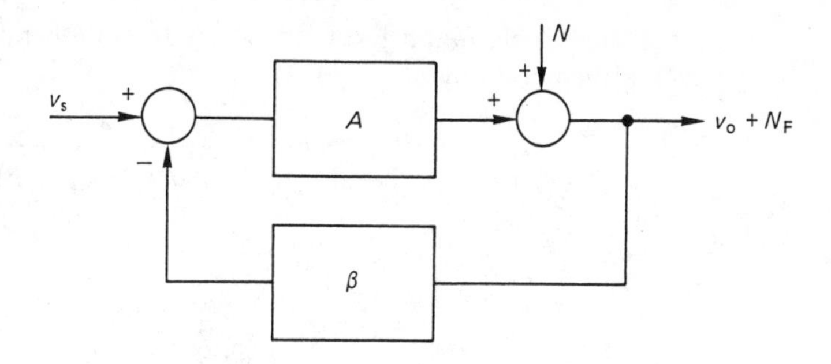

Figure 6.3 Noise reduction due to feedback

(d) Effects of Negative Feedback on Input and Output Resistance

This depends on the topology of the amplifier. As regards the input resistance, if the feedback is series-fed, then R_i is increased:

$$R_{iF} = R_i\,(1 + A\beta).$$

If the feedback is shunt-fed then R_i is decreased:

$$R_{iF} = \frac{R_i}{1 + A\beta}.$$

As regards the output resistance, if the feedback is voltage-derived then R_o tends to be decreased. In general,

$$R_{oF} = \frac{R_o}{1 + A\beta}.$$

If the feedback is current-derived then R_o tends to be increased. In general,

$$R_{oF} = R_o\,(1 + A\beta).$$

(e) Effect of Feedback on Amplifier Bandwidth

We have for an amplifier with feedback:

$$A_F = \frac{A}{1 + A\beta}.$$

To study the effect of feedback on bandwidth we need to be aware that A is frequency-dependent.

We have

$$A\,(\mathrm{j}f) = \frac{A_0}{1 + \mathrm{j}\left(\dfrac{f}{f_H}\right)}$$

for a single pole amplifier where A_0 is the mid-band gain. The gain width feedback is then given by

$$\therefore A_F(jf) = \frac{A_0}{1 + A_0\beta + j\left(\frac{f}{f_H}\right)}$$

$$= \frac{\frac{A_0}{1 + A_0\beta}}{1 + j\,\frac{f}{f_H\,(1 + A_0\beta)}}$$

$$= \frac{A_{0F}}{1 + j\left(\frac{f}{f_{HF}}\right)}$$

where A_{0F} is the mid-band gain with feedback and f_{HF} is the 3 dB frequency with feedback.

Thus the mid-band gain is given by

$$A_{0F} = \frac{A_0}{1 + A_0\beta}$$

and the 3 dB point is at

$$f_{HF} = f_H(1 + A_0\beta).$$

We may say that

$$f_H - f_L \approx f_H.$$

Thus the gain-bandwidth product with feedback is given by $A_0 f_H$ (i.e. the same as without feedback).

An idealised Bode plot of dB gain against log frequency is shown in Fig. 6.4, on which the bandwidth improvement may be seen.

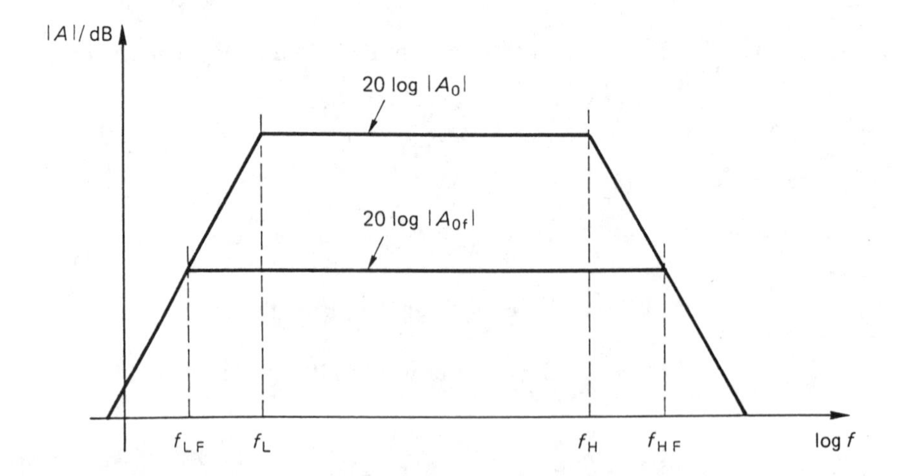

Figure 6.4 Bandwidth improvement using feedback

6.4 Cascaded Amplifiers with Feedback

When two amplifiers are cascaded together and overall negative feedback applied around the combination, then the overall frequency responses will be affected by the frequency response of both amplifiers. The cascaded amplifiers without feedback will typically have a double-pole transfer function of the form

$$A(j\omega) = \frac{A_0}{\left(1 + j\frac{\omega}{\omega_1}\right)\left(1 + j\frac{\omega}{\omega_2}\right)},$$

where A_0 is the low-frequency voltage gain and ω_1 and ω_2 are the individual cut-off frequencies.

If negative feedback is now applied to the combination

$$A_F(j\omega) = \frac{A}{1 + A\beta}$$

$$\therefore\ A_F(j\omega) = \frac{A_0}{\left(1 + j\frac{\omega}{\omega_1}\right)\left(1 + j\frac{\omega}{\omega_2}\right) + A_0\beta}.$$

Rearranging and letting

$$\omega_0^2 = \omega_1\omega_2(1 + A_0\beta) \qquad \text{and} \qquad Q = \frac{\omega_0}{\omega_1 + \omega_2},$$

where ω_0 = undamped resonant frequency, and
Q = quality factor at resonance

$$A_F(j\omega) = \frac{A_0}{1 + A_0\beta}\left(\frac{1}{1 + j\frac{\omega}{\omega_0}\frac{1}{Q} + \left(j\frac{\omega}{\omega_0}\right)^2}\right),$$

$$|A_F(j\omega)| = \frac{|A_F(0)|}{\sqrt{\left(1 - \frac{\omega^2}{\omega_0^2}\right)^2 + \frac{1}{Q^2}\left(\frac{\omega}{\omega_0}\right)^2}}.$$

The peak of this function may be found by taking the derivative of the term under the square root with respect to ω, giving a peak at

$$\omega = \omega_0\sqrt{1 - \frac{1}{2Q^2}},$$

where

$$\left|\frac{A_F(j\omega)}{A_F(0)}\right|_{\text{peak}} = \frac{Q}{\sqrt{1 - \frac{1}{4Q^2}}}.$$

Application of overall feedback over several amplifier stages connected in cascade gives much greater bandwidth improvement than separate feedback over each stage, provided that each configuration has been chosen to give the same mid-band gain.

6.5 Practical Analysis of Feedback Amplifiers

In analysing a multistage amplifier with feedback it is useful to separate the amplifier into the amplifier network without feedback and the feedback network. We may then apply the appropriate equations to calculate A_F, R_{iF} and R_{oF}.

To identify the amplifier input circuit:

(a) Short-circuit the output ($v_o = 0$) for voltage-derived feedback.
(b) Open-circuit the output ($i_o = 0$) for current-derived feedback.

To identify the amplifier output circuit:
(c) Short-circuit the input circuit ($v_i = 0$) for shunt-fed.
(d) Open-circuit the input ($i_i = 0$) for series-fed.

This ensures that the amplifier network without feedback retains the basic loading of the original amplifier.

The use of an unbypassed emitter resistor in the common-emitter voltage amplifier is an example of the use of negative feedback to reduce the voltage gain. This feedback may be described as current-derived, series-fed and has the effect of increasing R_{in} and increasing R_{out}.

The emitter follower and source follower are both examples of feedback that is voltage-derived and series-fed. It has the effect of increasing R_{in} and reducing R_{out}.

6.6 Amplifier Stability

Using the equation for the gain of an amplifier with negative feedback,

$$A_F = \frac{A}{1 + A\beta},$$

we can see that if the term $A\beta$ becomes equal to -1 then the amplifier will become unstable and break into oscillation. This will make it useless as an amplifier.

The term $A\beta$ is frequency-dependent and both the magnitude and phase vary with frequency. The system will be stable if the magnitude of the loop gain $|A\beta|$ is less than unity when the phase angle of $A\beta$ is 180°.

Two terms are used to describe how close an amplifier is to instability. They are easily understood by reference to the Bode plots of Fig. 6.5.

The gain margin is defined as the value of $|A\beta|$ in dB at the frequency at which the phase angle of $A\beta$ is 180°. For good stability the gain margin should be at least 10 dB.

The phase margin is 180° minus the phase angle of $A\beta$ at the frequency at which the value of $|A\beta|$ is unity (i.e. zero dB). For good stability the phase margin should be at least 50°.

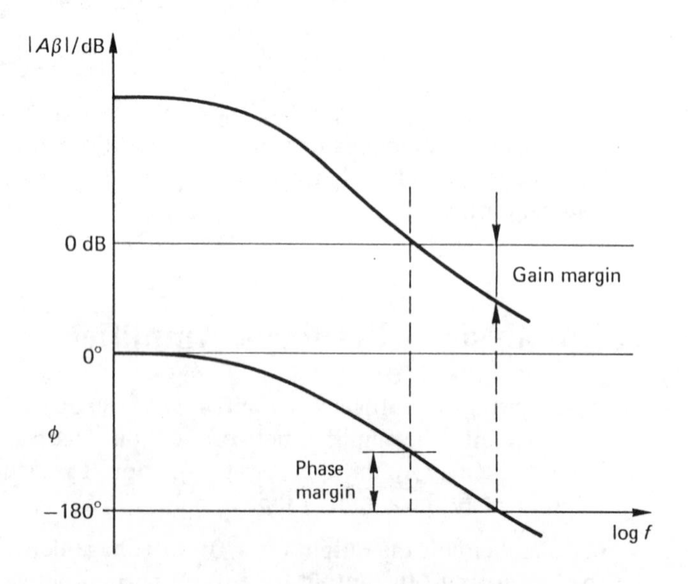

Figure 6.5 Gain and phase margins

A multistage amplifier will have multiple poles and hence a phase shift that may well exceed 180°. If excessive negative feedback is applied then the loop gain $A\beta$ could exceed unity at 180° phase shift, thus causing instability.

6.7 Worked Examples

Example 6.1

Two amplifiers, each with a gain of 20, are to be connected in cascade and negative feedback applied to the combination to return the overall gain to 20.

Calculate:

(a) the required feedback ratio;
(b) the percentage increase in overall gain with feedback if each amplifier suffers a 100 per cent increase in gain;
(c) the distortion with feedback if each amplifier has 10 per cent distortion.

Solution 6.1

(a)
$$A_F = \frac{A^2}{1 + \beta A^2} = \frac{20^2}{1 + \beta \times 20^2} = 20$$

$$\therefore \beta = \underline{0.0475}.$$

(b)
$$A_F = \frac{40^2}{1 + 0.0475 \times 40^2} = 20.78.$$

$$\therefore \text{Increase in overall gain with feedback} = \frac{0.78 \times 100}{20} \text{ per cent} = \underline{3.9 \text{ per cent}}.$$

(Notice how negative feedback reduces the sensitivity to gain variation.)

(c) For distortion in each amplifier of 10 per cent, the overall distortion is given by

$$D = (1 + D_1)(1 + D_2) - 1$$

$$= (1.1)^2 - 1 = 0.21 = 21 \text{ per cent}.$$

Distortion with feedback is given by

$$D_F = \frac{D}{1 + A\beta}$$

$$= \frac{21}{1 + 400 \times 0.0475} = \underline{1.05 \text{ per cent}}.$$

(Notice that negative feedback reduces distortion.)

Example 6.2

A voltage-series feedback amplifier is to have an output resistance of 100 Ω and a minimum voltage gain of 10. If the basic amplifier without feedback has an output resistance of 5 kΩ, determine the minimum gain required in the basic amplifier and the feedback factor to be used.

What will be the gain of the feedback amplifier if the basic amplifier has a gain double the minimum value calculated above?

Derive all formulae used.

Solution 6.2

To find the output resistance with negative feedback consider the equivalent circuit of the diagram, which shows the voltage series feedback as $v_f = \beta v_o$.

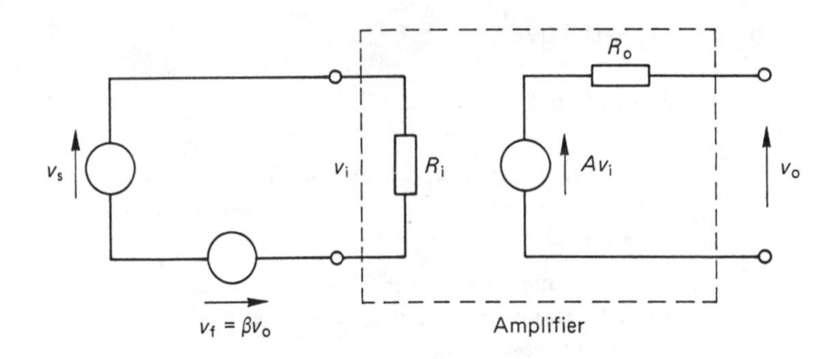

Now
$$R_{oF} = \frac{v_{o(oc)}}{i_{o(sc)}},$$

where
$$v_{o(oc)} = A\,(v_s - \beta v_{o(oc)})$$
$$= \frac{A\,v_s}{1 + A\beta}$$

and
$$i_{o(sc)} = \frac{Av_s}{R_o}.$$

$$\therefore\ R_{oF} = \frac{R_o}{1 + A\beta},$$

$$\therefore\ 100 = \frac{5 \times 10^3}{1 + A\beta},$$

$$\therefore\ A\beta = 49.$$

Now
$$A_F = \frac{A}{1 + A\beta} \quad \text{(see text for proof)}$$

$$\therefore A = A_F\,(1 + A\beta)$$
$$= 10\,(1 + 49) = \underline{500.}$$

$$\therefore\ \beta = \frac{49}{500} = \underline{0.098}.$$

If the gain of the basic amplifier is double, then

$$A'_F = \frac{1000}{1 + 0.098 \times 1000} = \underline{10.1}.$$

$\therefore$ Percentage increase with feedback = <u>1 per cent.</u>

Example 6.3

Produce a table to show the effects (i.e. increase or decrease) of the four feedback topologies (i.e. series/parallel voltage current) on the input and output impedances of an amplifier. For each case, state the type of gain stabilised by the feedback.

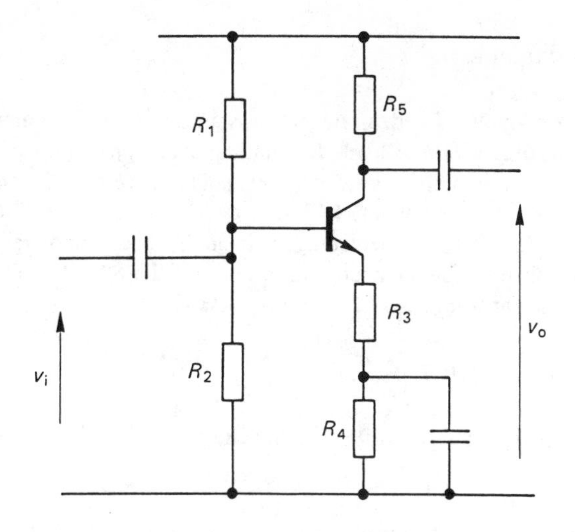

Identify the type of feedback, and the feedback factor, for the circuit in the diagram.
An operational amplifier has an open-loop gain defined by the equation

$$A(f) = \frac{10^4}{1 + j\dfrac{f}{10}}$$

and is connected in a non-inverting configuration with feedback factor $\beta = 0.01$. Deduce the transfer function for the feedback amplifier and hence determine the voltage gain at low frequencies, and the 3 dB frequency.

Solution 6.3

Input connection	Output derivation	Z_i	Z_o	A
series	voltage	↑	↓	V_o/V_i
series	current	↑	↑	I_o/V_i
parallel	voltage	↓	↓	V_o/I_i
parallel	current	↓	↑	I_o/I_i

The type of feedback in the diagram is current-derived, series-fed, with feedback factor $\beta = R_3$.

$$A_F = \frac{A}{1 + \beta A} = \frac{\dfrac{10^4}{1 + j\dfrac{f}{10}}}{1 + \dfrac{0.01 \times 10^4}{1 + j\dfrac{f}{10}}}$$

$$= \frac{10^4}{101 + j\dfrac{f}{10}}$$

$$= \frac{\dfrac{10^4}{101}}{1 + j\dfrac{f}{1010}} = \frac{A_0}{1 + j\dfrac{f}{f_H}}$$

where $A_0 = 99$ and $f_H = 1010$ Hz.
(Notice the increase in bandwidth due to negative feedback.)

Example 6.4

An amplifier with an output resistance of 600 Ω has an overall voltage gain of $10^4 \angle 180°$ when feeding a 600 Ω load. By simultaneous application of current and voltage feedback in series with the input, the overall voltage gain is to be reduced to $100 \angle 180°$, while the output resistance remains unaltered.

Calculate the percentage voltage feedback factor required and the value of current feedback resistor to be connected in series with the 600 Ω load. (You may assume that the value of the current feedback resistor is $\ll 600$ Ω.) (IEE)

Solution 6.4

In the circuit shown in the diagram,

$$v_f = \beta v_o + r i_o,$$

where β is the voltage feedback ratio and r is the current feedback resistor.

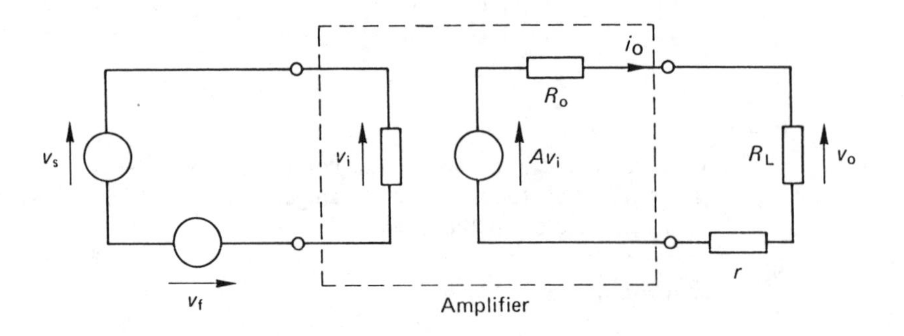

Also, $$v_i = v_s - v_f.$$

The output resistance with feedback is found from

$$R_{oF} = \frac{v_{o(oc)}}{i_{o(sc)}}.$$

Now $$v_{o(oc)} = Av_i = A\,(v_s - \beta\, v_{o(oc)})$$

since i_o is zero.

$$\therefore\ v_{o(oc)} = \frac{A\,v_s}{1 + A\beta}$$

$$i_{o(sc)} = \frac{A\,v_i}{R_o} \qquad \text{neglecting } r \text{ since } r \ll R_o$$

$$= \frac{A\,(v_s - r i_{o(sc)})}{R_o}$$

$$= \frac{Av_s}{R_o + Ar}$$

$$\therefore\ R_{oF} = \frac{R_o + Ar}{1 + A\beta}$$

Now the voltage gains with and without feedback are

$$A = 10^4 \qquad \text{and} \qquad A_F = 10^2.$$

Since v_o appears across R_L, where

$$R_L = R_o = 600\ \Omega,$$

then $\quad A' = 2 \times 10^4 \quad$ and $\quad A'_F = 2 \times 10^2.$

Now
$$A'_F = \frac{A'}{1 + A'\beta}$$

$$\therefore \beta = \frac{A' - A'_F}{A'A'_F} \approx \underline{0.005}.$$

Also,
$$R_{oF} = 600\ \Omega$$

$$\therefore 600 = \frac{600 + 2 \times 10^4 r}{1 + 2 \times 10^4 \times 0.005}$$

$$\therefore r = \underline{3\ \Omega}.$$

Example 6.5

Explain, using Nyquist or Bode diagrams, why a multistage amplifier can become unstable if excessive negative feedback is applied.

The open-loop voltage gain of an amplifier is represented by the expression

$$m = \frac{10^4}{\left(1 + \mathrm{j}\dfrac{f}{f_2}\right)^2 \left(1 + \mathrm{j}\dfrac{f}{f_1}\right)}$$

where $f_1 = 0.1$ MHz and $f_2 = 1$ MHz. Estimate, using a Bode diagram, the maximum amount of resistive negative voltage feedback that may be applied without causing instability. Suggest one method of compensation which would enable the closed-loop gain to be reduced to 10.

(CEI Part 2)

Solution 6.5

The means by which instability can occur is given in section 6.6. The Bode diagram for the expression given is shown in diagram (a).

To ensure stability the phase shift must be $< 180°$ lagging when the open-loop gain is unity.

From the diagram, at $\phi = -180°$, $|m| = 60$ dB $= 1000$.

When $|m\beta| = 1$

$$\beta = \frac{1}{1000} = \underline{0.001}.$$

Taking into account the fact that the Bode diagram (diagram (a)) is a straight-line approximation, at $f = 1$ MHz the breakpoint is in fact 6 dB below the point shown.

$$\therefore |m| = 54 \text{ dB} = 501 \text{ at } f = 1 \text{ MHz}$$

$$\therefore \beta_{max} \approx \underline{2 \times 10^{-3}}.$$

Also, since we normally require a phase margin of, say, 45°, then, at $\phi = -135°$, $|m| = 67$ dB $= 2239$,

$$\therefore \underline{\beta = 4.47 \times 10^{-4}}.$$

Now
$$A_F = \frac{A}{1 + A\beta} = 10.$$

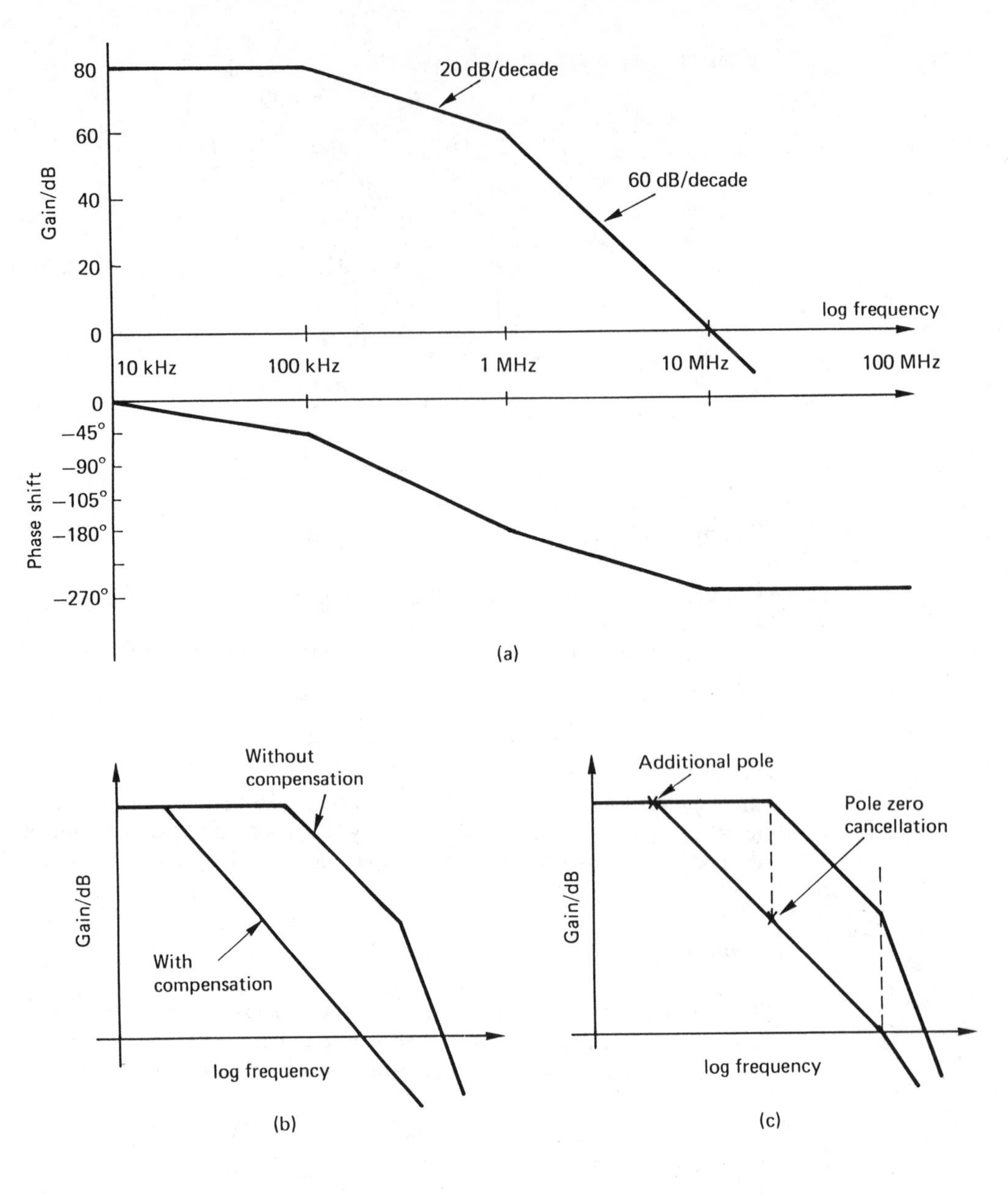

(a)

(b)

(c)

But $\quad A = 80 \text{ dB} = 10\,000$

$\therefore \ \beta = 0.1$ to give a closed-loop gain of 10.

This is acceptable provided some means is used to compensate for the instability that would occur. Possible methods of compensation are:

(i) Dominant pole or lag compensation, which consists of inserting an extra pole into the transfer function at a significantly lower frequency than the existing poles such as by connecting a capacitor between the output and ground. Suitable choice of this pole ensures that the open-loop gain–frequency curve crosses the 0 dB axis well before the phase shift approaches −180°. This is shown in diagram (b). The disadvantage of this method is severe loss of bandwidth.

(ii) Pole–zero or lag–lead compensation which adds both a pole and a zero to the transfer gain with the zero being chosen so as to effectively cancel the lowest pole, as shown in diagram (c). This method does not cause such severe bandwidth limitations.

Example 6.6

Comments on the effects of overall degenerative feedback on (a) frequency response, (b) distortion, (c) noise, (d) stability in a multistage broad band amplifier.

An amplifier has an open-loop mid-band voltage gain of 55 dB. The gain falls by 3 dB from this value at frequencies of 100 Hz and 20 kHz, each determined by a single time constant.

Determine the percentage of the output voltage which must be fed back in series opposition to the input to give a mid-band gain of 40 dB. What changes in open-loop gain are permissible if the closed-loop gain is not to change by more than ±0.5 dB?

Determine the frequencies at which the closed-loop gain is 1 dB below the mid-band gain.

(CEI Part 2)

Solution 6.6

The effects of negative feedback are described in the text.

The mid-band gain $A = 55$ dB $= 562$. The gain with feedback $A_F = 40$ dB $= 100$,

$$\therefore \beta = \frac{A - A_F}{A_F A} = \underline{8.2 \times 10^{-3}}.$$

$$A_{F(min)} = 39.5 \text{ dB} = 94.4, \qquad A_{F(max)} = 40.5 \text{ dB} = 105.9.$$

Now
$$A = \frac{A_F}{1 - A_F\beta},$$

$$\therefore A_{(min)} = 417.8 = 52 \text{ dB}, \qquad A_{(max)} = 804.6 = 58 \text{ dB}.$$

$$\therefore \underline{A = 55 \text{ dB} \pm 3 \text{ dB approx.}}$$

For the high-frequency break point f_H,

$$A(f) = \frac{A_0}{1 + j\dfrac{f}{f_H}} \quad \text{where } f_H = 20 \text{ kHz};$$

$$A_F(f) = \frac{A_{F0}}{1 + j\dfrac{f}{f_H(1 + A_0\beta)}}.$$

Now, when $A_F(f) = A_{F0} - 1$ dB,

$$A_F(f) = 39 \text{ dB} = 89.1.$$

$$\therefore 89.1 = \frac{100}{\sqrt{1 + \left(\dfrac{f}{f_H(1 + A_0\beta)}\right)^2}},$$

where $f_H = 20$ kHz, $A_0 = 562$ and $\beta = 8.2 \times 10^{-3}$,

leading to $f = \underline{57.1 \text{ kHz}}$ for 1 dB down.

For the low-frequency breakpoint f_L,

$$A(f) = \frac{A_0}{1 - j\dfrac{f_L}{f}}$$

$$A_F(f) = \frac{A_{F0}}{\sqrt{1 + \left(\dfrac{f_L}{f(1 + A_0\beta)}\right)^2}}$$

where $A_F(f) = 89.1$, $f_L = 100$ Hz, $A_0 = 562$ and $\beta = 8.2 \times 10^{-3}$,
leading to $f = \underline{83.5}$ Hz for 1 dB down.

Example 6.7

Comment briefly on the effect of degenerative feedback on the gain and phase shift characteristics of an amplifier.

The open-loop gain of an amplifier is represented by the relationship

$$m = \frac{1000}{\left(1 + \mathrm{j}\dfrac{f}{f_1}\right)^2},$$

where f is the frequency of the input signal.

Voltage feedback is applied in series with the input to reduce the d.c. gain to 100. Determine the resultant maximum gain and the frequency, in terms of f_1, at which it occurs.

Sketch the gain-frequency and the phase-frequency relationships for the amplifier with feedback.

Solution 6.7

See text for effects of feedback.

Writing $x = \dfrac{f}{f_1}$,

$$m_\mathrm{F} = \frac{\dfrac{1000}{(1 + \mathrm{j}x)^2}}{1 + \dfrac{1000\beta}{(1 + \mathrm{j}x)^2}} = \frac{1000}{(1 + \mathrm{j}x)^2 + 1000\beta}.$$

Also at $f = 0$,

$$m_\mathrm{F} = 100 = \frac{1000}{1 + 1000\beta}.$$

$$\therefore\ \beta = 0.009$$

$$\therefore\ 1 + 1000\beta = 10.$$

$$\therefore\ m_\mathrm{F} = \frac{1000}{\sqrt{[(10 - x^2)^2 + 4x^2]}}$$

$$= \frac{1000}{\sqrt{[100 - 16x^2 + x^4]}}.$$

m_F is maximum when the derivative of the square root term is minimum, which is when

$$4x^3 - 32x = 0$$

$$\therefore\ x = 0 \text{ or } 2.83;$$

$$\therefore\ f = \underline{2.83\, f_1}.$$

$$\therefore\ m_{\mathrm{F(max)}} = \frac{1000}{6} = \underline{166.7}.$$

To draw the gain and phase characteristics with frequency,

$$|m| = \frac{100}{\sqrt{(1 + x^2)}}, \qquad \angle m = -2 \arctan x,$$

$$|m_F| = \frac{1000}{\sqrt{[100 - 16x^2 + x^4]}}, \qquad \angle m_F = -\arctan\left(\frac{2x}{10 - x^2}\right).$$

The characteristics are shown in the diagram.

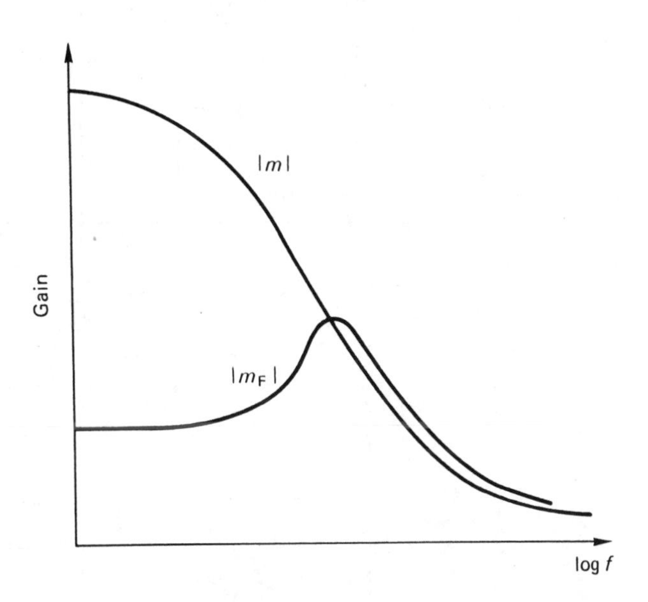

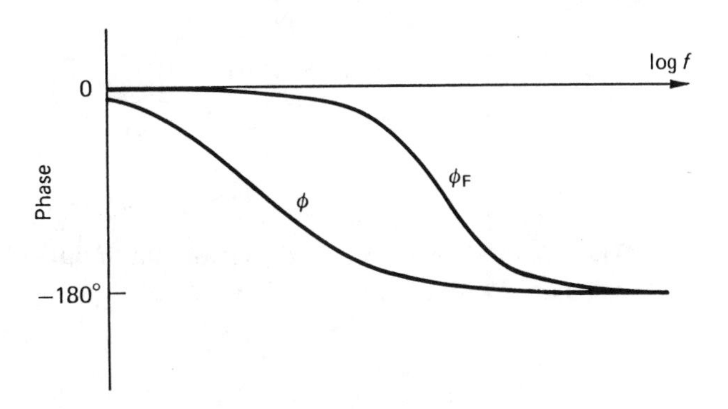

Example 6.8

Show that for a voltage-series feedback amplifier the input and output resistances are given by

$$R_{iF} = R_i\,(1 + A\beta) \qquad R_{oF} = \frac{R_o}{1 + A\beta},$$

where R_i and R_o are the values without feedback. The 2-stage amplifier shown in diagram (a) has $R_1 = 100\ \Omega$ and $R_2 = 3.3\ \text{k}\Omega$. Without feedback the amplifier parameters are

$$A_v = 200, \qquad R_i = 4\ \text{k}\Omega, \qquad R_o = 6\ \text{k}\Omega.$$

Calculate the corresponding values with feedback.

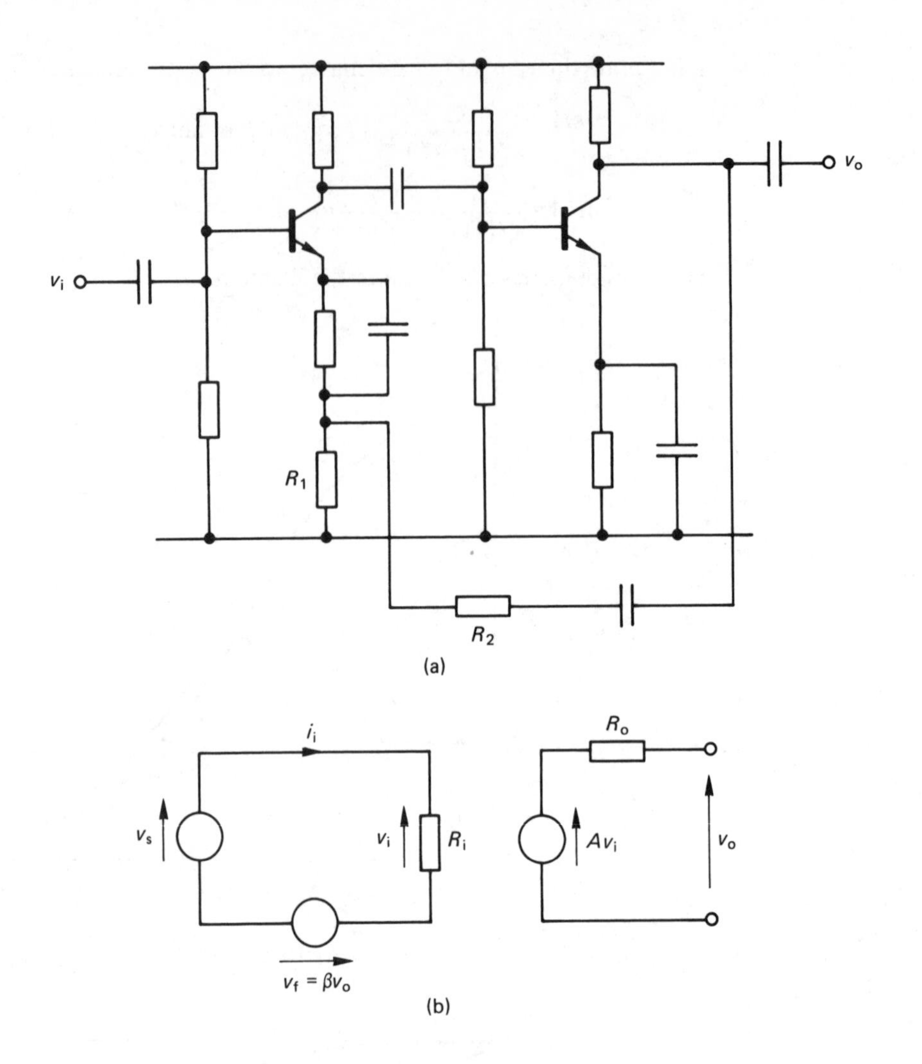

Solution 6.8

The effect of negative feedback on the input resistance of an amplifier may be calculated with reference to diagram (b).

$$
\begin{aligned}
v_s &= i_i R_i + v_f = i_i R_i + \beta v_o \\
&= i_i R_i + \beta A v_i \\
&= i_i R_i + \beta A i_i R_i \\
&= i_i R_i (1 + A\beta).
\end{aligned}
$$

$$\therefore\ R_{iF} = \frac{v_s}{i_i} = \underline{R_i\,(1 + A\beta).}$$

To calculate output resistance with feedback, use

$$R_{oF} = \frac{v_{o(oc)}}{i_{o(sc)}},$$

$$
\begin{aligned}
v_{o(oc)} &= A\,(v_s - \beta\, v_{o(oc)}) \\
&= \frac{A\, v_s}{1 + A\beta},
\end{aligned}
$$

and $$i_{o(sc)} = \frac{A\,v_s}{R_o}, \qquad \text{since } v_s = \beta v_o = 0.$$

$$\therefore\ R_{oF} = \frac{R_o}{1 + A\beta}.$$

In diagram (a) the feedback ratio is given by

$$\beta = \frac{R_1}{R_1 + R_2} = \frac{0.1}{0.1 + 3.3} = \frac{1}{34}.$$

$$\therefore\ A_F = \frac{A}{1 + A\beta} = 29,$$

$$R_{iF} = R_i\,(1 + A\beta) = \underline{27.5\ \text{k}\Omega},$$

$$R_{oF} = \frac{R_o}{1 + A\beta} = \underline{870\ \Omega}.$$

Example 6.9

For the voltage-series feedback amplifier shown in diagram (a) calculate:

(a) the voltage gain;
(b) the input resistance;
(c) the output resistance.

You may assume that $I_c = I_s \left[\exp\left(\frac{V_{BE}}{V_T} - 1\right)\right]$ and $h_{fe} = 50$.

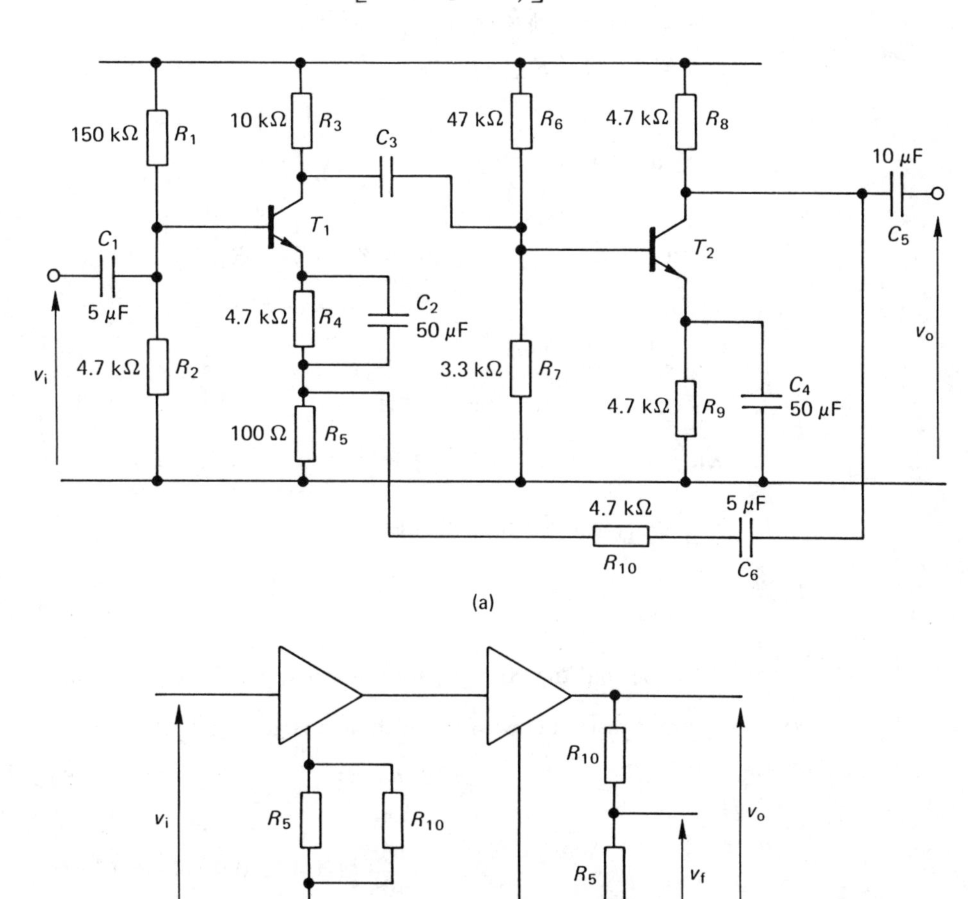

Solution 6.9

Using the technique described in section 6.5 of the text:

To identify the input circuit set $v_o = 0$, thus R_5 and R_{10} are in parallel. To identify the output circuit, open-circuit the input, thus R_5 and R_{10} are in series. The circuit may be represented as shown in diagram (b).

In this equivalent circuit the feedback ratio is given by

$$\beta = \frac{v_f}{v_o} = \frac{R_5}{R_5 + R_{10}}.$$

From the Ebers–Moll equation, the transistor transconductances are:

$$T_1: \quad g_{m1} \approx 40 \text{ mS} \quad \text{at } I_{C1} \approx 1 \text{ mA}$$

$$r_{i1} = \frac{h_{fe}}{g_{m1}} = \frac{50}{40 \times 10^{-3}} = 1.25 \text{ k}\Omega;$$

$$T_2: \quad g_{m2} \approx 80 \text{ mS} \quad \text{at } I_{C2} \approx 2 \text{ mA}$$

$$r_{i2} = \frac{h_{fe}}{g_{m2}} = \frac{50}{80 \times 10^{-3}} = 0.62 \text{ k}\Omega.$$

Voltage gain of stage 1:

$$A_{v1} = -\frac{g_{m1}R'_{C1}}{1 + g_{m1}R'_{E1}},$$

where $R'_{C1} = R_3 \parallel R_6 \parallel R_7 \parallel r_{i2} \approx 570\ \Omega$

and $R'_{E1} = R_5 \parallel R_{10} = 98\ \Omega$.

$$\therefore A_{v1} = -4.6.$$

Voltage gain of stage 2:

$$A_{v2} = -g_{m2}R'_{C2},$$

where $\quad R'_{C2} = R_8 \parallel (R_5 + R_{10}) = 2.4 \text{ k}\Omega.$

$$\therefore A_{v2} = -190.$$

Overall voltage gain:

$$A_v = A_{v1} \times A_{v2} = 4.6 \times 190 = 874$$

Also, $\quad \beta = \dfrac{R_5}{R_5 + R_{10}} = \dfrac{1}{48} = 0.02.$

$\therefore$ Voltage gain with feedback:

$$A_F = \frac{A_v}{1 + A_v\beta} = \underline{47.3}.$$

(Notice that the voltage gain is given approximately by $1/\beta = \dfrac{R_5 + R_{10}}{R_5} = 48.$)

Input resistance without feedback is given by:

$$R_i = \frac{h_{fe}}{g_{m1}}(1 + g_{m1}R'_{E1})$$

$$= \frac{50}{40}(1 + 40 \times 0.098) = 6.15 \text{ k}\Omega.$$

Input resistance with feedback:

$$R_{iF} = R_i (1 + A\beta)$$
$$= 6.15 (1 + 17.5) = \underline{114 \text{ k}\Omega}.$$

Output resistance without feedback:

$$R_o = R'_{C2} = 2.4 \text{ k}\Omega.$$

Output resistance with feedback:

$$R_{oF} = \frac{R_0}{1 + A\beta} = \frac{2.4}{1 + 17.5} = \underline{130 \ \Omega}.$$

Example 6.10

Explain why the gain of the series feedback pair circuit of diagram (a) may be approximated to

$$\frac{R_4 + R_5}{R_4}.$$

Use a more detailed analysis to establish a more accurate result and calculate the gain with the component values shown. (Assume $h_{fe} = 50$.)

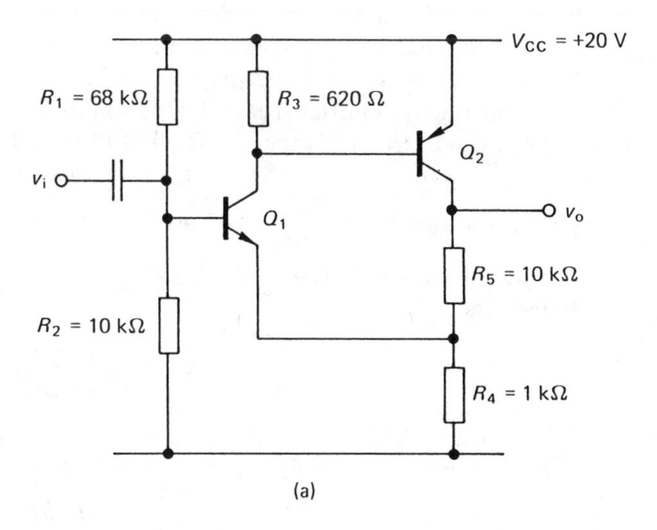

(a)

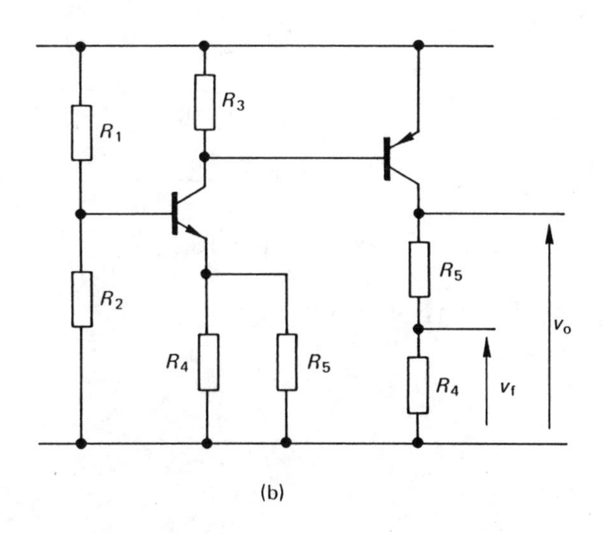

(b)

Solution 6.10

For the series feedback pair the feedback is voltage-derived and series-fed, for which

$$A_F = \frac{A}{1 + A\beta} .$$

Now when A is large

$$A_F \approx \frac{1}{\beta} ,$$

where $\beta = \dfrac{R_4}{R_4 + R_5}$.

$$\therefore \; A_F \approx \frac{R_4 + R_5}{R_4} .$$

A more detailed analysis could consider the overall gain of the amplifier without feedback and then, given the feedback ratio, calculate the resultant gain.

The method is described in section 6.5. Referring to the modified circuit of diagram (b) proceed as follows:

(i) First find the effective input circuit without feedback by short-circuiting the output; thus R_5 appears in parallel with R_4.

(ii) Then find the effective output circuit without feedback by open-circuiting the input stage (i.e. holding the current in Q_1 constant.) This is effectively like disconnecting Q_1 emitter from the junction of R_4 and R_5.

The overall circuit without feedback is thus as shown in diagram (b).

We may now calculate the voltage gain without feedback A, and then use the feedback ratio $\beta = \dfrac{R_4}{R_4 + R_5}$ to find the gain with feedback.

Since the currents in Q_1 and Q_2 are about 1 mA then $g_m \approx 40$ mS for both transistors.

$$\therefore \; A_{v1} \approx \frac{g_m R_L}{1 + g_m R_E} ,$$

where $R_L = R_3 \parallel R_{i2} = R_3 \parallel \dfrac{h_{fe}}{g_m}$ and $R_E = R_4 \parallel R_5$.

$$\therefore \; A_{v1} \approx 0.45.$$

Also, $A_{v2} \approx g_m \, (R_4 + R_5)$

≈ 440

$$\therefore A_v \approx 198.$$

$$\therefore A_F = \frac{A_v}{1 + A_v\beta} , \qquad \text{where } \beta = \frac{1}{11} ,$$

$$\therefore A_F \approx 10.4.$$

Using $A_F \approx \dfrac{1}{\beta}$ gives $A_F = 11$.

6.8 Unworked Problems

Problem 6.1

Show that when negative voltage feedback is applied in series with the input of an amplifier of gain A, the gain is modified to $G = A/(1 + A\beta)$, where β is the feedback ratio.

The value of the gain A varies from its nominal value of 60 dB. Determine the feedback ratio β and the permissible range of values of A (in dB) if the gain G is to be 40 ± 0.5 dB.

The magnitude of the open-loop gain A is found to fall by 3 dB from its mid-band value at a frequency of 200 kHz. At what frequency will G fall by 3 dB from its mid-band value? Calculate the magnitude and angle of G at five times this frequency. Assume that the open-loop gain A is dominated by a single time-constant, and the feedback β is independent of frequency.

(CEI Part 2)

Problem 6.2

The measured response for a multistage amplifier without feedback is given in the table.

A fraction $0.002\angle 0°$ of the output signal is fed back in series with the input. Calculate the gain with feedback and sketch graphs of the gain with and without feedback to a common base of log(frequency). Determine the bandwidths between the 3 dB points and comment on the results.

Frequency (kHz)	0.02	0.05	0.1	0.2	0.5	1.0	20	50	100	200	500	1000
Voltage gain:												
magnitude	180	430	690	860	940	1000	1000	940	820	580	350	200
phase angle (°)	−70	−102	−135	−155	−167	−180	+165	+143	+120	+80	+34	−40

(CEI Part 2)

Problem 6.3

A fraction $\beta\angle 0°$ of the output voltage of an amplifier having a gain $-m\angle\theta$ is fed back degeneratively in series with the input. Show that the overall phase shift, θ, of the amplifier is reduced by

$$\phi = \arctan\left[\frac{\beta m \sin\theta}{1 + \beta m \cos\theta}\right].$$

An amplifier has a mid-band voltage gain of $-500\angle\phi°$ and its upper 3 dB frequency of 100 kHz is determined by a single time constant. Sketch the gain magnitude and phase shift to a logarithmic base of frequency.

Evaluate the gain magnitude and phase shift at 100 kHz if 1 per cent of the output is fed back degeneratively in series with the input.

(CEI Part 2)

Problem 6.4

Derive an expression for the gain v_o/v_i of the amplifier shown. State any simplifying assumptions made.

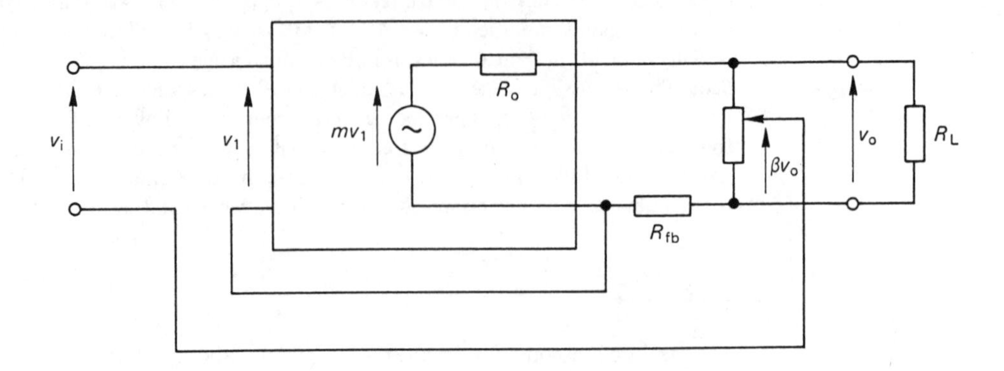

An amplifier without feedback has output resistance of 600 Ω and a voltage gain of 60 dB when supplying a 600 Ω resistive load R_L. Determine the value of β and R_{fb} such that the gain with feedback is reduced to 30 dB without changing the output resistance of the amplifier.

(CEI Part 2)

Problem 6.5

An amplifier stage has a mid-band voltage gain of 30 dB. The gain falls by 3 dB from this value at frequencies of 200 Hz and 50 kHz, each determined by a single time-constant.

If 3 per cent of the output voltage is fed back in series opposition to the input at mid-band frequencies, determine the new mid-band gain in decibels and the new '3 dB' frequencies.

Two of the amplifier stages without feedback are now connected in cascade. Overall feedback is then applied to give a mid-band gain of 40 dB. Calculate the resulting '3 dB' frequencies and comment on the resulting frequency response.

(CEI Part 2)

7 Operational Amplifiers

7.1 The Ideal Op-amp

An operational amplifier (op-amp) is an integrated-circuit amplifier with a very large voltage gain (10^5 to 10^6), a high input impedance (usually several megohms) and a low output impedance (several tens of ohms).

Op-amps are available in hundreds of different types, the most general group being the d.c.-coupled differential-voltage amplifiers, with an inverting and a non-inverting input and a single-ended output as represented by the symbolic diagram of Fig. 7.1. The power supply connections are not normally shown.

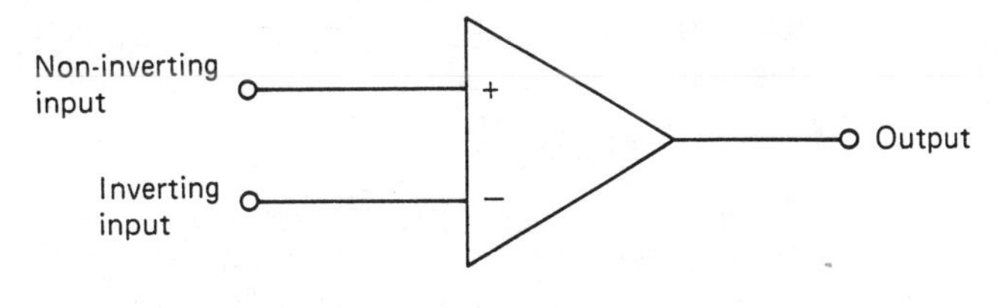

Figure 7.1

The output is an amplified version of the difference between the two inputs (the output goes positive when the non-inverting input is greater than the inverting input).

Only a very small difference in potential is required between the two inputs to cause a large output-voltage swing.

The voltage gain is large and unpredictable, and the device is never used as an amplifier without negative feedback being applied.

A typical op-amp is the 741, available from a large number of manufacturers. Many op-amps use integrated bipolar transistors in their design, but op-amps are available that use an FET input stage and thus give a higher input impedance (BiFETs).

The ideal op-amp has infinite input impedance, zero output impedance, infinite voltage gain, infinite bandwidth and zero output voltage when the two input voltages are equal. Unfortunately this device would also have infinite cost!

7.2 Practical Limitations of Op-amps

True op-amps differ from the ideal in terms of the following parameters:

(a) Input Bias Current (I_B)

I_B is the average of the currents into the two input terminals with the output at zero volts. In the 741, I_B is typically 80 nA. This current causes a voltage drop across the equivalent source impedance seen by the op-amp input (see Examples 7.1 and 7.2).

(b) Input Offset Current (I_{os})

I_{os} is the difference between the currents into the two input terminals with the output at zero volts. This means that even when driven from identical source impedance the op-amp inputs 'see' unequal voltage drops and hence there is a voltage difference between the two inputs (see Example 7.2). In the 741 I_{os} is typically 20 nA.

(c) Input Offset Voltage (V_{os})

V_{os} is the d.c. voltage that must be applied between the input terminals to force the quiescent d.c. output voltage to zero. It is due to imbalance in the input stages and in the case of the 741 is typically 1 mV. Offset compensation may be achieved for the 741 op-amp using the circuit shown in Fig. 7.2. An alternative arrangement is to provide an adjustable offset compensating voltage as shown in Fig. 7.3.

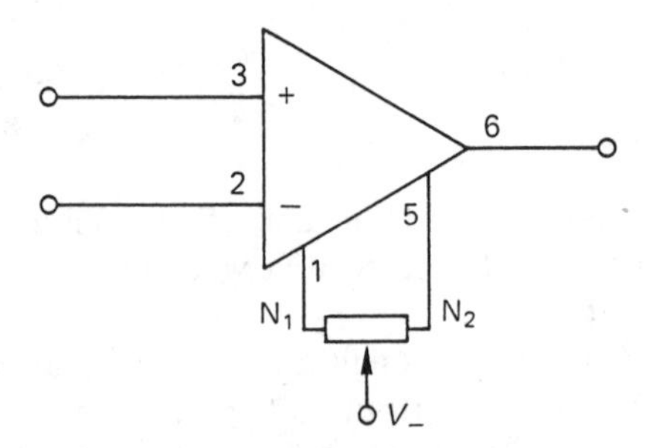

Figure 7.2 N_1 and N_2 are the 'offset null' pins. $V_- = -V_{CC}$

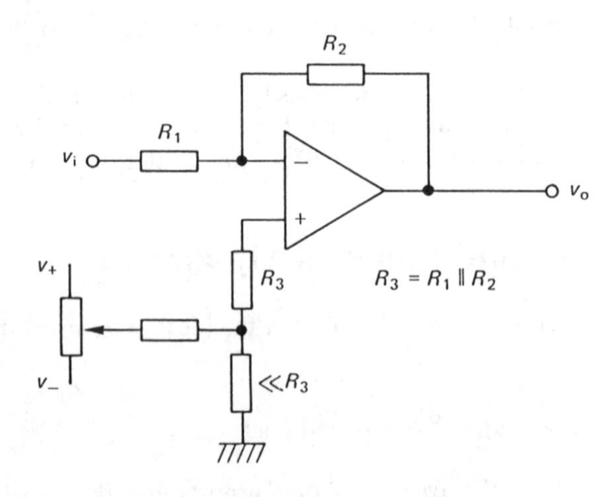

Figure 7.3 Adjustable offset compensation

(d) Input Resistance (R_i)

R_i is the resistance between the input terminals with one input grounded. For the 741 it is typically 2 MΩ but with negative feedback it is raised to a very high value and is not often as important a parameter as input bias current.

(e) Output Resistance (R_o)

R_o is the op-amp output impedance without feedback. For the 741 it is typically 75 Ω, but with negative feedback reduces to a negligible value. Most op-amps currently manufactured have output short-circuit protection. The output current is limited during a short circuit by a low-value resistor in series with the output. It does not significantly effect the output impedance in normal operation.

(f) Voltage Gain (A_v)

A_v is the ratio of the output voltage to input voltage under the stated conditions for source resistance and load resistance.

(g) Difference and Common-mode Gain

Considering the two input signals as v_1 and v_2 and the output signal as v_o, each measured with respect to ground, then in the ideal differential amplifier,

$$v_o = A_d (v_1 - v_2)$$

where A_d is the difference gain.

In a practical differential amplifier, the output depends not only on the difference signal, v_d, but also on the average level, called the common-mode signal, v_c, where

$$v_d = v_1 - v_2 \qquad \text{and} \qquad v_c = 0.5(v_1 + v_2).$$

The output signal is thus given by:

$$v_o = A_d v_d + A_c v_c,$$

where A_c is the common-mode gain.

(h) Common-mode Rejection Ratio (CMRR)

This is the ratio of differential-voltage amplification A_d to common-mode amplification A_c. It is the ability of an op-amp to amplify the differential input signal while rejecting the common mode input signal. CMRR = A_d/A_c. In the 741 the CMRR is typically 90 dB (about 30 000 : 1).

(i) Bandwidth

The open-loop voltage gain of an op-amp falls off to unity gain at a frequency typically between 1 mHz and 10 MHz. A graph of the open-loop large-signal differential voltage gain vs. frequency is shown for the 741 in Fig. 7.4. The bandwidth is generally defined as the frequency at which the gain has fallen to $1/\sqrt{2}$ of the d.c. gain (the 3 dB point). The gain–bandwidth product is generally defined as the frequency at which the gain has fallen to unity (1 MHz in the case shown in Fig. 7.4). The application of negative feedback reduces the gain but correspondingly increases the bandwidth. The gain–bandwidth product is the gain x the bandwidth.

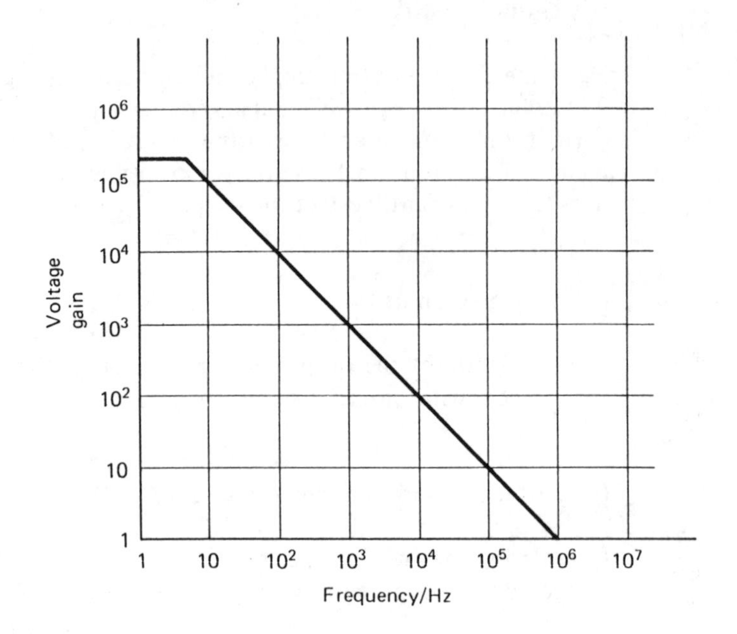

Figure 7.4

Because the open-loop gain of an op-amp is very high, there is always a possibility of instability (self-oscillation) unless the amplifier is frequency-compensated (see Chapter 6). This requires the connection of external components, possibly just a capacitor. Many amplifiers (including the 741) are internally compensated, and thus do not require external compensating components. Prevention of oscillation is generally the largest single problem engineers have with op-amps.

(j) Slew Rate (S)

This is the rate of change of output voltage with a large-amplitude step function applied to the input. It is due to the internal compensation capacitor and the small internal drive currents. The slew rate limits the amplitude of a sine wave output above a certain frequency and for large output signals will be the limiting factor on maximum operating frequency. A sine wave frequency f (Hz) and peak amplitude V_p (volts) requires (for an undistorted output) a minimum slew rate given by $S = 2\pi f\ V_p$ volts/second. (See Example 7.4.)

(k) Temperature Dependence

All of the parameters mentioned are somewhat temperature-dependent. The two parameters that have most effect on performance with temperature are input offset voltage and input offset current (see Example 7.3). For the 741A, maximum parameters are as follows:

input offset voltage drift: 15 μV/°C;
input offset current drift: 0.5 nA/°C.

Op-amps are available with improved parameters and facilities such as high input impedance ($I_B < 25$ nA), low drift (ΔV_{os}/temp < 10 μV/°C), high slew rate ($S > 10$ V/μS), low power consumption, single supply, high voltage ($V_{CC} >$ +25 V), buffers, programmable, high output current ($I_o > 200$ mA), compensated, uncompensated, special instrumentation amplifiers and comparators.

7.3 Basic Op-amp Circuits

(a) Inverting Amplifier

The basic inverting op-amp circuit is shown in Fig. 7.5. The disadvantage is that the input impedance is low and equal to the input resistance R_1, since the non-inverting input point is a virtual earth. R_3 is made equal to $R_1 \parallel R_2$ so as to minimise the effect of input bias current. The output offset voltage due to V_{os} and I_{os} is

$$V_o = \pm V_{os}\left(1 + \frac{R_2}{R_1}\right) \pm I_{os}R_2,$$

where the ± signs are necessary due to the polarities of the offsets being unpredictable.

The term $\pm I_{os}R_2$ in the output offset equation imposes a limit to the range of usable resistance values. Very high values of R_2 (necessary for high gain or high input impedance in the case of the inverting amp) produce a large offset $I_{os}R_2$.

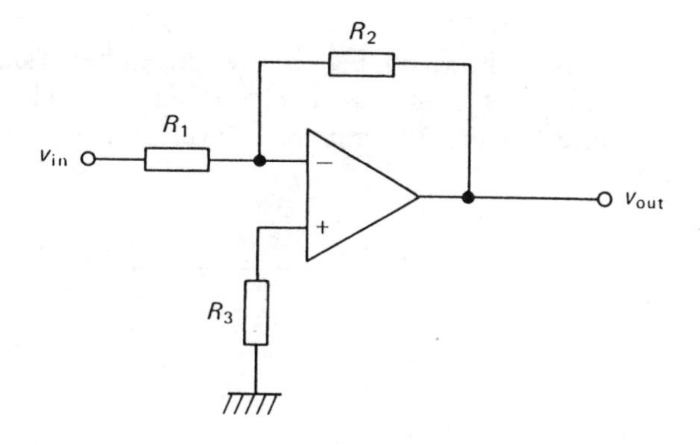

Figure 7.5 Basic inverting op-amp circuit

$$A_v = \frac{v_{out}}{v_{in}} = -\frac{R_2}{R_1}$$

$Z_{in} = R_1$ $\quad$ Z_{out} = fraction of an ohm

$R_3 = R_1 \parallel R_2$ (minimises the effect of input bias current)

While this could be nulled out, long-term compensation would be impossible due to the temperature sensitivity of I_{os}. For general-purpose bipolar op-amps, resistance values greater than 1 MΩ are rarely used for this reason. FET input op-amps may be used with much higher resistance values owing to their lower I_{os}.

(b) Non-inverting Amplifier

A high-input-impedance, non-inverting amplifier circuit is shown in Fig. 7.6.

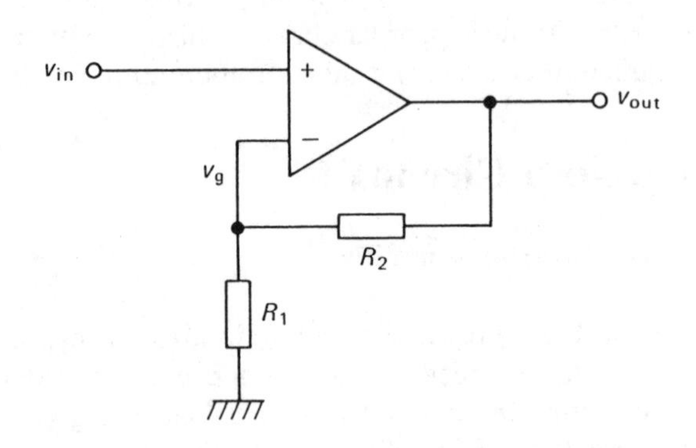

Figure 7.6 High-input-impedance non-inverting amplifier

$$A_v = \frac{v_{out}}{v_{in}} = 1 + \frac{R_2}{R_1}$$

$Z_{in} \approx \infty$ (at least 100 MΩ)

Z_{out} = fraction of an ohm

(c) Unity-gain Buffer Amplifier (Voltage Follower)

This is shown in Fig. 7.7, where quite often R_1 is simply replaced by a short circuit. If used, R_1 should be made equal to the source resistance at the non-inverting input to minimise the error due to input bias current.

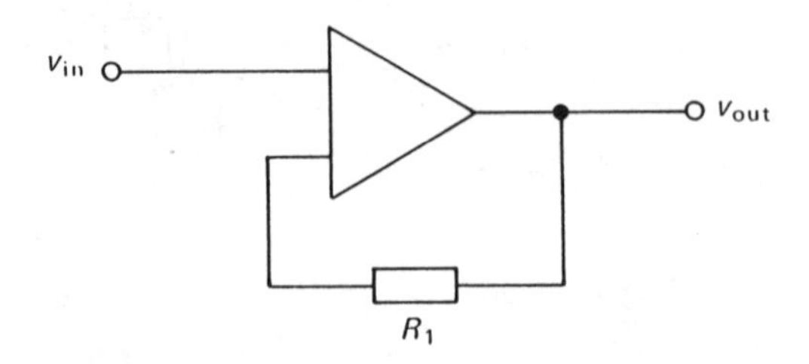

Figure 7.7 Unity-gain buffer amplifier (voltage follower)

$$A_v = \frac{v_{out}}{v_{in}} \approx 1$$

$Z_{in} \approx \infty$ $\qquad Z_{out}$ = fraction of an ohm

$R_1 = R_{source}$ for minimum error due to input bias current

(d) Differential Amplifier

The differential amplifier is shown in Fig. 7.8 and gives effective minimisation of common-mode signals. The disadvantages are that the input impedance is comparatively low, that the input impedances are different for each input, and that the CMRR is affected greatly by the source impedances due to mismatch.

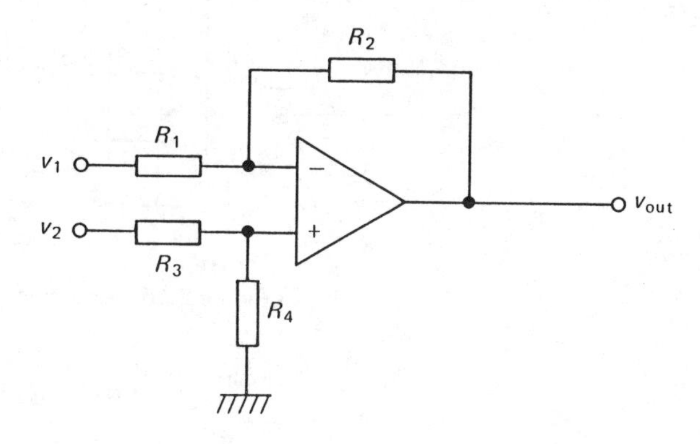

Figure 7.8 Differential amplifier

$$v_{out} = \frac{R_2}{R_1}(v_2 - v_1)$$

with $R_1 = R_3$ and $R_2 = R_4$

Impedance to ground should be matched by ensuring $R_1 \| R_2 = R_3 \| R_4$

(e) Integrating Amplifier

The integrating amplifier is shown in Fig. 7.9. The output is the integral of the input waveform, provided the amplifier does not go into saturation. The disadvantage is that the output tends to drift because of op-amp offset and bias currents. The frequency response of the integrating amplifier falls off linearly at 20 dB/decade, the output amplitude thus dropping off rapidly with frequency.

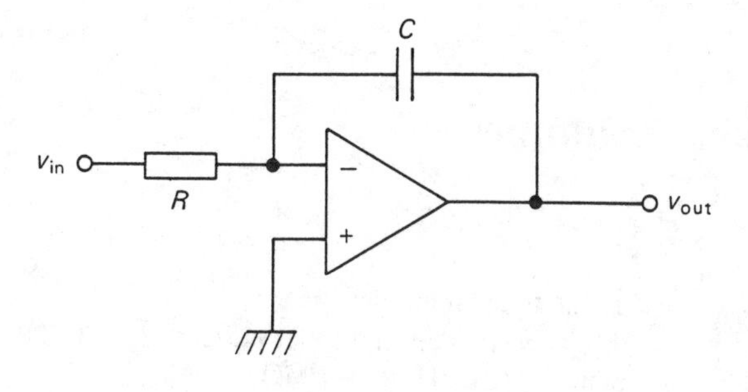

Figure 7.9 Integrating amplifier

$$v_o = -\frac{1}{CR}\int v_i \, dt$$

(f) Summing Amplifier

The summing amplifier is shown in Fig. 7.10. The output is the sum of the input signals.

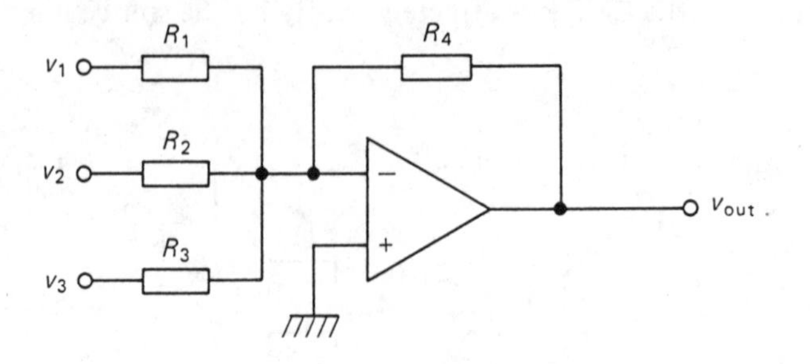

Figure 7.10 Summing amplifier

$$v_{out} = -R_4\left(\frac{v_1}{R_1} + \frac{v_2}{R_2} + \frac{v_3}{R_3}\right)$$

(g) Comparator

The simplest form of comparator is a high-gain differential amplifier using an op-amp as shown in Fig. 7.11. The op-amp output goes into positive or negative saturation, depending on the difference between the input voltages. It may be used to detect which of the two voltages is the larger or when a signal exceeds a certain level. Special integrated circuits intended for use as comparators (such as the LM311) provide very fast slew rate and often have an open collector output that may be connected via a resistor to, say, +5 V, and are thus ideal for driving TTL logic gates.

Other applications of op-amps, such as multivibrators and voltage stabilisers, will be considered in the appropriate chapters.

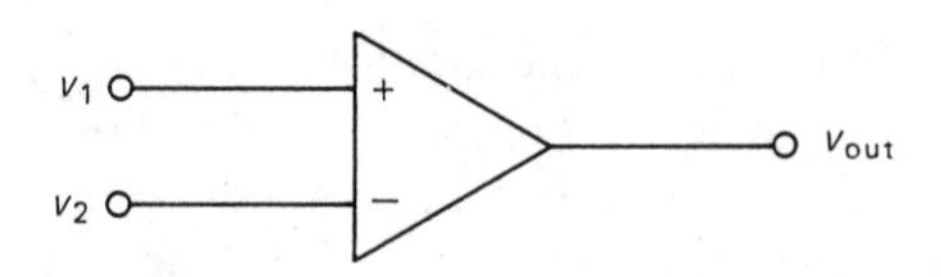

Figure 7.11 Voltage comparator

7.4 Worked Examples

Example 7.1

Define the term 'input bias current'.

An op-amp has an input bias current (I_B) of 200 nA at 20 °C. In the circuit shown in the diagram: R_1 = 10 kΩ, R_2 = 1 MΩ.

(a) Calculate the voltage gain.
(b) Calculate the output offset voltage due to the input bias current (I_B), assuming R_3 to be zero.
(c) Calculate the value of R_3 to minimise the effect of input bias current.

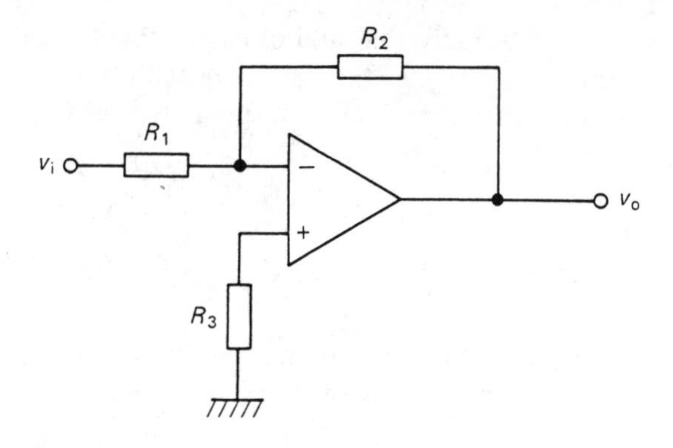

Solution 7.1

See text for definition of terms.

(a) Voltage gain $A_v = -\dfrac{R_2}{R_1} = -\dfrac{1\ \text{M}\Omega}{10\ \text{k}\Omega} = -100.$

(b) Offset voltage at input due to I_B is given by

$$V_{os} = I_B\left(\frac{R_1 R_2}{R_1 + R_2}\right)$$
$$= 200 \times 10^{-9} \times 9.9 \times 10^3$$
$$= \underline{1.98\ \text{mV}}.$$

(c) To minimise the effects of input bias current, both inputs should 'see' the same driving resistance and therefore the value of $\underline{9.9\ \text{k}\Omega}$ is chosen for R_3.

Example 7.2

A non-inverting amplifier is arranged as shown in the diagram. Calculate the voltage gain, noting that $A_F = A/(1 + \beta A)$ and β in this case is $R_1/(R_1 + R_2)$. Assume that the amplifier input current is negligible. Explain why R_3 is necessary and estimate a suitable value, assuming that it is driven from a signal source of resistance R_s.

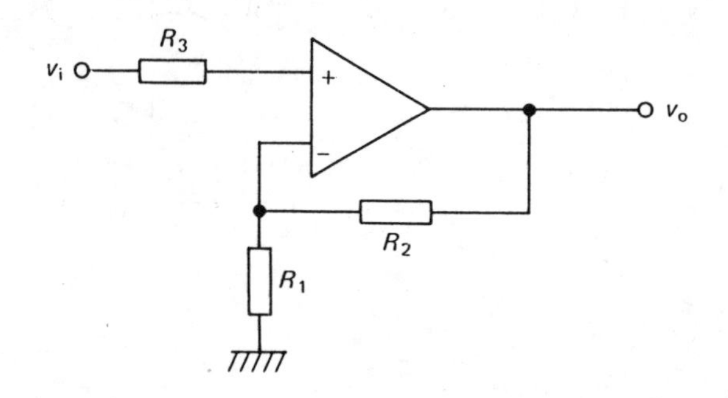

Solution 7.2

$$A_F = \frac{A}{1 + A\beta}$$

Now, if A is large then

$$A_F \approx \frac{1}{\beta} = \frac{R_1 + R_2}{R_1}.$$

An alternative method of calculating the gain would be to assume that, in linear amplifier circuits, the output attempts to do whatever is necessary to make the voltage difference between the inputs zero.

$$\therefore v_i = \frac{R_1}{R_1 + R_2} v_o,$$

$$\therefore \frac{v_o}{v_i} = \frac{R_1 + R_2}{R_1}.$$

This rule is often useful in establishing op-amp behaviour.

The purpose of R_3 is to reduce the input offset voltage drift. We require that

$$R_3 + R_s = \frac{R_1 R_2}{R_1 + R_2},$$

so that both inputs 'see' the same driving resistance.

Example 7.3

Define the terms (i) input offset current, (ii) input offset voltage, (iii) offset null control, (iv) input offset voltage drift.

An op-amp has the following parameters at 20 °C:

input offset voltage (V_{os}) = 6 mV;
input offset current (I_{os}) = 200 nA.

In the circuit shown in the diagram, R_2 = 100 kΩ, R_1 = 10 kΩ and the effect of input bias current is minimised by making $R_3 = R_1 \| R_2$.

(a) Calculate the output voltage offset due to V_{os} and I_{os}.
(b) If the output is adjusted to zero at 20 °C using the 'offset null' control, calculate the worst-case output offset voltage at 70 °C;
Input offset current drift = 0.5 nA/°C max;
Input offset voltage drift = 10 μV/°C max.

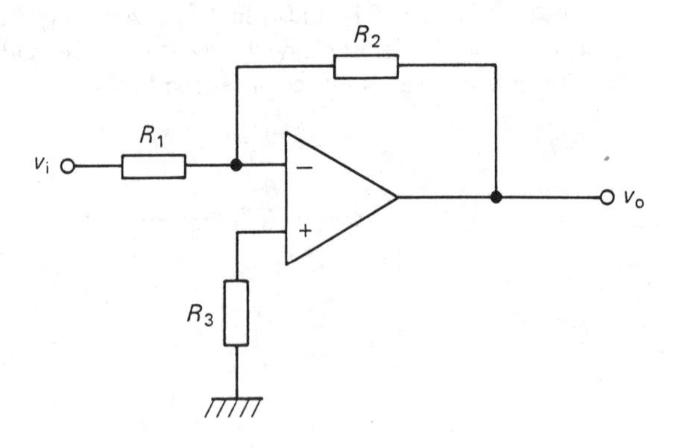

Solution 7.3

See text for definition of terms.

(a) Output offset voltage due to V_{os} is given by

$$V_o = V_{os}\left(\frac{R_1 + R_2}{R_1}\right)$$

(it acts like a non-inverting amplifier with V_i held constant and V_{os} appearing between the non-inverting and the inverting inputs).

Since the inverting input is a virtual earth, all of the offset current is supplied via R_2 and the output offset voltage due to I_{os} is given by

$$V_o = I_{os}R_2.$$

The total output offset voltage is given by

$$V_o = V_{os}\left(1 + \frac{R_2}{R_1}\right) + I_{os}R_2$$

$$= 6 \times 10^{-3}\ (1 + 10) + 200 \times 10^{-9} \times 100 \times 10^3$$

$$= 66\ \text{mV} + 20\ \text{mV} = \underline{86\ \text{mV}}.$$

(b) Worst-case input offset voltage drift at 70 °C is given by

$$\text{drift} = 10\ \mu\text{V} \times 50 = 500\ \mu\text{V}.$$

Worst-case input offset current drift at 70 °C is given by

$$\text{drift} = 0.5\ \text{nA} \times 50 = 25\ \text{nA}.$$

∴ Worst-case output offset voltage drift is given by

$$\text{drift} = 500 \times 10^{-6}\left(1 + \frac{R_2}{R_1}\right) + 25 \times 10^{-9} \times R_2$$

$$= 5.5\ \text{mV} + 2.5\ \text{mV}$$

$$= \underline{8\ \text{mV}}.$$

Example 7.4

An op-amp has a slew rate $S = 20\ \text{V}/\mu\text{s}$. Determine from first principles the maximum frequency at which it can give an undistorted sinusoidal output of 20 V, peak-to-peak. An op-amp has a rated full-power bandwidth of 50 kHz for a sinusoidal output of peak amplitude 10 V. Describe the response of the circuit shown in the diagram to a step input of +2 V and calculate the rise time (10 per cent to 90 per cent).

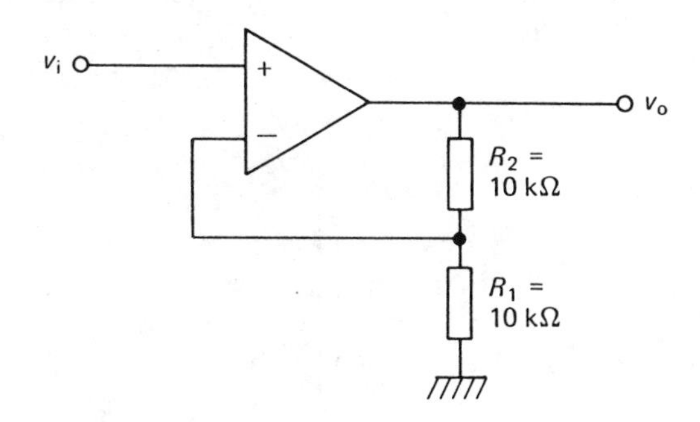

Solution 7.4

For a sinusoidal voltage waveform, $v = V_m \sin 2\pi ft$.

The rate of change of voltage with time is given by:

$$\frac{dv}{dt} = 2\pi f\ V_m \cos 2\pi ft.$$

The maximum rate of change is given by

$$\left.\frac{dv}{dt}\right|_{max} = 2\pi f\ V_m.$$

Now, the maximum possible rate of change of voltage is equal to the slew rate,

$$\therefore\ 2\pi f\, V_m = S.$$

∴ Maximum frequency f_{max} for an undistorted output is

$$f_{max} = \frac{S}{2\pi\, V_m}\,.$$

(a) $$f_{max} = \frac{20 \times 10^6}{2\pi \times 10} = \underline{318\ \text{kHz}}.$$

(b) $$S = 2\pi f\, V_m\,,$$

where $f = 50$ kHz and $V_m = 10$ V.

$$\therefore\ S = 2\pi \times 50 \times 10^3 \times 10 = 3.14\ \text{V}/\mu\text{s}.$$

The amplifier gain is

$$A = 1 + \frac{R_2}{R_1} = 2.$$

The response to a 2 V step is an approximately linear rise to 4 V with a rise time (10 per cent to 90 per cent) given by

$$0.8 \times \frac{4}{3.14} \approx \underline{1\ \mu\text{s}}.$$

Example 7.5

An op-amp has a gain frequency open-loop response as shown in the diagram.

(a) Estimate from this curve the gain-bandwidth product and the gain-frequency roll-off rate.
(b) Calculate the bandwidth if the amplifier is used with negative feedback to give a gain of 200.
(c) Calculate the maximum output voltage obtainable at this frequency if the slew rate is 0.5 V/μs.

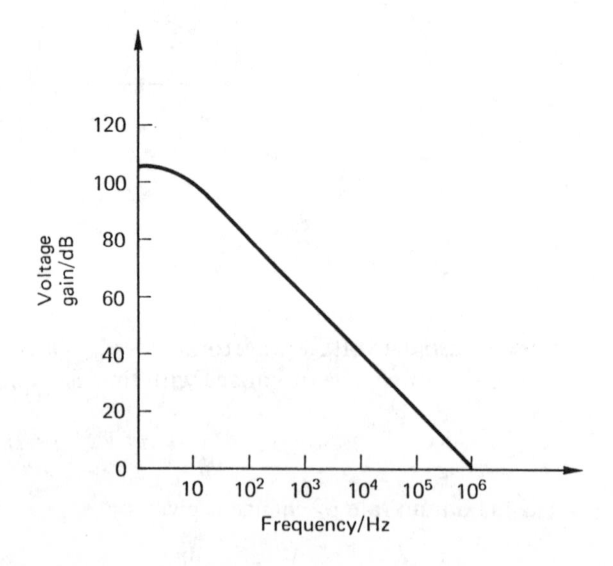

Solution 7.5

(a) From the curve, the gain–bandwidth product is 1 MHz (i.e. when the gain has fallen to 0 dB the frequency is 1 MHz).

Also, roll-off rate is 20 dB/decade.

(b) Gain x bandwidth = 10^6.

At a gain of 200, bandwidth = 5 kHz.

(c) $S = 2\pi f \; V_m$

$$\therefore \; V_m = \frac{0.5 \times 10^6}{2\pi \times 5 \times 10^3} = 15.92 \text{ V peak.}$$

Example 7.6

For the circuit shown in the diagram, assuming that the input current is negligible, show that $v_o/v_i = f(R_2, R_1, A)$ and that it may be approximated to $v_o/v_i = -(R_2/R_1)$. Explain the purpose of R_3 and calculate an approximate value for R_3 such that it matches the source resistance seen by the inverting input.

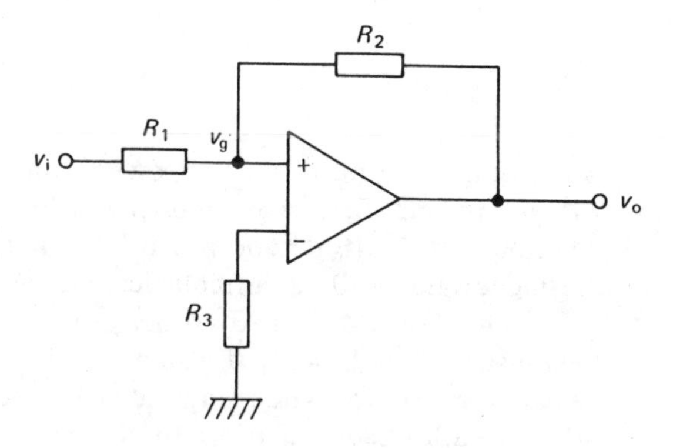

Solution 7.6

Let the non-inverting input be v_g, current in R_1 be i_1, and current in R_2 be i_2;

$$\therefore \; i_1 = \frac{v_i - v_g}{R_1} \quad \text{and} \quad i_2 = \frac{v_g - v_o}{R_2}.$$

But since amplifier input current is zero, $i_1 = i_2$.

$$\therefore \; \frac{v_i - v_g}{R_1} = \frac{v_g - v_o}{R_2}.$$

Now $v_o = -A\, v_g$, where A is the open-loop gain,

$$\therefore \; \frac{v_i + \dfrac{v_o}{A}}{R_1} = -\frac{\left(\dfrac{v_o}{A} + v_o\right)}{R_2},$$

$$\therefore \; \frac{v_o}{v_i} = -\frac{R_2}{R_1}\left(\frac{1}{1 + \dfrac{R_1 + R_2}{AR_1}}\right).$$

For large values of A,

$$\frac{v_o}{v_i} \approx -\frac{R_2}{R_1}.$$

The purpose of R_3 is to reduce the effect of input offset voltage due to input bias current.

$$\therefore\ R_3 = \frac{R_1 R_2}{R_1 + R_2}.$$

Example 7.7

Explain why it is often necessary to connect frequency compensation components around an op-amp, but not in the case of the 741 op-amp. Explain how external frequency compensation can be achieved.

An op-amp has a low-frequency, open-loop voltage gain of 10^5, unity gain at 1 MHz and a slope of 20 dB/decade.

(a) Calculate the open-loop break frequency.

(b) Calculate the break frequency if negative feedback reduces the gain to 100.

Solution 7.7

An op-amp gain will roll off at some frequency due to internal capacitive effects within the amplifier, each producing a low-pass filter. Each RC low-pass filter contributes a 20 dB/decade roll off beyond the break point, and a phase shift moving towards $-90°$ at a frequency of about ten times the break frequency.

If a phase shift of $> 180°$ is achieved while the gain is still greater than unity then instability will occur. Diagram (a) shows typical responses for both uncompensated and compensated amplifiers. An uncompensated amplifier generally rolls off at 20 dB/decade, then 40 dB/decade and even 60 dB/decade as various frequency-limiting effects within the amplifier come into play. If a loop with this type of response is closed then it will oscillate. This may be overcome by keeping the open-loop phase shift much less than $180°$ at a frequency where the gain is greater than unity. This is achieved by adding a compensating capacitor such that the roll off rate is 20 dB/decade at all frequencies where the gain is greater than unity, as shown in the compensated curve of diagram (a).

The 741 op-amp has this compensation capacitor included internally.

(a) The open-loop response for the amplifier specified is drawn in diagram (b), from which it may be seen that the open-loop break frequency is 10 Hz.

(b) The open-loop gain is given by

$$A(j\omega) = \frac{10^5}{1 + j(f/10)}.$$

The closed-loop gain is given by

$$A_F(j\omega) = \frac{A}{1 + A\beta}\left(\frac{1}{1 + j\dfrac{f}{10\,(1 + A\beta)}}\right).$$

But $|A_F| = 100$ and $|A| = 10^5$,

$$\therefore\ 1 + A\beta = 10^3.$$

$\therefore$ Closed-loop break frequency is

$$f_H = 10\,(1 + A\beta) = \underline{10\ \text{kHz}}.$$

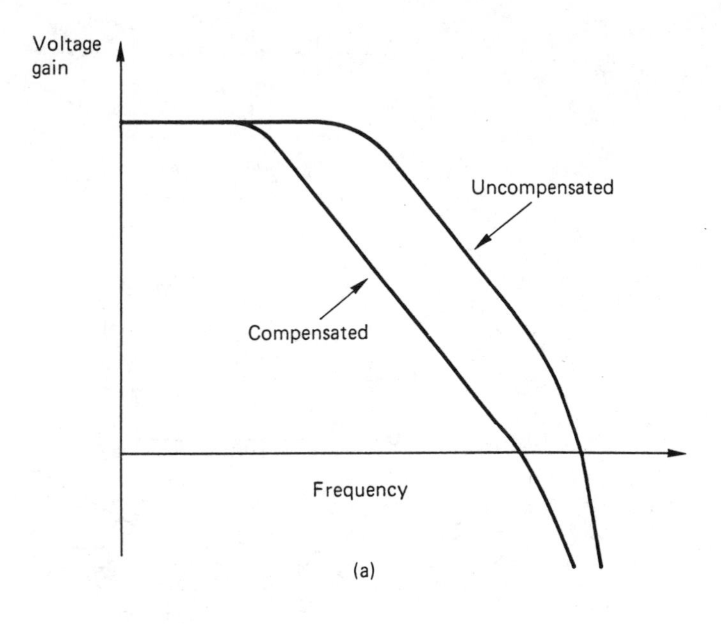

(a)

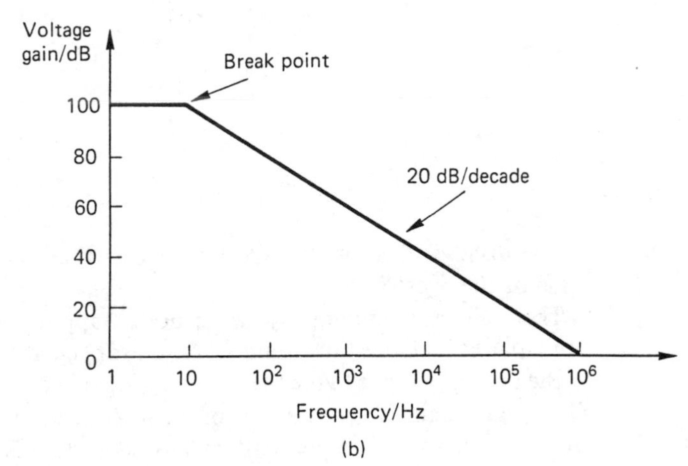

(b)

Example 7.8

Explain what is meant by 'bootstrapping' and how it can increase the input resistance of an amplifier. Suggest an alternative method of ensuring a high input resistance. In the a.c. amplifier shown in diagram (a) state the reason for the inclusion of the resistor R_3, and estimate a suitable value. State the disadvantages of including this resistor and explain with the aid of a circuit diagram how this disadvantage could be overcome using bootstrapping.

Solution 7.8

Bootstrapping is a technique of increasing an amplifier input resistance by arranging for the voltage change across the input bias circuit to be very small, thus giving the effect of a high input resistance. (An alternative might be to precede the amplifier with a high-input-impedance pre-amplifier.)

The purpose of R_3 in the a.c. amplifier of diagram (a) is to provide a d.c. path so as to reduce the effect of input offset current.

$$R_3 = \frac{R_1 R_2}{R_1 + R_2}$$

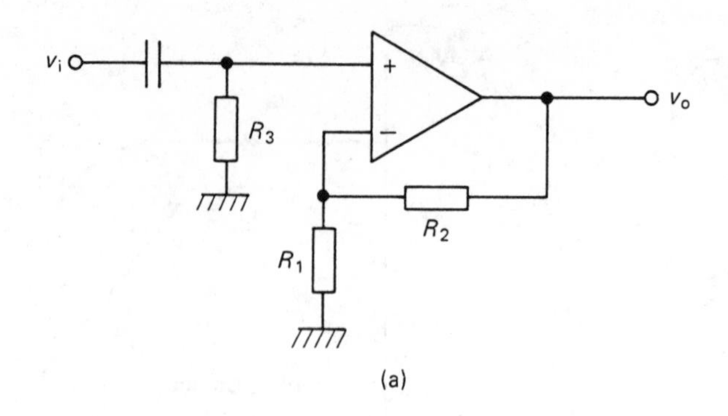

(a)

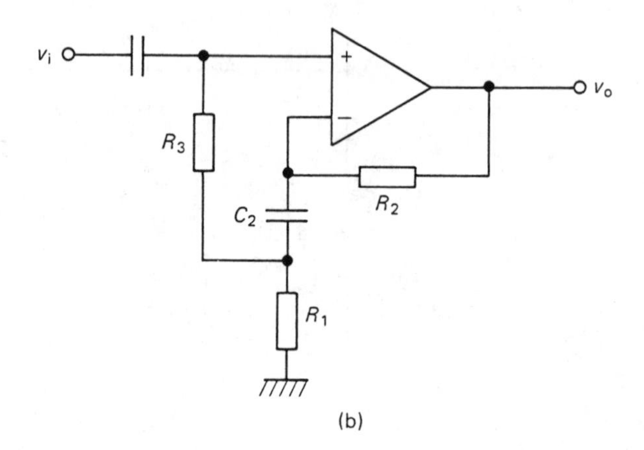

(b)

The disadvantage of this addition is that the input impedance is reduced to the value of the resistor R_3.

This can be overcome by using bootstrapping as shown in diagram (b). The principle is that the lower end of R_3 is, for signal frequencies, connected (via C_2) to the inverting input, which is itself at a signal voltage equal to that at the output. The voltage gain is very nearly unity, so that the voltages at the two ends of R_3 are very nearly the same. This means that the current flow through R_3 is very small, and the effective resistance is in fact equivalent to a resistance $R_3\ (1 + A)$ connected between the input and common.

Example 7.9

Define the terms (a) differential voltage gain, (b) common-mode voltage gain, (c) common-mode rejection ratio.

A differential amplifier with a differential gain of 200 and with a CMRR of 90 dB is fed from a balanced source that provides a difference mode signal of 40 mV. On each input there is noise pickup of 400 mV. Calculate the desired output signal and the noise output signal and the ratio of signal to noise at the input and at the output.

Solution 7.9

See text for definition of terms.

$$\text{CMRR} = \frac{A_d}{A_c} = 90\ \text{dB} \approx 30\,000,$$

$$\therefore A_c = \frac{A_d}{30\,000} = \frac{200}{30\,000} = 6.67 \times 10^{-3}.$$

The desired output voltage signal is given by

$$v_o = A_d v_s = 200 \times 40 \text{ mV}$$
$$= \underline{8 \text{ V}}.$$

The noise output signal is given by

$$v_o = A_c v_n = 6.67 \times 10^{-3} \times 400 \text{ mV}$$
$$= \underline{2.7 \text{ mV}}.$$

The signal-to-noise ratio at the input is

$$\frac{40 \text{ mV}}{400 \text{ mV}} = 0.1 = \underline{-20 \text{ dB}}$$

The signal-to-noise ratio at the output is

$$\frac{8 \text{ V}}{2.7 \text{ mV}} = \underline{70 \text{ dB}}.$$

Notice that this result could have been calculated directly:

$$\begin{matrix}\text{signal-to-noise ratio} \\ \text{at output (dB)}\end{matrix} = \begin{matrix}\text{signal-to-noise ratio} \\ \text{at input (dB)}\end{matrix} + \text{CMRR (dB)} = -20 \text{ dB} + 90 \text{ dB}$$
$$= \underline{70 \text{ dB}}.$$

Example 7.10

Draw the equivalent circuit of an op-amp.

An op-amp has a differential input resistance R_i, negligible output resistance, and voltage gain A_v. Obtain an expression for the gain of the feedback amplifier (v_o/v_i) when used as an inverting amplifier as shown in diagram (a). Determine the minimum value of differential input resistance that will cause a 1 per cent error in the definition of 'ideal' voltage gain ($-R_2/R_1$), given that $A_v = 10\,000, R_1 = R_2 = 10$ MΩ. (CEI Part 2)

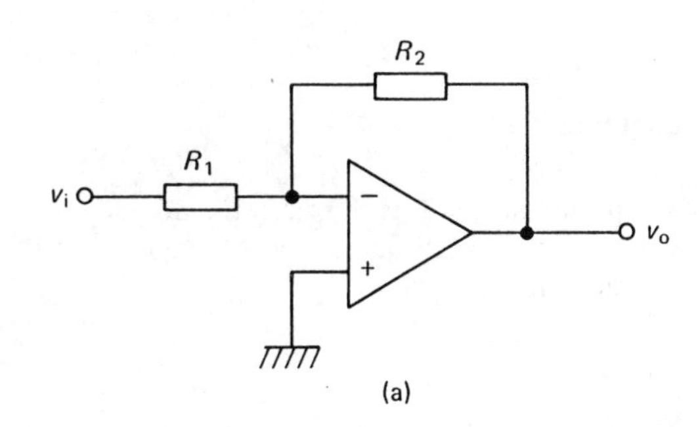

(a)

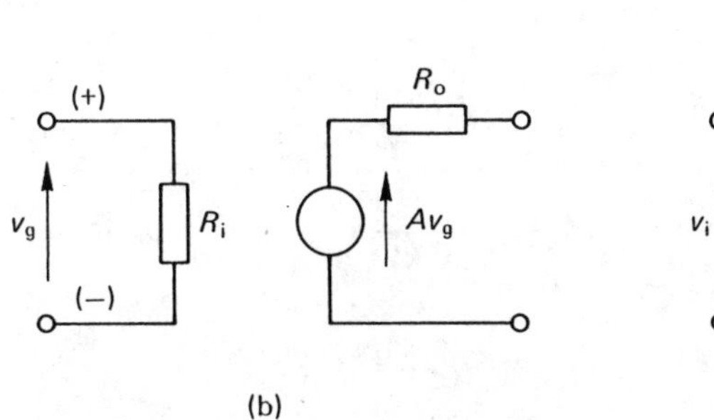

(b)

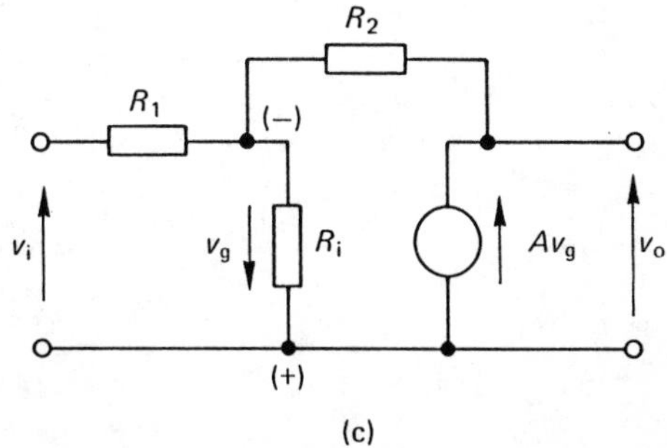

(c)

Solution 7.10

The op-amp equivalent circuit is drawn in diagram (b). Substituting this equivalent circuit into the configuration of diagram (a) with $R_o = 0$ produces the circuit of diagram (c). Summing the currents at the junction of the resistors, and noting that the signal is applied to the inverting input, then

$$\frac{v_i - (-v_g)}{R_1} = -\frac{v_g}{R_i} + \frac{(-v_g) - Av_g}{R_2},$$

$$\therefore \frac{v_i}{R_1} = -\frac{v_g}{R_1} - \frac{v_g}{R_i} - \frac{v_g}{R_2} - \frac{Av_g}{R_2},$$

$$= -v_g\left[\frac{1}{R_1} + \frac{1}{R_i} + \frac{1+A}{R_2}\right].$$

But $v_o = Av_g$,

$$\therefore \frac{v_i}{R_1} = -\frac{v_o}{A}\left[\frac{1}{R_1} + \frac{1}{R_i} + \frac{1+A}{R_2}\right],$$

$$\therefore \frac{v_o}{v_i} = -\frac{R_2}{R_1}\left(\frac{1}{1 + \dfrac{R_1R_2 + R_1R_i + R_iR_2}{AR_1R_i}}\right).$$

For $A_v = 10^4$ and $R_1 = R_2 = 10$ MΩ,

$$\frac{v_o}{v_i} = -\frac{1}{1 + \dfrac{2R_i + 10^7}{10^4 R_i}}$$

An error of 1 per cent in the definition of the ideal voltage gain requires

$$\frac{1}{1 + \dfrac{2R_i + 10^7}{10^4 R_i}} = 0.99 \text{ or } 1.01$$

$$\therefore R_i \geqslant 10^5\ \Omega = \underline{100\ \text{k}\Omega}.$$

Example 7.11

An op-amp integrator is shown in the diagram, where $R = 10$ kΩ and $C = 0.1$ μF. If the input waveform is a 2 kHz square wave of amplitude 2 V, sketch the output waveform and calculate its amplitude.

Solution 7.11

$$v_o = -\frac{1}{RC}\int v_i\,dt$$

$$= -\frac{1}{10\ \text{k}\Omega \times 0.1\ \mu\text{F}}\int v_i\,dt$$

$$= -1000\int v_i\,dt.$$

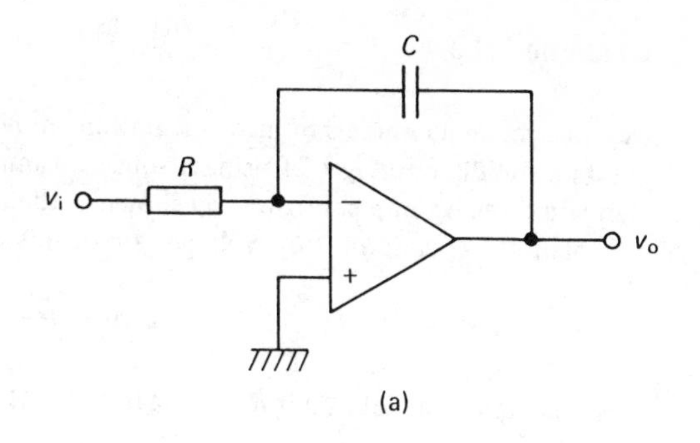

(a)

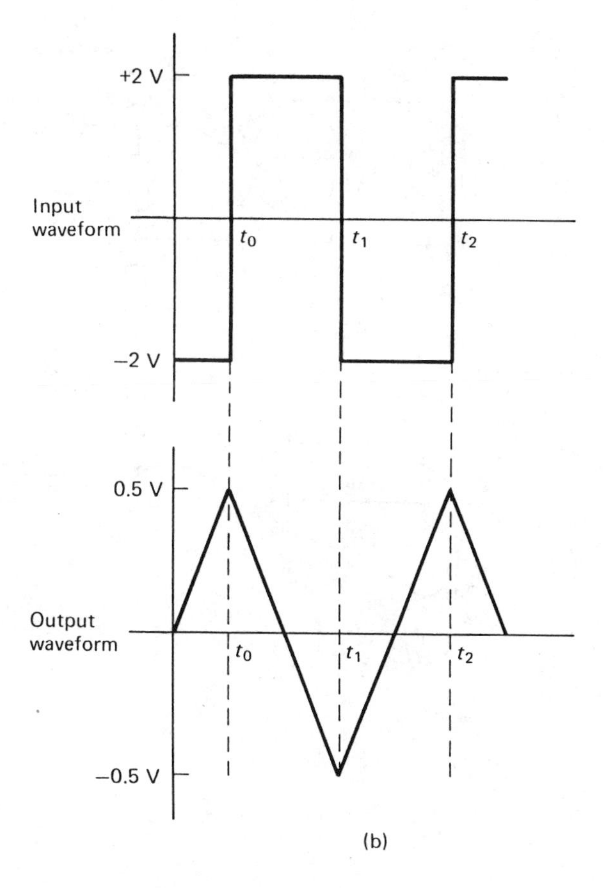

(b)

For an input step, the output will be a linear ramp. The resultant waveform will be as shown in diagram (b). For t_0 to t_1

$$v_o = -1000 \int v_i \, dt .$$

For a step change from −2 V to +2 V (i.e. 4 V change), the change in output voltage will be

$$\Delta V_o = -1000 \times 4 \times 0.25 \times 10$$

$$= \underline{-1 \text{ V}}.$$

A similar change will take place between t_1 and t_2 but in the opposite direction, thus giving an output waveform as shown in diagram (b).

Example 7.12

Describe the requirements of instrumentation amplifiers.

State the disadvantages of using a single op-amp in such an application and explain why the standard three-op-amp configuration shown in the diagram is used in preference.

Obtain the equation for the voltage gain of this circuit as

$$A = \frac{v_o}{e_2 - e_1}.$$

Calculate the actual gain if $R_3 = R_4, R_1 = 50\ \Omega$ and $R_2 = 20\ \text{k}\Omega$.

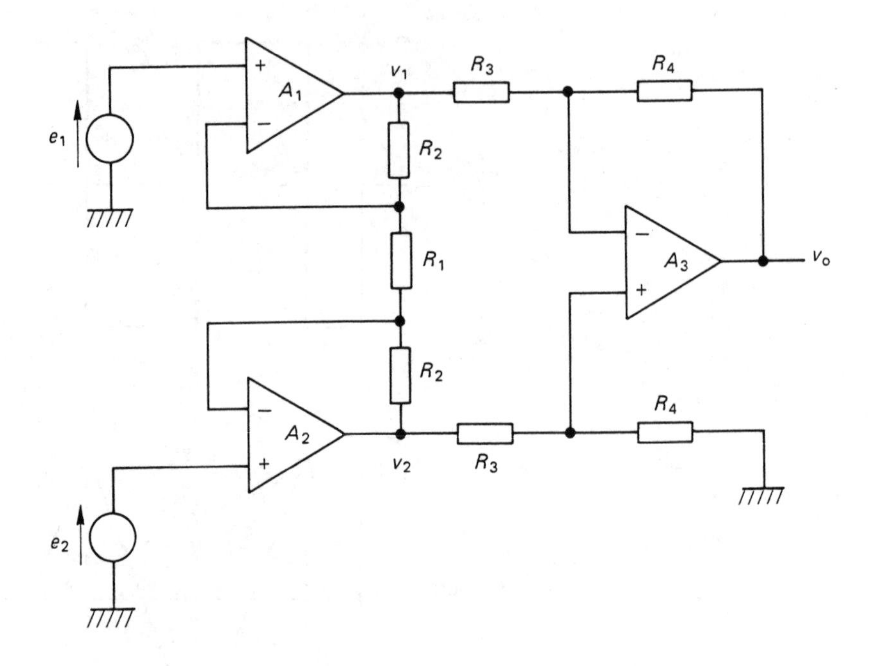

Solution 7.12

Instrumentation amplifiers are high-gain, d.c.-coupled, differential amplifiers with a single-ended output, high input impedance and high CMRR. They are used to amplify *small* difference signals produced by transducers, in which there may be a large common-mode signal or d.c. level. A CMRR of 100 dB is a typical requirement, and bias currents and offsets are important considerations.

The problem with instrumentation amplifiers that use only one op-amp is gain accuracy, which is often required to be 0.1 per cent. This puts restrictions on source impedances such that they must be kept very low and well matched. This difficulty is overcome by using the three op-amp configuration of the diagram in which op-amp followers achieve very high input impedance.

The gain of the difference amplifier A_3 can be evaluated by considering it as a combination of inverting and non-inverting amplifiers. Using the superposition principle, the output voltage due to v_1 and v_2 can be obtained by adding the effects due to v_1 and v_2 individually.

With $v_2 = 0$, $$v_{o1} = -\frac{R_4}{R_3} v_1.$$

With $v_1 = 0$, $$v_2 \frac{R_4}{R_4 + R_3} = \frac{R_3}{R_3 + R_4} v_{o2},$$

$$\therefore v_{o2} = \frac{R_4}{R_3} v_2,$$

$$\therefore v_o = \frac{R_4}{R_3} (v_2 - v_1).$$

For the unity-gain amplifiers A_1 and A_2 the inverting inputs must also be at e_1 and e_2.

$$\therefore \frac{v_2 - v_1}{2R_2 + R_1} = \frac{e_2 - e_1}{R_1},$$

$$\therefore \frac{v_o}{e_2 - e_1} = \frac{R_4}{R_3} \left(\frac{2R_2 + R_1}{R_1}\right).$$

$$\therefore A = \frac{2 \times 20 \times 10^3 + 50}{50} = 801.$$

7.5 Unworked Problems

Problem 7.1

Define the following terms relating to an operational amplifier: input offset current, input offset voltage, output offset voltage. How can the effects of these offsets be minimised?

Calculate the voltage gain and input impedance of the amplifier shown. The operational amplifier has an open-circuit gain $A = 10^5$. $R_{cm} = 25\ \text{M}\Omega$, $R_i = 100\ \text{k}\Omega$, $R_1 = 2\ \text{k}\Omega$, $R_2 = 100\ \text{k}\Omega$.

(CEI Part 2)

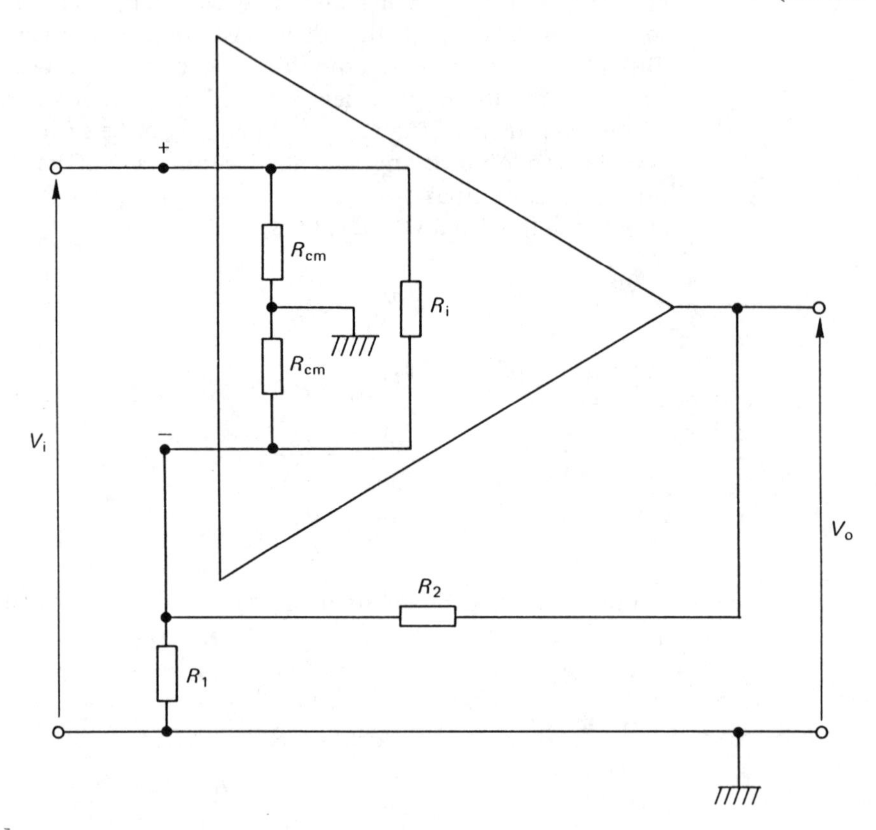

Problem 7.2

A unity-gain buffer amplifier (voltage follower) is shown in the diagram. Draw the equivalent circuit, neglecting the effect of output resistance, and derive the relationship for the closed-loop voltage gain (v_o/v_i). Show that the closed-loop voltage gain is very close to unity using the conditions $A = 100\,000$, $R_i = 10\ \text{k}\Omega$ and $R_s = 10\ \text{k}\Omega$. Calculate the input resistance of the circuit. What statement can you make about the effect of source and input resistance on the closed loop voltage gain?

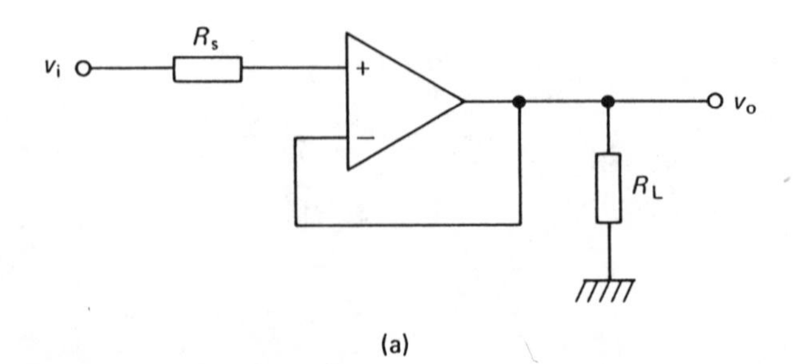

(a)

Problem 7.3

For the op-amp integrator of the diagram, derive the equation that relates v_o and v_i. State the disadvantage of this circuit, and explain how the addition of a high-value resistor across the capacitor can overcome the disadvantage. An integrator is made up as shown in the diagram, but with the inclusion of a resistor R_2 across the capacitor. If the desired response of the integrator is to be $v_o = -2000 \int v_i \, dt$ and the accuracy should be within 2 per cent for a 20 kHz sinewave, calculate values for R_1 and R_2, assuming that $C = 0.1$ μF.

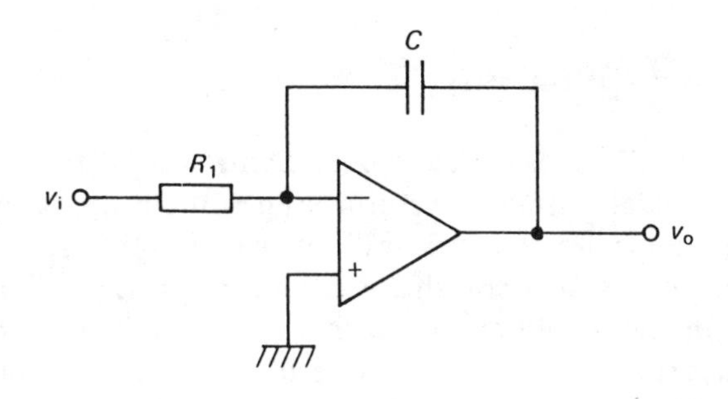

8 Sinusoidal Oscillators

8.1 Principle of Operation

An oscillator is an electronic circuit whose function is to deliver an essentially sinusoidal output waveform even without any input excitation. There are many different circuit configurations, all of which rely on the application of positive feedback to a circuit that is capable of providing amplification.

In the amplifier shown in Fig. 8.1, if it could be arranged that v_f is equal in magnitude to v_s but opposite in sign, then v_s could be removed and the output signal would be sustained with no input. This requires that:

(a) $$|A\beta| = 1;$$

(b) the loop phase shift is zero;
i.e. $$\angle A\beta = -180°$$

In fact, the frequency of oscillation will be that at which the loop phase shift is zero.

From Fig. 8.1, it would appear that if $|A\beta|$ is slightly greater than unity then the amplitude of the oscillations would increase without limit. However, non-linearity in the amplifier becomes more marked at larger amplitudes and this onset of non-linearity limiting the amplitude of oscillations is an essential feature of the operation of all practical oscillators. In every practical oscillator the loop gain is slightly larger than unity and the amplitude of the oscillations is limited by the onset of non-linearity.

The frequency-determining network may consist of an *LC* circuit, an *RC* circuit or a piezo-electric crystal. The amplifying device may be a bipolar-junction transistor (BJT), an FET or an op-amp.

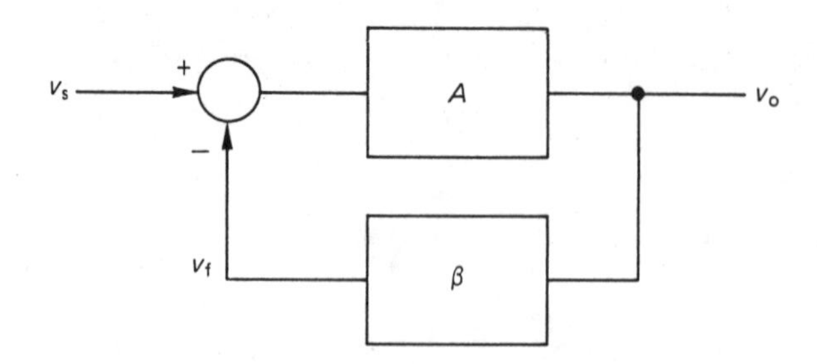

Figure 8.1 Amplifier with feedback

8.2 Phase Shift Oscillators

The phase shift oscillator using an FET as the active device shown in Fig. 8.2 exemplifies the principles of an oscillator. The FET amplifier has a phase shift of

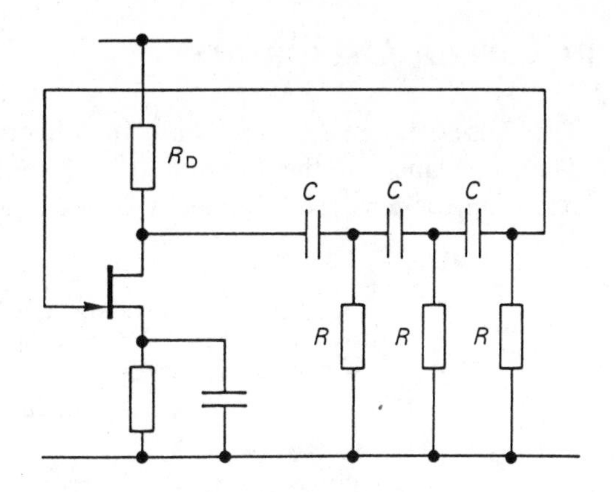

Figure 8.2 Phase shift oscillator using FET

180 °. Also at some frequency the RC network has a phase shift of 180 °. At this frequency the loop phase shift ϕ is zero. This is the frequency at which the circuit will oscillate provided that the loop gain is sufficiently large. For the phase shift network, the phase shift is 180 ° for

$$f = \frac{1}{2\pi CR\sqrt{6}}$$

and the voltage gain is $\beta = \frac{1}{29}$.

Therefore $|A\beta| \geqslant 1$ requires that $A \geqslant 29$.

In the case of a bipolar transistor being used as the active device, the output of the feedback network is shunted by the relatively low input resistance of the transistor. Voltage-shunt feedback is therefore used instead of voltage-series feedback. The arrangement is shown in Fig. 8.3 and leads to

$$f = \frac{1}{2\pi RC\sqrt{6 + \frac{4R_c}{R}}} .$$

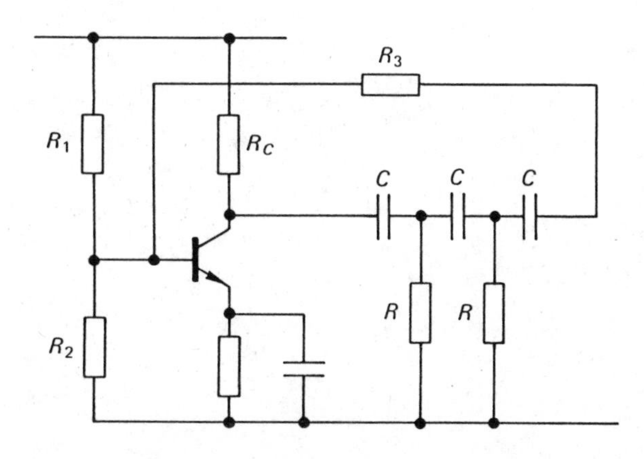

Figure 8.3 Phase shift oscillator using BJT

8.3 Resonant Circuit Oscillators

These use a tuned LC circuit as the collector or drain load as shown in Fig. 8.4. The impedance of the LC circuit at resonance is purely resistive, and the loop phase shift is zero. The frequency of oscillation is given approximately by

$$\omega = \frac{1}{\sqrt{LC}} .$$

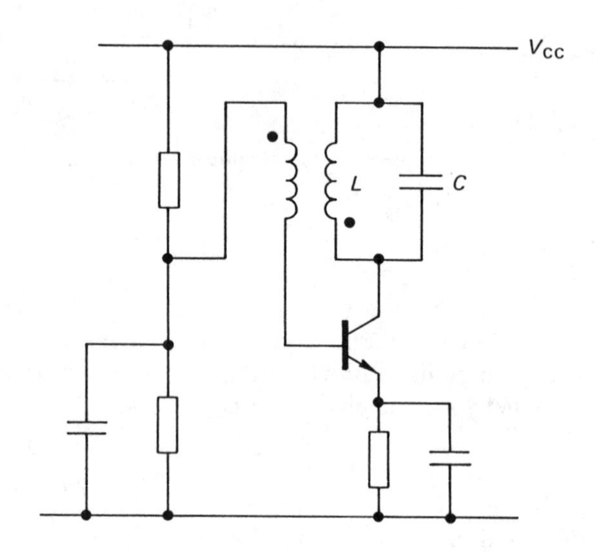

Figure 8.4 Tuned LC oscillator

8.4 Wien Oscillator

The Wien bridge oscillator uses a balanced bridge as the feedback network as shown in Fig. 8.5. In the case shown, with both capacitors and resistors made equal, the oscillation frequency is given by

$$f = \frac{1}{2\pi CR}$$

and the required amplifier gain by

$$A_v = 3.$$

If the bipolar or field effect transistor is used to provide the amplification, then two stages are required to provide the necessary 360° phase shift. Also, some means of stabilising the gain is required.

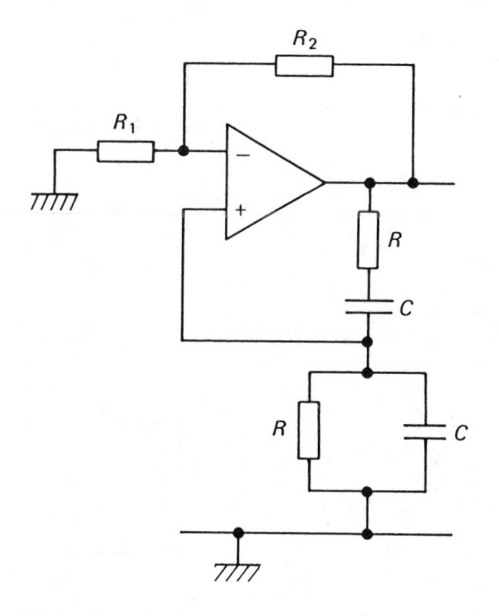

Figure 8.5 Wien bridge oscillator using an op-amp

8.5 Oscillator Design Criteria

The analysis of the criteria for oscillation for any given design of oscillator involves lengthy algebra and provides two conditions:

(a) the voltage gain necessary to maintain oscillation;
(b) the circuit conditions that control the oscillation frequency.

These two conditions are generally found by solving for the open-loop gain and setting it equal to unity. The real part of the result gives the gain requirements of the active device, while the imaginary part gives the frequency of oscillation.

Other oscillator designs often used are shown in Fig. 8.6.

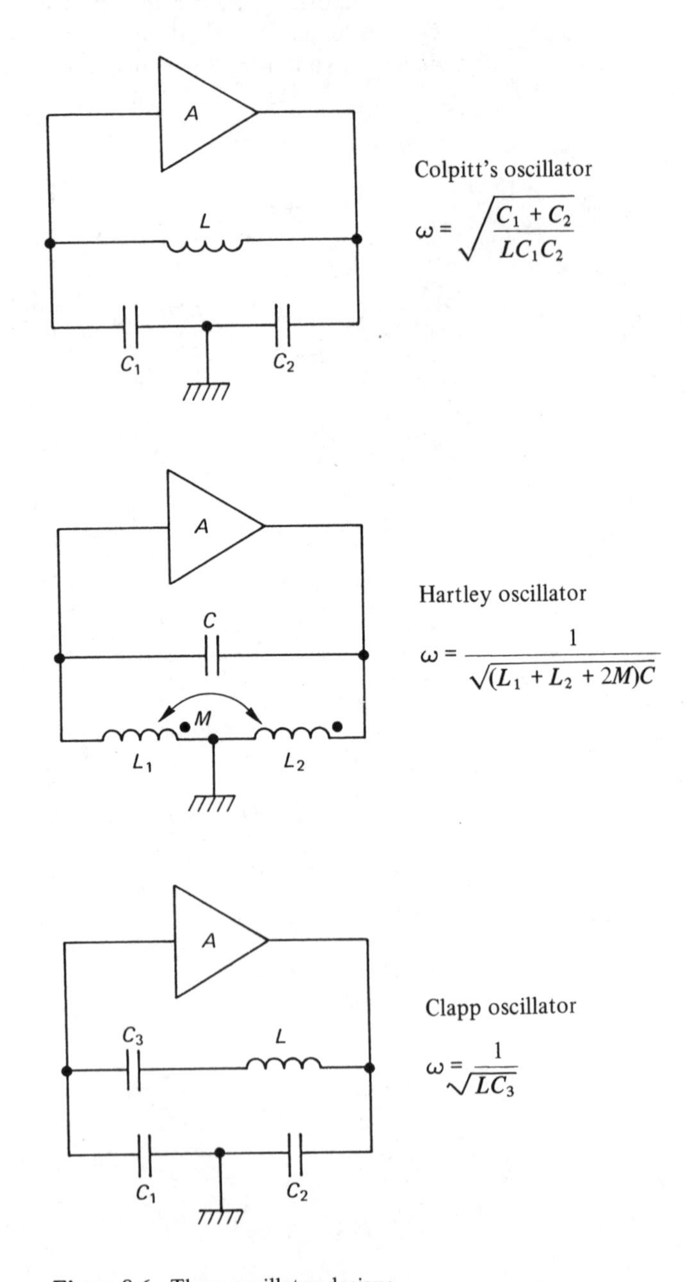

Figure 8.6 Three oscillator designs

8.6 Crystal Oscillators

A crystal oscillator is one in which the frequency-determining network is provided by a piezo-electric crystal. Crystals are avilable with fundamental natural frequencies from about 4 kHz to 10 MHz (see Example 8.8).

8.7 Worked Examples

Example 8.1

For the network shown in diagram (a), calculate v_f/v_o, assuming negligible source impedance and infinite load impedance.

Show how this network may be used as the feedback network around an FET amplifier, and show that for the circuit to oscillate it is required that

$$g_m \frac{r_d R_d}{r_d + R_d} \geqslant 29,$$

where the variables have their usual meaning.

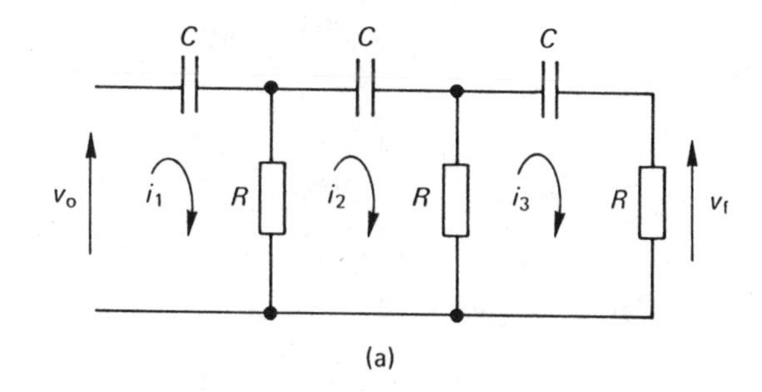

(a)

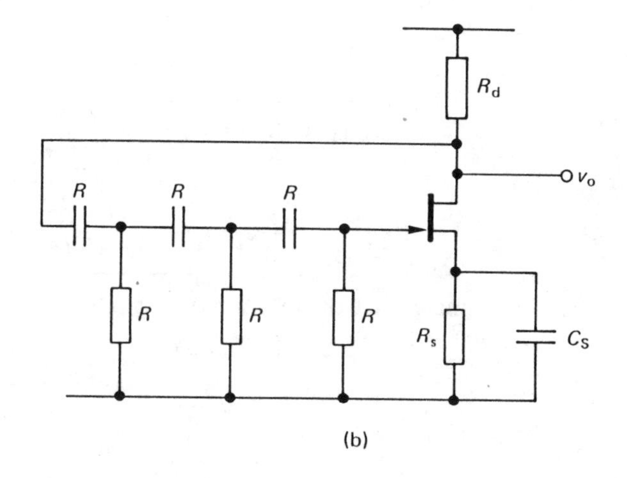

(b)

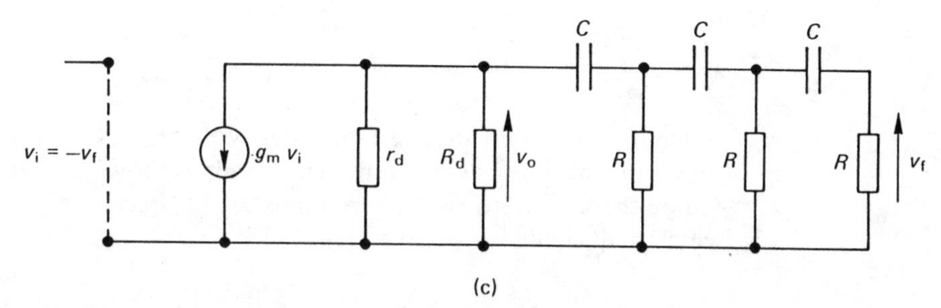

(c)

Solution 8.1

Using mesh analysis and letting the currents in the meshes of diagram (a) be i_1, i_2 and i_3, then

$$v_o = i_1 (R - jX) - i_2 R,$$
$$0 = - i_1 R + i_2 (2R - jX) - i_3 R$$
$$= - i_2 R + i_3 (2R - jX).$$

$$\therefore \begin{bmatrix} v_o \\ 0 \\ 0 \end{bmatrix} = \begin{bmatrix} R - jX & -R & 0 \\ -R & 2R - jX & -R \\ 0 & -R & 2R - jX \end{bmatrix} \begin{bmatrix} i_1 \\ i_2 \\ i_3 \end{bmatrix}.$$

Let $$\alpha = \frac{1}{\omega CR}.$$

$$\therefore \begin{bmatrix} v_o \\ 0 \\ 0 \end{bmatrix} = R^3 \begin{bmatrix} 1 - j\alpha & -1 & 0 \\ -1 & 2 - j\alpha & -1 \\ 0 & -1 & 2 - j\alpha \end{bmatrix} \begin{bmatrix} i_1 \\ i_2 \\ i_3 \end{bmatrix}.$$

The determinant Δ is given by

$$\Delta = R^3 \begin{vmatrix} 1 - j\alpha & -1 & 0 \\ -1 & 2 - j\alpha & -1 \\ 0 & -1 & 2 - j\alpha \end{vmatrix}$$
$$= R^3 [(1 - j\alpha)(2 - j\alpha)^2 - (1 - j\alpha) - (2 - j\alpha)]$$
$$= R^3 [1 - 5\alpha^2 + j(\alpha^3 - 6\alpha)].$$

Using Cramer's rule,

$$\Delta_3 = \begin{vmatrix} R - jX & -R & v_o \\ -R & 2R - jX & 0 \\ 0 & -R & 0 \end{vmatrix}$$
$$= v_o R^2.$$

$$\therefore i_3 = \frac{\Delta_3}{\Delta} = \frac{R^2 v_o}{R^3 [1 - 5\alpha^2 + j(\alpha^3 - 6\alpha)]},$$

$$\therefore \frac{v_f}{v_o} = \frac{i_3 R}{v_o} = \frac{1}{1 - 5\alpha^2 + j(\alpha^3 - 6\alpha)}.$$

For 180° phase shift the j-term = 0.

$$\therefore \alpha^3 - 6\alpha = 0,$$
$$\therefore \alpha = \sqrt{6},$$
$$\therefore f = \frac{1}{2\pi CR\sqrt{6}},$$

also $$\frac{v_f}{v_o} = \frac{1}{1 - 5\alpha^2} = \frac{1}{1 - 30} = -\frac{1}{29}.$$

The network may be used together with a JFET to produce an oscillator as shown in diagram (b) with equivalent circuit as shown in diagram (c).

For the circuit to oscillate requires a 180° phase shift between v_o and v_f, at which frequency

$$\left|\frac{v_f}{v_o}\right| = \frac{1}{29}.$$

$\therefore$ required FET voltage gain is 29 to give a loop gain of unity for oscillation.

$$\therefore A = g_m \frac{r_d R_d}{r_d + R_d} \geqslant 29.$$

Example 8.2

A transistor phase shift oscillator is shown in diagram (a), where $R_3 = R - R_i$, R_i being the input resistance of the transistor. Draw the equivalent circuit and show that the oscillation frequency is given by

$$f = \frac{1}{2\pi CR\sqrt{6 + \frac{4R_c}{R}}}.$$

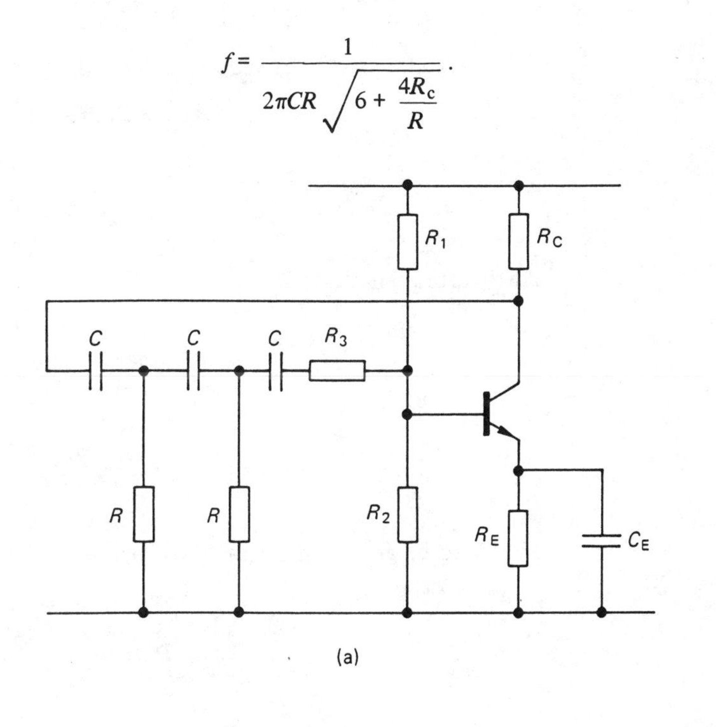

(a)

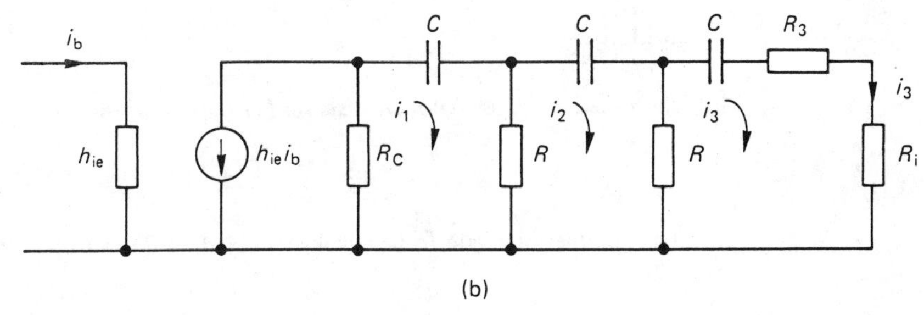

(b)

State the limitations on the transistor current gain to ensure oscillation.

Solution 8.2

The equivalent circuit is shown in diagram (b) where $R_i \approx h_{ie}$. For the circuit to oscillate there should be zero phase shift between i_3 and i_b.

Replacing the current generator by its Thévenin equivalent voltage generator,

$$-h_{fe}\, i_b\, R_C = i_1\,(R_C + R - jX) - i_2 R,$$

$$0 = -i_1 R + i_2\,(2R - jX) - i_3 R$$

$$0 = -i_2 R + i_3\,(2R - jX);$$

noting that $R_3 + R_i = R$.

$$\begin{bmatrix} -h_{fe}\, i_b\, R_C \\ 0 \\ 0 \end{bmatrix} = \begin{bmatrix} R_C + R - jX & -R & 0 \\ -R & 2R - jX & -R \\ 0 & -R & 2R - jX \end{bmatrix} \begin{bmatrix} i_1 \\ i_2 \\ i_3 \end{bmatrix}$$

By Cramer's rule, $i_3 = \dfrac{\Delta_3}{\Delta}$.

∴ By applying a similar analysis to Example 8.1,

$$i_3 = \frac{-h_{fe}\, i_b\, R_C R^2}{3R_C R^2 + R^3 - 5RX^2 - R_C X^2 + jX(-6R^2 - 4R_C R + X^2)}$$

For zero phase shift between i_3 and i_b, the j-term is zero,

$$\therefore\ 6R^2 + 4R_C R - X^2 = 0,$$

$$\therefore\ \omega = \frac{1}{CR\sqrt{6 + 4\dfrac{R_C}{R}}}.$$

At this frequency,

$$\frac{i_3}{i_b} = \frac{-h_{fe}\, R_C R^2}{3R_C R^2 + R^3 - 5RX^2 - R_C X^2}.$$

Now $X^2 = 6R^2 + 4RR_C$,

$$\therefore\ \frac{i_3}{i_b} = \frac{h_{fe}\, R_C R^2}{23R^2 R_C + 29R^3 + 4RR_C{}^2}.$$

For $\dfrac{i_3}{i_b}$ to be greater than unity to ensure oscillation requires

$$h_{fe} > 4\left(\frac{R_C}{R}\right) + 23 + 29\left(\frac{R}{R_C}\right).$$

Example 8.3

For the tuned-drain oscillator of diagram (a) show that the frequency of oscillation is given by

$$\omega^2 = \frac{1}{LC}\left(1 + \frac{r}{r_d}\right),$$

and that for the loop gain to be equal to unity the FET must have a mutual conductance given by

$$g_m = \frac{\mu r C}{\mu M - L},$$

where r is the equivalent series resistance of the coil primary, μ is the amplification factor and M is the mutual inductance of the coil.

Solution 8.3

The equivalent circuit is shown in diagram (b).

For unity loop gain, $v_2 = v_g$.

Now $v_2 = -j\omega MI = \dfrac{-j\omega M v_1}{r + j\omega L}$.

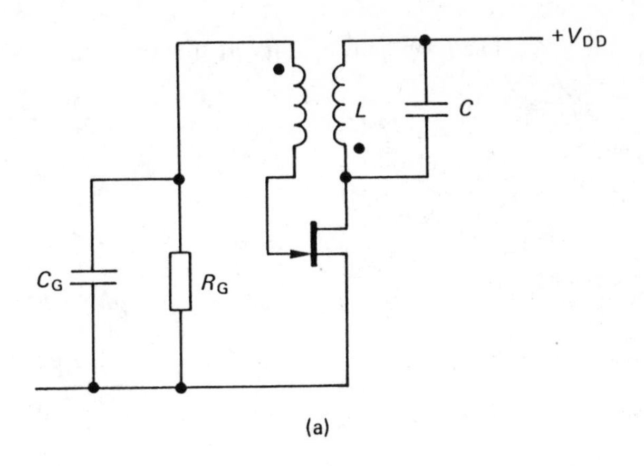

(a)

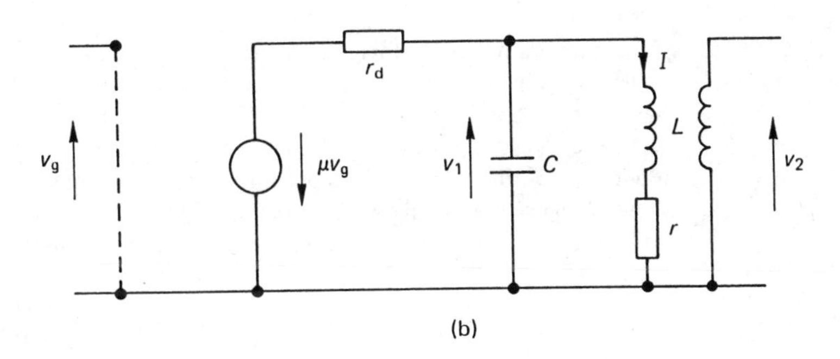

(b)

Also, $v_1 = \dfrac{-\mu v_g Z}{Z + r_d}$,

where Z is the impedance of the tuned circuit.

$$\therefore Z = \frac{(r + j\omega L)\, 1/j\omega C}{r + j\omega L + 1/j\omega C}$$

$$= \frac{r + j\omega L}{1 + j\omega C\,(r + j\omega L)} .$$

$$\therefore\ v_1 = \frac{-\mu\, v_g\,(r + j\omega L)}{(r + j\omega L) + r_d\,(1 + j\omega C\,(r + j\omega L))}$$

$$= \frac{-\mu\, v_g\,(r + j\omega L)}{r + j\omega L + j\omega C\, r_d\,\left(r + j\left(\omega L - \dfrac{1}{\omega C}\right)\right)}$$

Now the loop gain is given by

$$\frac{v_2}{v_g} = \frac{j\omega M\mu\,(r + j\omega L)}{(r + j\omega L)\left[r + j\omega L + j\omega C\, r_d\left(r + j\left(\omega L - \dfrac{1}{\omega C}\right)\right)\right]}$$

$$= \frac{\mu\omega M}{\omega\,(L + r\, r_d C) + j\,(\omega^2 r_d\, LC - r_d - r)} .$$

The loop phase shift is zero when

$$\omega^2 = \frac{r_d + r}{r_d\, LC}$$

$$\therefore\ \omega^2 = \frac{1}{LC}\left(1 + \frac{r}{r_d}\right) .$$

The loop gain is unity at this frequency when

$$\frac{\mu\omega M}{\omega\,(L + r\,r_d C)} = 1.$$

$$\therefore\ \mu M = L + r\,r_d C.$$

Now

$$\mu = g_m\ r_d\,;$$

$$\therefore\ \mu M - L = \frac{r\,\mu C}{g_m}\,,$$

$$\therefore\ g_m = \frac{\mu\,rC}{\mu M - L}$$

Example 8.4

Why are laboratory oscillators covering the frequency range up to 1 MHz normally of *CR* type?

Derive the conditions for the output voltage of the two-port network in diagram (a) to be in phase with its input.

Draw a circuit showing how this two-port network may be connected with a differential amplifier and other essential components to produce sinusoidal oscillations at 250 Hz. Determine suitable values for all significant components.

Comment briefly on the effects of the driving source resistance and load resistance across v_o.

(CEI Part 2)

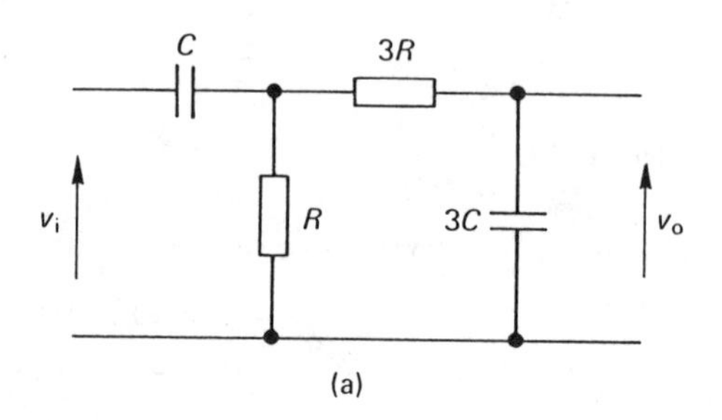

(a)

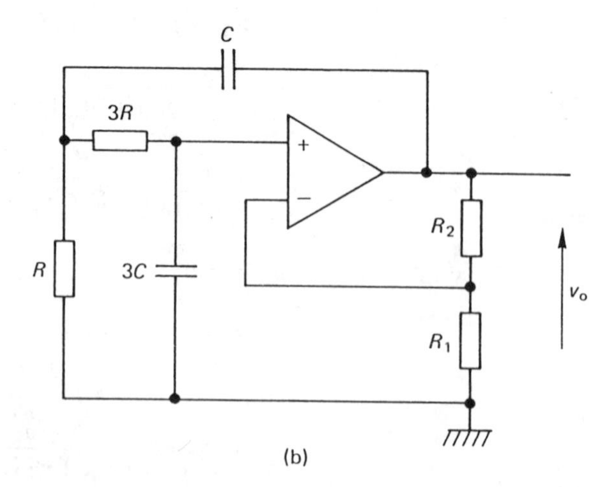

(b)

Solution 8.4

Laboratory oscillators in the low to medium frequency range are normally *CR* type since the values of *L* required for inductive oscillators would be inconveniently large. Also, a wider range of variation is possible using variable C's or R's than is possible with LC oscillators, since for CR network oscillators

$$f \propto \frac{1}{CR},$$

whereas for LC network oscillators

$$f \propto \frac{1}{\sqrt{LC}}.$$

Letting the mesh currents in diagram (a) be i_1 and i_2,

$$\begin{bmatrix} v_i \\ 0 \end{bmatrix} = \begin{bmatrix} R + \frac{1}{sC} & -R \\ -R & 4R + \frac{1}{3sC} \end{bmatrix} \begin{bmatrix} i_1 \\ i_2 \end{bmatrix}$$

$$v_o = i_2/3sC$$

Using Cramer's rule leads to

$$\frac{v_o}{v_i} = \frac{1}{13 + j\left(9\omega CR - \frac{1}{\omega CR}\right)}.$$

For zero phase shift,

$$9\omega CR = \frac{1}{\omega CR},$$

$$\underline{\omega = \frac{1}{3CR}}.$$

At this frequency, $\frac{v_o}{v_i} = \frac{1}{13}$.

If this network is used as shown in diagram (b) to provide an oscillator, then the amplifier should be non-inverting with a gain of 13.

$$\therefore 1 + \frac{R_2}{R_1} = 13,$$

$$R_2 = 12R_1.$$

For an oscillation frequency of 250 Hz,

$$2\pi \times 250 = \frac{1}{3CR}.$$

Choose R_1 = <u>10 kΩ</u>, R_2 = <u>120 kΩ</u>.

To minimise bias current effects, $4R = R_1$.

$\therefore R$ = <u>2.5 kΩ</u>, $3R$ = <u>7.5 kΩ</u>, $\therefore C$ = <u>85 nF</u>, $3C$ = <u>255 nF</u>.

Example 8.5

The circuit of diagram (a) shows a Wien oscillator in which the frequency is controlled by a single variable resistor. Determine the frequency range obtainable, describe the action of the thermistor, and calculate the range of variation of thermistor resistance required as the oscillator frequency is varied over its range. Comment on the required amplifier stability.

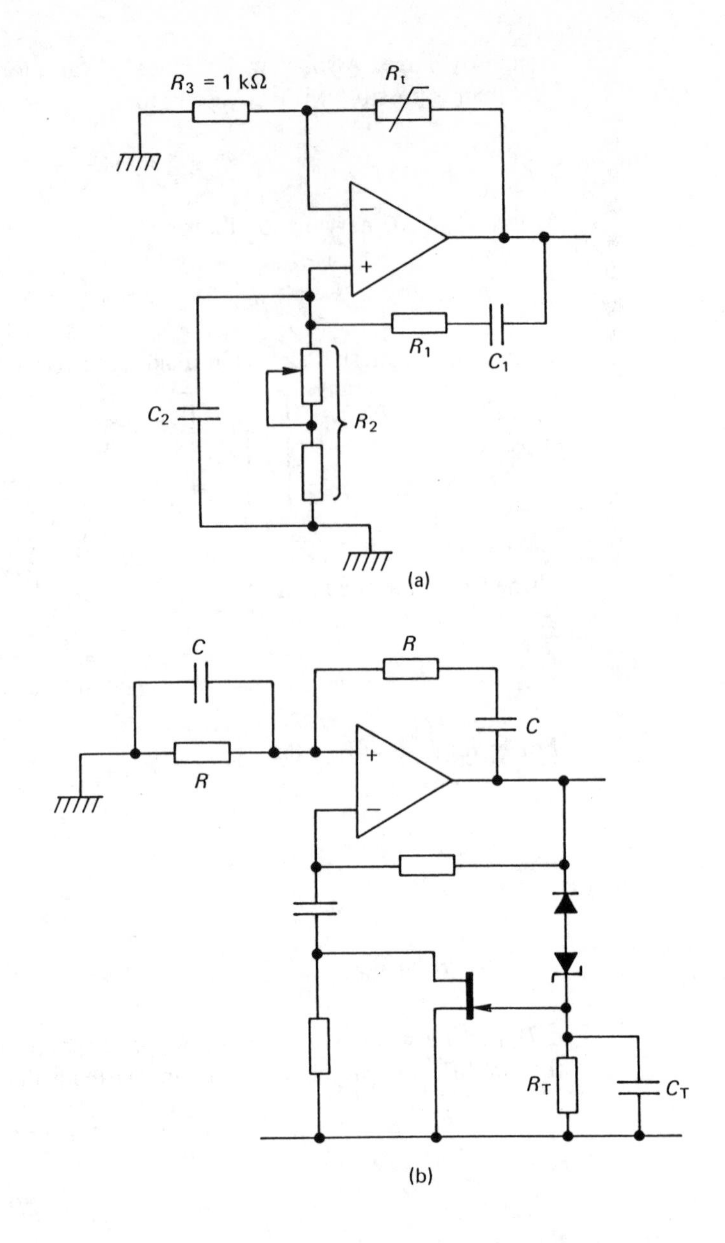

Sketch an alternative amplitude stabilisation arrangement using a FET, and describe its operation.

The component values are:

R_1 = 1kΩ, R_2 = 1 kΩ + 100 kΩ potentiometer,
C_1 = 33 nF, C_2 = 330 nF.

Solution 8.5

The standard analysis of a Wien network oscillator gives an oscillation frequency of

$$f = \frac{1}{2\pi \sqrt{C_1 C_2 R_1 R_2}}$$

and the required gain of the amplifier as

$$A = 1 + \frac{R_1}{R_2} + \frac{C_2}{C_1} = 1 + \frac{R_t}{R_3},$$

where R_t = thermistor resistance,

$$\therefore\ A_{min} = 1 + \frac{1}{100} + 10 = 11.01,$$

$$A_{max} = 1 + 1 + 10 = 12,$$

giving a frequency range of 152 Hz to 1525 Hz.

The corresponding thermistor resistance range is 10.01 kΩ to 11 kΩ.

The purpose of the thermistor is amplitude stabilisation. If the loop gain of the amplifier is not *exactly* unity, then either the oscillations will cease or the output will saturate. Using a thermistor as part of the gain-determining network, if the output increases, the thermistor resistance decreases, thus stabilising the output amplitude.

Any method of gain-setting feedback with a long time constant is suitable. An alternative is the network of diagram (b).

The FET acts as a voltage variable resistor for small applied voltages. The a.c. gain is controlled by the FET resistance variation. The $R_T C_T$ time constant is made long to avoid output distortion.

Example 8.6

The circuit shown in diagram (a) includes a negative resistance. Derive the critical conditions for oscillation and an expression for the oscillation frequency.

Show that the circuit diagram (b) is a negative-impedance converter whose input impedance is given by

$$Z_i = -R_x.$$

Solution 8.6

The network makes use of a negative resistance that is achieved using an op-amp connected to provide a 'negative-impedance converter' (NIC). For the circuit to oscillate requires that the negative resistance must cancel out the circuit losses.

The equivalent parallel representation of the LCr circuit is shown in diagram (c), where

$$R_0 = \frac{L}{Cr} \quad \text{is the dynamic impedance.}$$

Thus the critical conditions for oscillation are that the negative resistance is equal and opposite to the dynamic impedance R_0.

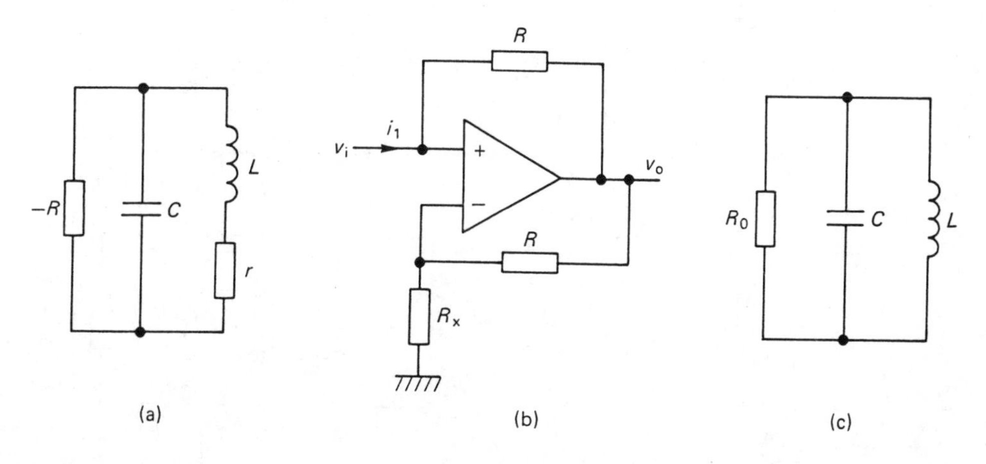

The oscillation frequency is given by

$$f = \frac{1}{2\pi}\sqrt{\frac{1}{LC} - \frac{r^2}{L^2}}.$$

For the NIC,

$$v_i = \frac{R_x}{R + R_x} v_o.$$

Also,
$$i_i = \frac{v_i - v_o}{R},$$

$$\therefore \frac{v_i}{i_i} = \frac{R}{1 - \frac{R + R_x}{R_x}} = -R_x.$$

Example 8.7

For the Clapp oscillator shown in diagram (a), obtain the oscillation condition, in terms of the circuit elements and the FET parameters. Hence evaluate the oscillation frequency if

$L = 200\ \mu H$, $r = 5\ \Omega$, $R_L = 2\ k\Omega$;
$C_1 = 0.15\ \mu F$, $C_2 = 0.01\ \mu F$, $C_3 = 100$ pF.

Assume that r_d is negligibly large, and ignore the effect of unlabelled components.
Discuss the frequency stability of the oscillator compared with that of the Colpitts oscillator.

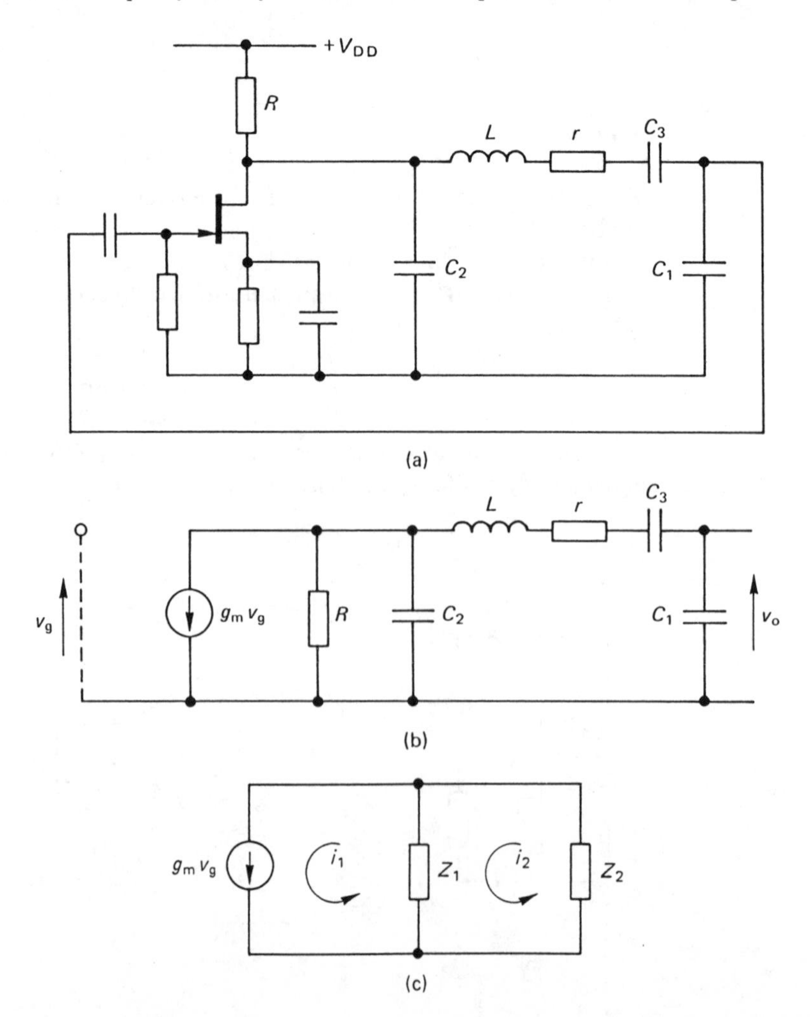

Solution 8.7

The equivalent circuit is shown in diagram (b). We may consider the model as made up of two impedances Z_1 and Z_2 as shown in diagram (c), where

$$Z_1 = \frac{R}{1 + j\omega C_2 R},$$

$$Z_2 = r + j\left(\omega L - \frac{1}{\omega C_1} - \frac{1}{\omega C_2}\right).$$

$$\therefore\ v_o = j\frac{i_2}{\omega C_1} = \frac{j g_m v_g Z_1}{\omega C_1 (Z_1 + Z_2)}$$

$$= \frac{j g_m v_g R}{\omega C_1 \left[R + r(1 + j\omega C_2 R) + j(1 + j\omega C_2 R)\left(\omega L - \frac{1}{\omega C_1} - \frac{1}{\omega C_2}\right)\right]}.$$

Now for oscillation we require unity loop gain with $v_o = v_g$.

$$\therefore\ R + r + j\omega C_2 Rr + j\left(\omega L - \frac{1}{\omega C_1} - \frac{1}{\omega C_3}\right) - \omega^2 L C_2 R + C_2 R\left(\frac{1}{C_1} + \frac{1}{C_3}\right) = \frac{j g_m R}{\omega C_1}.$$

Equating real terms,

$$\omega^2 L C_2 R = R\left(\frac{C_2}{C_1} + \frac{C_2}{C_3} + 1\right) + r,$$

$$\therefore\ \omega^2 = \frac{1}{L}\left(\frac{1}{C_1} + \frac{1}{C_2} + \frac{1}{C_3}\right) + \frac{r}{L C_2 R}.$$

The first term defines the resonant frequency in terms of L and C_1, C_2 and C_3 in series. The second term is a 'correction' term often negligible in practice.

Inserting the values given,

$$f_{osc} = \underline{1.13\ \text{MHz}}.$$

The effects of transistor junction capacitance are 'swamped' by the high values of C_1 and C_2. The oscillation frequency is mainly determined by the high-Q network LC_3.

Example 8.8

Diagram (a) shows the equivalent circuit of a quartz crystal resonator. Show that the ratio between the two resonant frequencies of the circuit is given by

$$\sqrt{1 + (C_1/C_0)}.$$

Give typical values for the elements of the circuit and hence explain why such a resonator is widely used in high-stability oscillators.

Explain the operation of, and give circuits for, crystal oscillators that use the crystal as (a) an inductive and (b) a resistive circuit element. (CEI Part 2)

Solution 8.8

In the equivalent circuit of diagram (a), r, L and C_1 represent the equivalent series components of the crystal and C_0 the capacitance of the holder in which it is mounted, where

$$C_0 \gg C_1,$$

$$\therefore\ \omega_0 \approx \frac{1}{\sqrt{LC_1}},$$

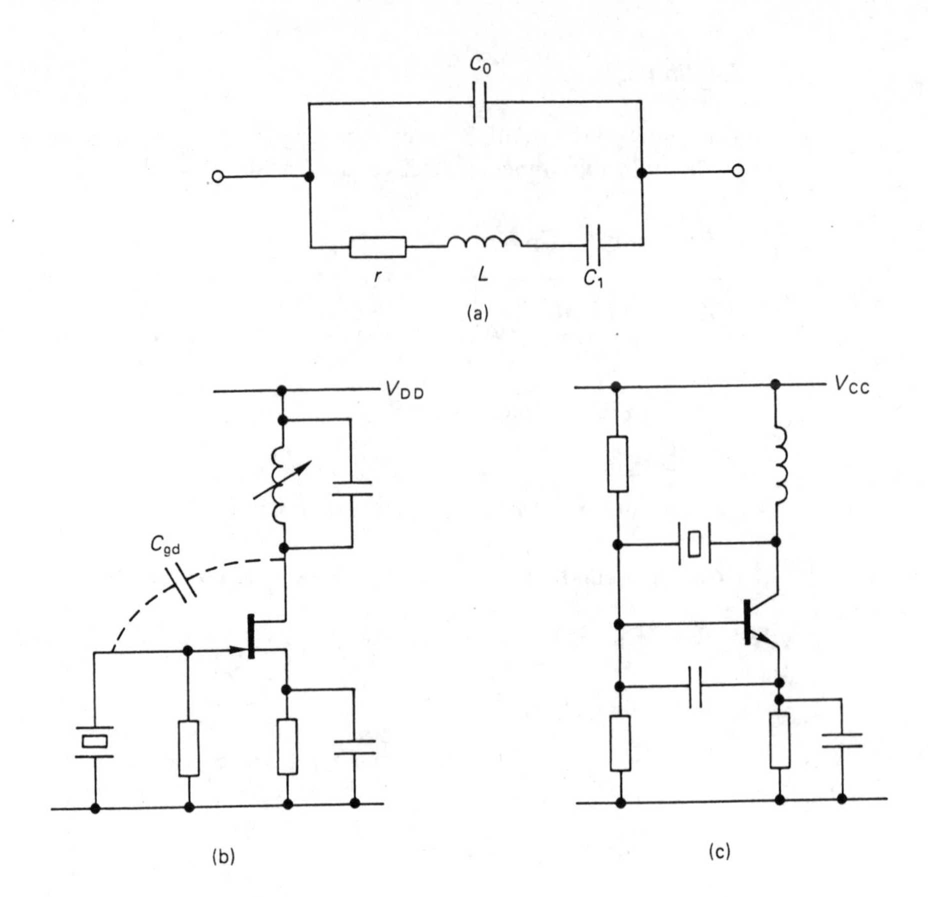

and the bandwidth is given by

$$B \approx \frac{1}{2\pi R C_1}.$$

Typical values for, say, a 400 kHz crystal are

$$L = 3.2 \text{ H}, \qquad C_1 = 0.05 \text{ pF},$$
$$r = 4 \text{ k}\Omega, \qquad C_0 = 6 \text{ pF}.$$

∴ The LC_1r branch of the equivalent circuit is series-resonant with

$$\omega_s L = \frac{1}{\omega_s C_1}$$

$$\therefore \omega_s = \frac{1}{\sqrt{LC_1}}.$$

At higher frequencies the effective reactance of the series branch is inductive. Parallel resonance occurs at a frequency where the effective inductive reactance of the series branch has the same magnitude as the reactance of the parallel capacitor C_0

$$\therefore \omega_p L - \frac{1}{\omega_p C_1} = \frac{1}{\omega_p C_0},$$

$$\therefore \omega_p^2 = \frac{1}{LC_1}\left(1 + \frac{C_1}{C_0}\right).$$

Thus $$\omega_p = \omega_s \sqrt{1 + \frac{C_1}{C_0}}.$$

Since $C_0 \gg C_1$, there is very little difference between the series and parallel resonant frequencies and

$$f_p \approx 1.01 f_s.$$

Crystal oscillators are widely used for the following reasons:

(a) Since the inductive effect is large, the Q-factor is large (typically 10 000).
(b) There is good frequency stability (better than 1 in 10^6).
(c) The external circuit capacitance does not significantly effect the resonant frequency.
(d) The temperature coefficient is of the order of 1 p.p.m./°C.

An oscillator using a crystal as an inductive element is shown in diagram (b). It uses a tuned-collector load, and the crystal acts as an LC network with effective feedback via the drain–gate capacitance.

The circuit of diagram (c) uses the crystal as a feedback network.

Example 8.9

The general form of an oscillator using an LC network is shown in diagram (a), where the amplifier may be considered to have an infinite input resistance, a voltage gain A_v and an output resistance R_o. Ideal lossless reactances may be assumed. Show that for oscillation to take place,

(a) X_1 and X_2 must have the same sign (i.e. both reactances either inductive or capacitive);
(b) X_3 must be of the opposite sign;

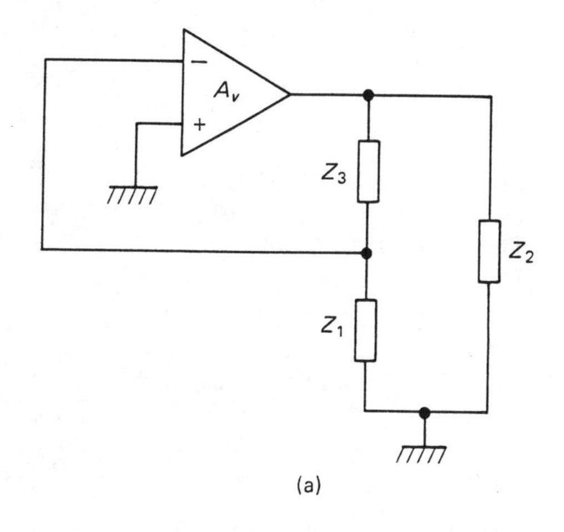

(a)

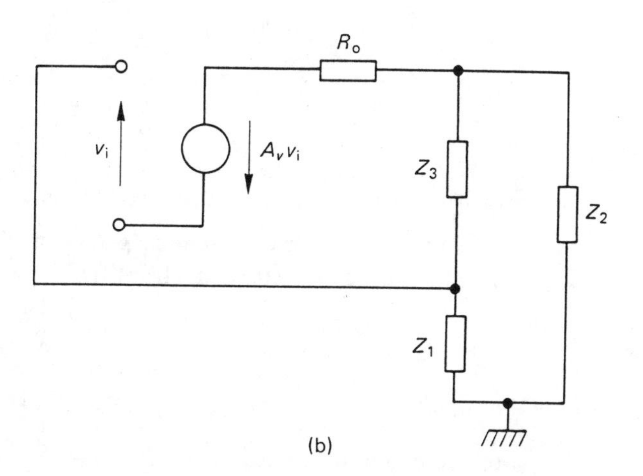

(b)

(c) $$A_v = \frac{X_2}{X_1}.$$

Hence derive an equation for the oscillation frequency of a Colpitts oscillator, where

$$Z_1 = \frac{1}{j\omega C_1}, \qquad Z_2 = \frac{1}{j\omega C_2}, \qquad Z_3 = j\omega L.$$

Solution 8.9

The equivalent circuit is shown in diagram (b).

Consider the loop gain $A\beta$. The gain without feedback is given by

$$A = -\frac{A_v Z_L}{Z_L + R_o},$$

where $$Z_L = Z_2 \parallel (Z_1 + Z_3) = \frac{Z_2 (Z_1 + Z_3)}{Z_1 + Z_2 + Z_3}.$$

The feedback factor is given by

$$\beta = -\frac{Z_1}{Z_1 + Z_3},$$

$$\therefore A\beta = \frac{-A_v Z_1 Z_2}{R_o (Z_1 + Z_2 + Z_3) + Z_2 (Z_1 + Z_3)}.$$

If ideal lossless reactances are assumed, then

$$Z_1 = jX_1, \qquad Z_2 = jX_2, \qquad Z_3 = jX_3.$$

$$\therefore A\beta = \frac{A_v (jX_1)(jX_2)}{jR_o (X_1 + X_2 + X_3) + jX_2 (jX_1 + jX_3)}$$

$$= \frac{-A_v X_1 X_2}{jR_o (X_1 + X_2 + X_3) - X_2 (X_1 + X_3)}.$$

The requirements for oscillation are that the loop gain $A\beta = -1$.

The phase condition is met if

$$X_1 + X_2 + X_3 = 0 \qquad \text{...(i)}$$

The magnitude condition is met if

$$\frac{-A_v X_1 X_2}{-X_2 (X_1 + X_3)} = -1.$$

But $X_1 + X_3 = -X_2$ from (i)

$$\therefore \frac{A_v X_1 X_2}{X_2^2} = 1,$$

$$\therefore A_v = \frac{X_2}{X_1} \qquad \text{...(ii)}$$

Equations (i) and (ii) lead to the following interpretations:

(a) X_1 and X_2 must have the same sign (from (ii)).

(b) X_3 must be of opposite sign (from (i)).

(c) $$A_v = \frac{X_2}{X_1}.$$

In practice, $$A_v > \frac{X_2}{X_1}$$

to overcome the circuit losses.

For the Colpitts oscillator,

$$X_1 + X_2 + X_3 = 0,$$

where $X_1 = -\dfrac{1}{\omega C_1}$. $\quad X_2 = -\dfrac{1}{\omega C_2}$, $\quad$ and $X_3 = \omega L$.

$$\therefore \quad -\frac{1}{\omega C_1} - \frac{1}{\omega C_2} + \omega L = 0,$$

$$\omega L = \frac{C_1 + C_2}{\omega C_1 C_2},$$

$$\therefore \quad \omega^2 = \frac{C_1 + C_2}{L C_1 C_2}.$$

8.8 Unworked Problems

Problem 8.1

A varactor diode having a capacitance $C = 65/\sqrt{V}$ pF, where V is in volts, is connected across the frequency-determining circuit of an oscillator. The inductance in the tuned circuit is 600 μH.

(a) With a 3.5 V bias across the diode, determine the capacitance that, connected in parallel with the diode and inductance, will give an oscillation frequency of 500 kHz.
(b) With the above value of C, what bias voltage on the diode would be required to give an oscillation frequency of 520 kHz? (EC Part 1)

Problem 8.2

In the circuit shown, R_1 is slowly decreased from its maximum value of 10 kΩ until oscillation occurs. Find the corresponding value of R_1 and the frequency of oscillation. Derive the equations used and state any assumptions made.

What factors limit the amplitude of the output v_o? Describe briefly the effect of reducing R_1 further, e.g. to 1 kΩ. (EC Part 1)

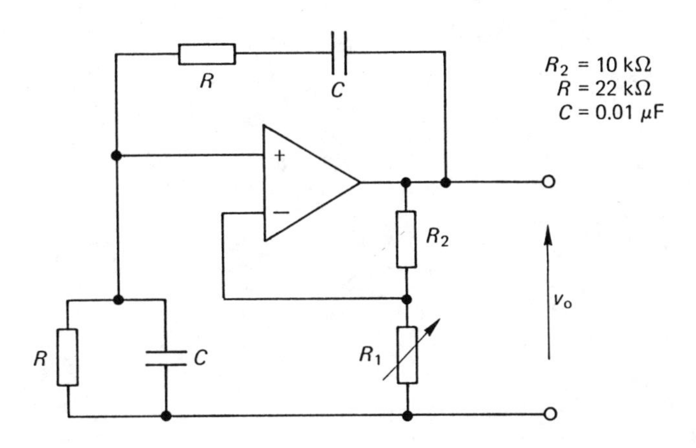

Problem 8.3

Comment briefly on the factors that influence the choice of a frequency-determining network for an oscillator.

Calculate the frequency of oscillation for the circuit shown for (a) $R_L = \infty$, (b) $R_L = 2$ kΩ. The transistor T_1 has $h_{ie} = 1.5$ kΩ, $h_{fe} = 80$, $h_{oe} = 120$ μS and negligible h_{re} and the effects of the unlabelled capacitors may be neglected. What assumptions have been made in your calculations? How could the circuit be modified to give greater frequency stability? (CEI Part 2)

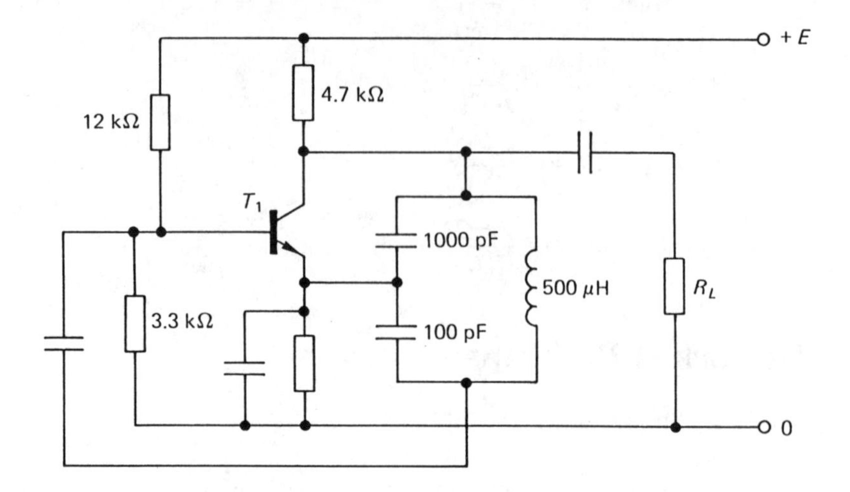

9 Non-linear Analogue Systems

9.1 Introduction

There is a wide range of non-linear analogue systems that may be designed using discrete components, op-amps, or specialised integrated circuits.

Basic types of non-linear system include square-wave generators, triangular-waveform generators, monostables (one-shots), Schmitt triggers, comparators, analogue multipliers, logarithmic amplifiers, amplitude modulators, sample and hold circuits, digital-to-analogue converters and analogue-to-digital converters.

Some of these systems are considered in the following examples.

9.2 Worked Examples

Example 9.1

Describe the operation of the Schmitt trigger circuit shown in diagram (a), and discuss the uses of the circuit. Determine the output waveform when the input is the triangular wave shown in diagram (b). The amplifier output has limits of ± 10 V.

Suggest a modification to the circuit that would limit the output voltage swing to ± 5 V.

(CEI Part 2)

Solution 9.1

The output voltage of a Schmitt trigger is either high or low, depending on whether the input is above or below some threshold level. The circuit in the diagram is regenerative since the feedback is to the non-inverting input, thus causing the output to switch suddenly as the input voltage passes the threshold voltage.

Uses:

(a) reconstituting pulses;
(b) voltage sensing;
(c) squaring waveforms;
(d) driving relays where hysteresis avoids jitter.

The voltage at the inverting input is

$$\frac{R_3}{R_3 + R_4} \times 15 \text{ V} = +1.5 \text{ V}.$$

Remembering that the Schmitt trigger acts as a saturating switch where the output can only be high or low, for v_i less than 1.5 V, then

$$v_o = -10 \text{ V}.$$

$\therefore$ Voltage at non-inverting input is

$$v = -10 + \left(\frac{R_2}{R_1 + R_2}\right)(v_i + 10).$$

When this voltage exceeds 1.5 V then the output suddenly switches to +10 V. Just prior to switching,

$$v_i = (1.5 + 10)\left(\frac{R_1 + R_2}{R_2}\right) - 10$$

$$= 1.73 \text{ V}.$$

For v_i greater than 1.5 V, then $v_o = +10$ V.
$\therefore$ Voltage at non-inverting input is

$$v = +10 + \frac{R_2}{R_1 + R_2}(v_i - 10).$$

When this voltage falls below 1.5 V then the output suddenly switches to −10 V. Just prior to switching,

$$v_i = (1.5 - 10)\left(\frac{R_1 + R_2}{R_2}\right) + 10$$

$$= 1.33 \text{ V}.$$

The input hysteresis is thus

$$1.73 - 1.33 = 0.4 \text{ V about } 1.53 \text{ V}.$$

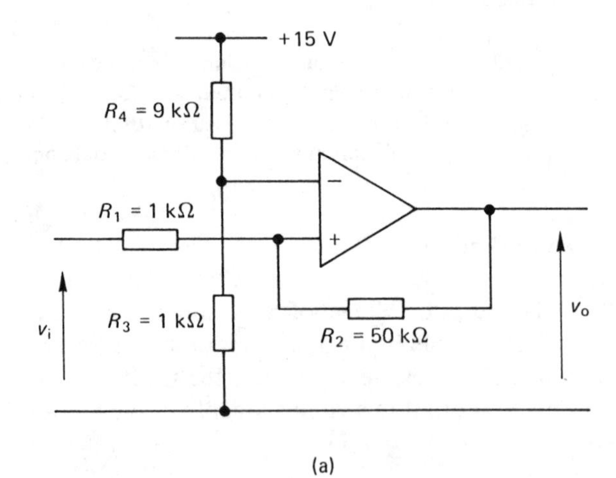

(a)

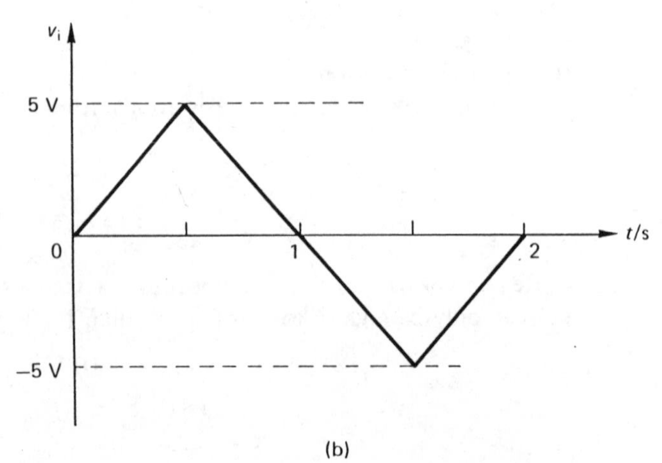

(b)

The output waveform is a square wave with the period of the mark given by

$$\left(\frac{5 - 1.73}{5}\right) 0.5 + \left(\frac{5 - 1.33}{5}\right) 0.5 = 0.694 \text{ seconds.}$$

The space period is given by

$$2 - 0.694 = 1.306 \text{ s.}$$

$$\therefore \text{ Mark–space ratio} = 0.53.$$

To limit the output voltage to ±5 V use zeners or clipping diodes.

Example 9.2

Define the term slew rate as it relates to an operational amplifier. Show that signals of large amplitude will be distorted at high frequencies, and derive an expression relating the maximum undistorted signal to the slew rate. Describe a method of measuring the slew rate.

Determine the frequency and waveshape of the output v_o of the circuit shown in diagram (a). The amplifier output may be assumed to saturate at ± 10 V. What factors limit the performance of this circuit at high frequencies? (CEI Part 2)

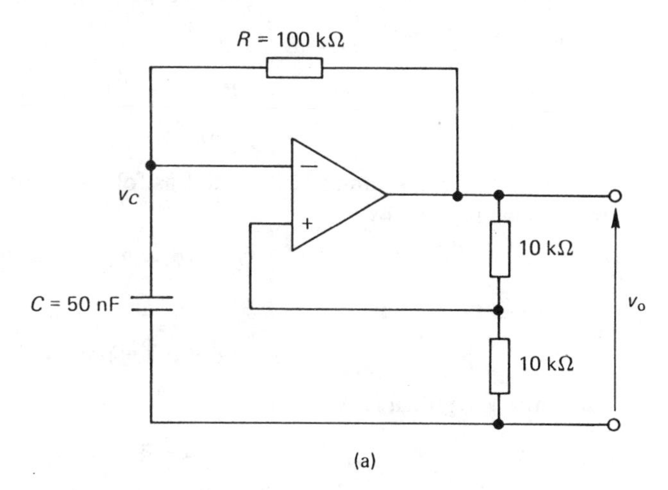

(a)

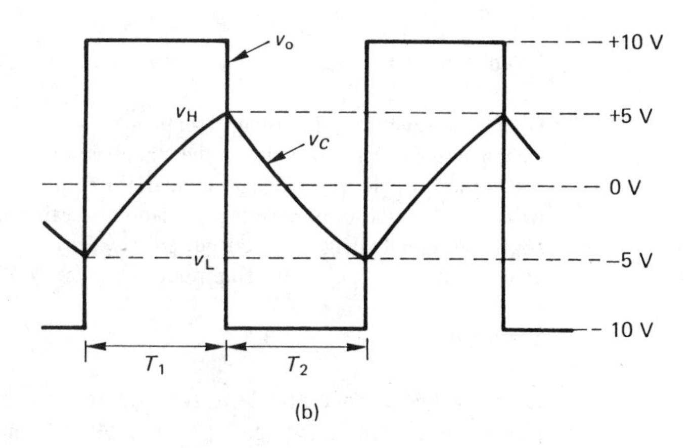

(b)

Solution 9.2

Slew rate is discussed in Chapter 7. Slew rate may be measured using an oscilloscope, and applying a voltage to the input so that a complete swing is achieved at the output from one saturating level to the other.

The op-amp and the 10 kΩ resistors together form an inverting comparator. The addition of the *CR* network converts the arrangement into an astable multivibrator.

The output waveform is a square wave of amplitude ± 10 V. The waveform at the inverting input is an exponential rise and decay between two switching thresholds as shown in diagram (b). The time constant is determined by the *CR* network and the exponential waveforms are 'aiming' at + 10 V or − 10 V.

The upper switching threshold is at

$$v_o = v_{o(max)}$$

$$\therefore\ v_H = v_r + \beta\,[v_{o(max)} - v_r]$$

$$= v_r\,(1 - \beta) + \beta v_{o(max)},$$

where β = feedback factor = 0.5 and v_r = reference voltage connected to bottom end of resistor pair (0 V in this case).

The lower switching threshold is when $v_o = v_{o(min)}$.

$$\therefore\ v_L = v_r + \beta\,[v_{o(min)} - v_r]$$

$$= v_r\,(1 - \beta) + \beta v_{o(min)}.$$

Inserting the values gives

$$v_H = 0.5 \times 10\text{ V} = 5\text{ V},$$

$$v_L = 0.5 \times (-10\text{ V}) = -5\text{ V}.$$

The waveform timing is calculated as follows. During the period T_1 the capacitor voltage is given by

$$v_C = -5\mathrm{e}^{-t/CR} + 10\,(1 - \mathrm{e}^{-t/CR})$$

$$= +5 \qquad \text{at } t = T_1.$$

$$\therefore\ T_1 = CR\ \ln\left(\tfrac{15}{5}\right) = \underline{5.49\text{ ms}}.$$

Since the waveform is symmetrical,

$$T_2 = T_1 = \underline{5.49\text{ ms}}.$$

The factors that limit the high-frequency performance of this circuit are op-amp slew rate and frequency response.

Example 9.3

Define the functions of a voltage comparator, and compare the performances of an op-amp and a comparator integrated circuit in these applications.

Diagram (a) shows a voltage comparator connected as a monostable multivibrator. Sketch waveforms of the output voltage and non-inverting input voltage after the application of a trigger pulse, and calculate the output pulse width. Assume that the comparator output saturates at + 5 V and 0 V. Specify the amplitude and polarity of the trigger pulse required.

Solution 9.3

Using a Thévenin equivalent for the resistor network R_1 and R_2, it may be replaced by a source voltage of − 1.5 V with output resistance R = 14.85 kΩ. The comparator output normally resides at 0 V. The application of a negative-going trigger pulse at the input causes the output to switch to + 5 V. The voltage at the non-inverting input thus rises by 5 V and then decays exponentially back toward − 1.5 V as shown in diagram (b), giving an output waveform as shown.

Now $v_C = 5\mathrm{e}^{-t/CR} - 1.5$.

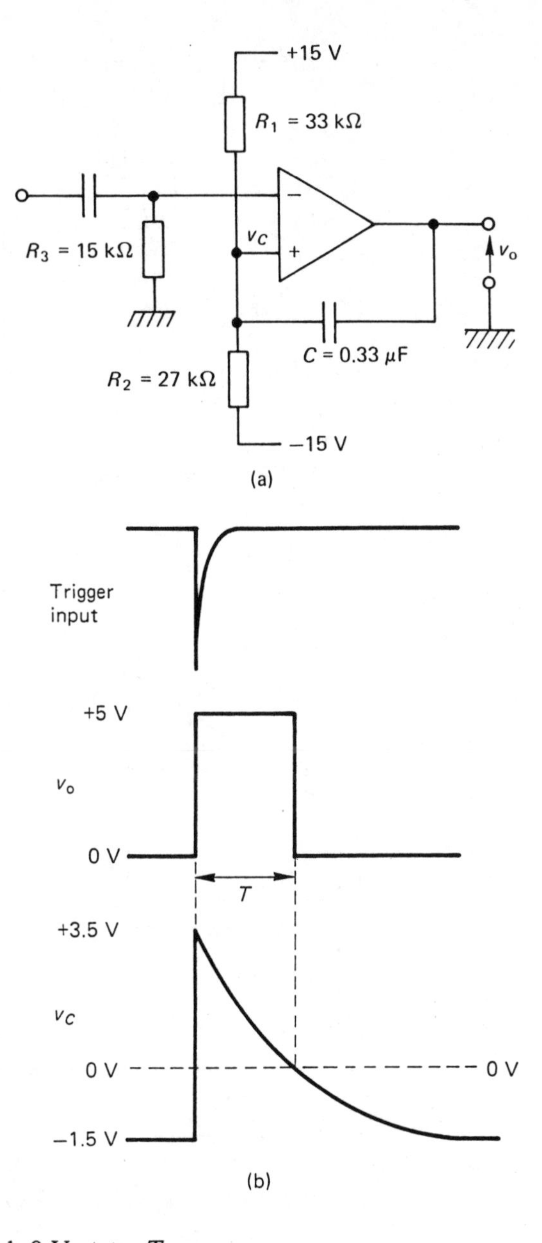

This passes through 0 V at $t = T$.

$$\therefore\ 0 = 5e^{-T/CR} - 1.5$$

$$T = CR \ln 5/1.5 = \underline{5.9 \text{ ms}}.$$

The negative trigger pulse must be sufficiently large to go below the − 1.5 V at the non-inverting input.

Example 9.4

Explain the action of the circuit shown in diagram (a), when v_i is increased linearly with time from 0 V to + 10 V and similarly decreased back to 0 V.

Sketch to scale v_i and v_o to a common base of time.

Reasonable approximations may be made. Describe a use for this circuit in practice. Explain why the capacitor C is included in the circuit, and state what factors influence the choice of its value. (CEI Part 2)

R_1 = 2.2 kΩ, R_2 = 820 Ω, R_3 = 47 kΩ,
R_4 = 470 Ω, R_5 = 68 kΩ; β = 70 for each transistor.

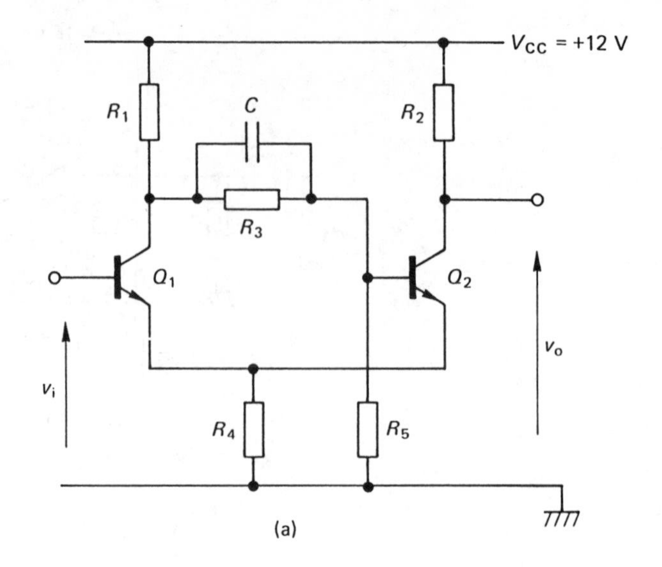

(a)

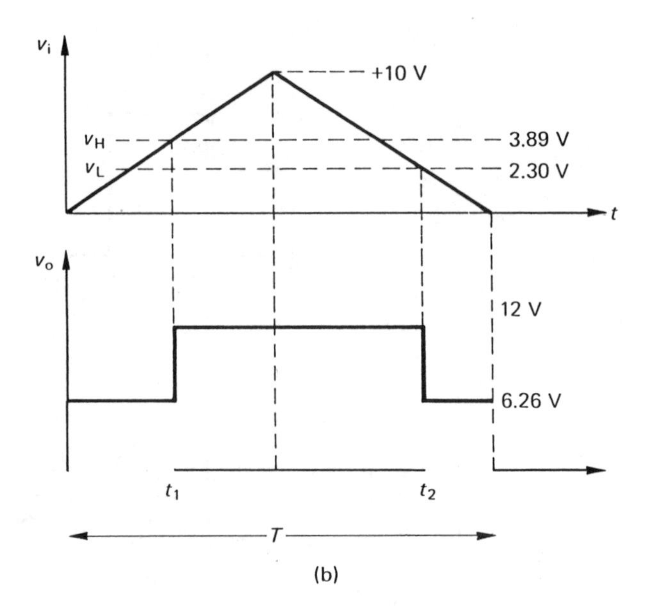

(b)

Solution 9.4

The circuit is a Schmitt trigger. The output voltage will be either high or low, depending on whether the input is above or below some threshold value.

With v_i at 0 V, Q_1 is off and Q_2 is biased on by the divider chain $R_1 R_3 R_5$. The emitter current in Q_2 develops a voltage across R_4 so that the emitter of Q_1 is positive, and holds Q_1 cut-off.

When the input goes positive, then at some threshold value Q_1 switches on, thus pulling its collector potential low, and switching Q_2 off.

The threshold levels of the input are determined by considering the values of Q_2 base potential in the two states. This analysis is an approximation due to the slight difference in V_{BE} between a transistor that is fully conducting and one that is at the switching threshold.

Let the switch-on threshold = v_H,

Let the switch-off threshold = v_L.

The hysteresis $H = v_H - v_L$.

Analysis with Q_2 on and Q_1 off:
Using Thévenin's rule,

$$v_H = V_{B2(on)} = \frac{R_5 V_{CC}}{R_1 + R_3 + R_5} - I_{B2} \frac{R_5 (R_1 + R_3)}{R_1 + R_3 + R_5},$$

where $I_{B2} = \dfrac{I_{E2}}{1 + h_{fe}}$ at $v_i = v_H$

$$= \frac{v_H - V_{BE}}{(1 + h_{fe}) R_4}.$$

$$\therefore\ v_H = \frac{R_5 V_{CC}}{R_1 + R_3 + R_5} - \frac{(v_H - V_{BE})}{(1 + h_{fe}) R_4} \frac{R_5 (R_1 + R_3)}{R_1 + R_3 + R_5}.$$

Using $V_{BE} = 0.6$ V,

$$v_H = \underline{3.89 \text{ V}}$$

Analysis with Q_2 off and Q_1 on:
Using Thévenin's rule,

$$v_L = V_{B2(off)} = \frac{R_5 V_{CC}}{R_1 + R_3 + R_5} - I_{C1} \frac{R_1 R_5}{R_1 + R_3 + R_5},$$

where $I_{C1} \approx I_{E1}$ at $v_i = v_L$,

$$= \frac{v_L - V_{BE}}{R_4}.$$

$$\therefore\ v_L = \frac{R_5 V_{CC}}{R_1 + R_3 + R_5} - \frac{(v_L - V_{BE})}{R_4} \times \frac{R_1 R_5}{R_1 + R_3 + R_5}$$

$$= \underline{2.30 \text{ V}}.$$

The output voltages are:

with Q_2 off, $v_{oH} = \underline{12 \text{ V}}$;

with Q_2 on, $v_{oL} = 12 - I_{C2} R_2$,

where $I_{C2} = \dfrac{v_H - V_{BE}}{R_4} = 7$ mA,

$$\therefore\ v_{oL} = \underline{6.26 \text{ V}}.$$

With the input waveform specified, the output waveform is as shown in diagram (b), where

$$t_1 = 0.19T \qquad \text{and} \qquad t_2 = 0.88T.$$

The uses include reconstituting pulses, voltage sensing, squaring waveforms and driving relays where the hysteresis avoids jitter.

C is a speed-up capacitor to increase the rate of switching. A typical value would be 150 pF.

9.3 Unworked Problems

Problem 9.1

A sinusoidal input voltage of peak amplitude 5 V is applied to the input terminals of the circuit shown. Explain the operation of the circuit and sketch the input and output voltage waveforms.

Determine approximately the significant voltages and phase angles of the input sinusoid. Indicate where approximations have been made.

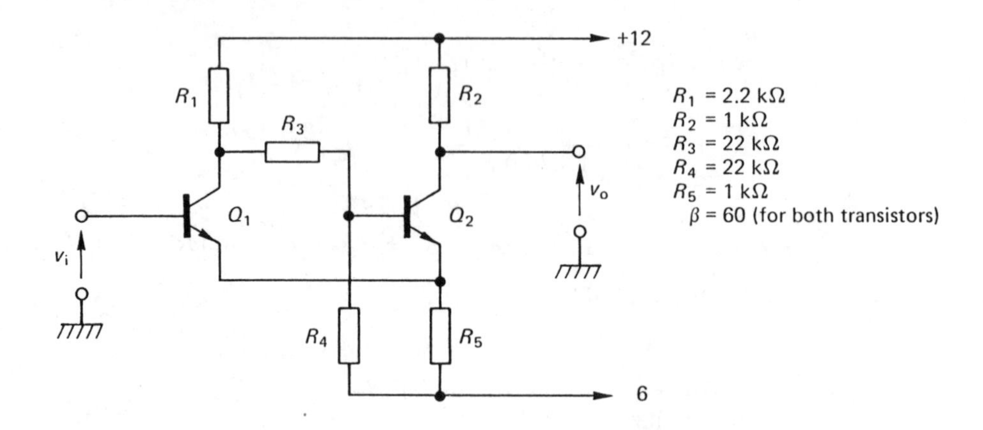

Problem 9.2

State the function of a voltage comparator and describe the action of the circuit shown in the diagram. Suggest suitable resistor values to give a hysteresis of 0.2 V if the ideal operational amplifier output saturates at ± 10 V.

Calculate the mark-space ratio of the output voltage waveform if the input signal is $V = 2 + 8 \sin \omega t$ volts. Explain the deficiencies of operational amplifier comparators as opposed to comparator integrated circuits.

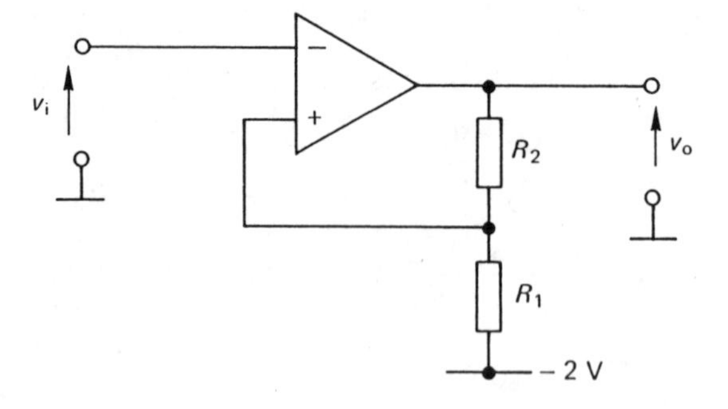

Problem 9.3

What features of an oscillator circuit determine
(a) the frequency of oscillation, and
(b) whether it operates as a sinusoidal or a relaxation oscillator?

Explain the operation of the oscillator circuit shown. Sketch the waveforms of the voltages v_+ and v_- that appear at the inputs of the amplifier and that of v_o, the output voltage. Assuming identical saturation voltages and an ideal amplifier, derive an expression for the frequency of oscillation. (CEI Part 2)

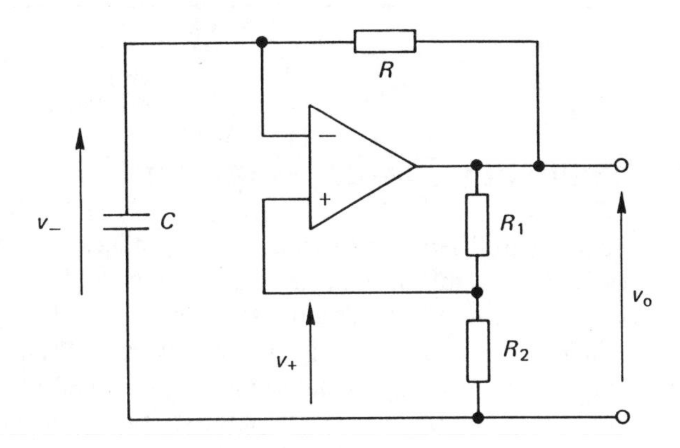

Problem 9.4

Calculate the frequency of oscillation of the multivibrator shown in the diagram. Sketch, to scale and to a common base of time, the collector and base voltage waveforms for each transistor.

Assume that the two transistors are identical and that collector current flows for $V_{BE} \geqslant 0.6$ V. State clearly any assumptions made in the calculation.

Modify the circuit to provide monostable operation with an output pulse duration 1.75 ms. Show on the diagram the location, the polarity and the approximate minimum amplitude of the input signal to trigger the circuit. (CEI Part 2)

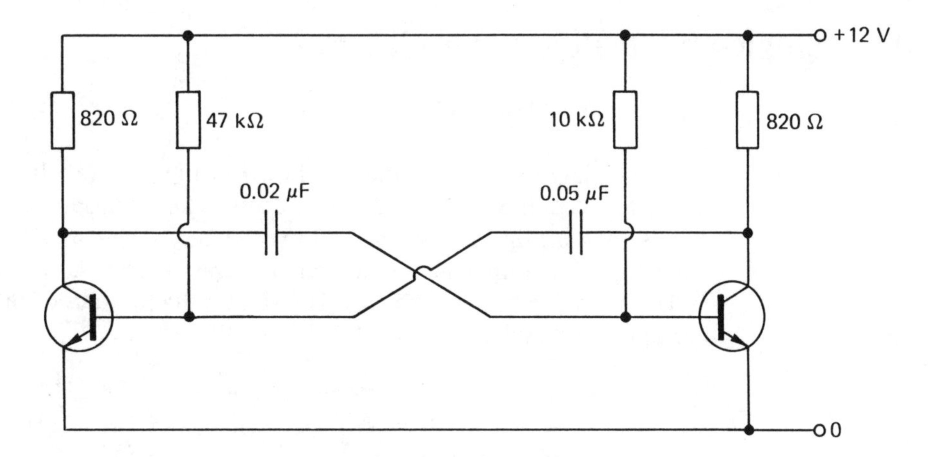

10 Power Amplifiers

10.1 Audio-frequency Power Amplifiers

Power amplifiers are used as output stages to drive loudspeakers, servo-motors, cathode ray tubes, etc., and need to be able to deliver large voltage or current swings or appreciable amounts of power.

The simplified small-signal transistor model is not valid for use in analysing amplifiers with large signal swings, and instead the analysis is generally carried out graphically on the device characteristics.

We are mainly concerned with aspects such as output power, power conversion efficiency and percentage distortion.

The transistor must be biased so that its maximum current, voltage and power ratings are not exceeded.

Transistor power amplifiers are either single-ended (a single-output transistor; see section 10.2) or push–pull (using two transistors connected in push–pull) (see section 10.3).

10.2 Single-ended Power Amplifiers

(a) Direct-coupled Load

A class-A large-signal amplifier is shown in Fig. 10.1(a). It is a simple series-fed transistor amplifier with a resistive load. The corresponding transistor output curves and load line are as shown in Fig. 10.1(b). They are shown with equidistant collector current increments for equal increments of base current.

The output power to the load is given by the product of the r.m.s. voltage and current as follows:

$$P_L = \frac{V_m}{\sqrt{2}}\frac{I_m}{\sqrt{2}} = \frac{V_m I_m}{2} = \frac{I_m^2 R_L}{2} = \frac{V_m^2}{2R_L}.$$

This may also be written in the form

$$P_L = \frac{(V_{max} - V_{min})(I_{max} - I_{min})}{8}.$$

Thus the output power may be calculated simply by plotting the load line on the output characteristics.

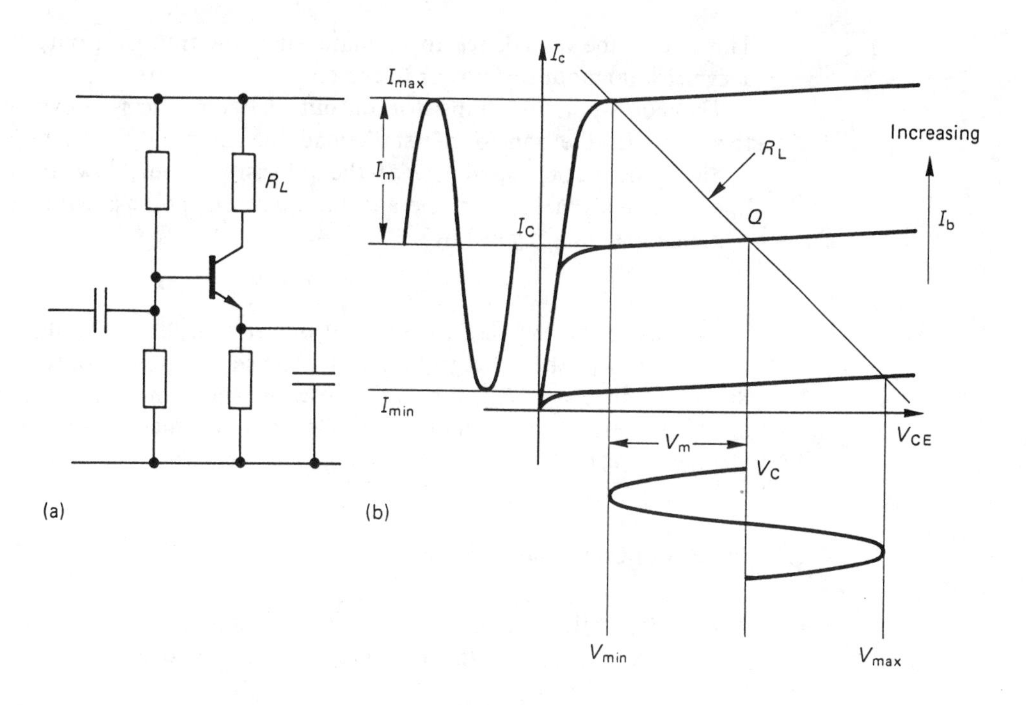

Figure 10.1

(b) Transformer-coupled load

In the case of direct-coupled loads, the quiescent current passes through the load resistance. This current represents a waste of power. Also, the actual load impedance is rarely the optimum value for maximum power into the load.

A better method is to use a transformer as shown in Fig. 10.2(a). An example might be the transformer coupling of the voice coil of a loudspeaker in an audio power amplifier. The effective resistance seen by the transistor is given by

$$R_L' = n^2 R_L .$$

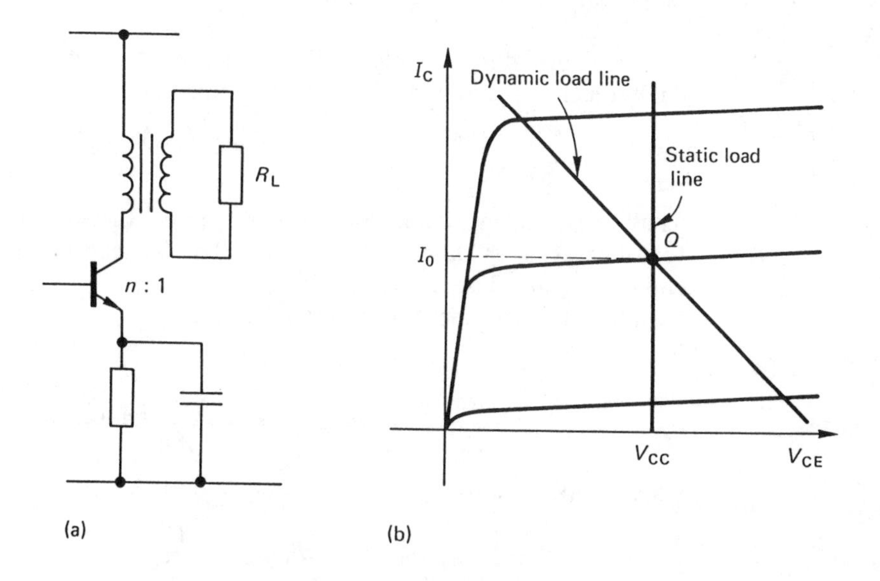

Figure 10.2

This allows the impedance to be matched to the transistor output so as to deliver a significant amount of power to the coil.

The corresponding transistor output characteristic is shown in Fig. 10.2(b), together with the transformer static load line and the dynamic resistance.

Since the static impedance of the transistor is very low, then the static load line is virtually a straight line and the quiescent point is with $V_{CE} \approx V_{CC}$. The dynamic load line is given by

$$R_L' = n^2 R_L .$$

In considering the case for optimal power transfer, it should be noted that for very small R_L' the voltage swing and hence the power output approach zero. Similarly, for large R_L' the current swing and hence the power output again approach zero. The maximum power output is achieved when R_L' is such that a voltage swing up to $2V_{CC}$ is possible.

(c) Power Conversion Efficiency

Conversion efficiency (collector circuit efficiency) η is the ability of a transistor to convert d.c. power into a.c. signal power to the load.

$$\therefore \eta = \frac{\text{a.c. signal power output to load}}{\text{average d.c. power taken from power supply}} = \frac{P_L}{P_{dc}} \times 100 \text{ per cent.}$$

Now average d.c. power

$$P_{dc} = V_{CC} I_0 ,$$

and signal power delivered to load

$$P_L = \frac{V_m I_m}{2} .$$

$$\therefore \eta = \frac{V_m I_m}{2V_{CC} I_0} \times 100 \text{ per cent.}$$

(d) Maximum Value of Efficiency

The case for a direct-coupled load is represented in Fig. 10.3, which shows a load line for a class-A amplifier. The load power is given by

$$P_L = \frac{1}{2} \frac{(V_{max} - V_{min})}{2} \frac{(I_{max} - I_{min})}{2} = \frac{(V_{max} - V_{min})^2}{8R_L} .$$

Thus to obtain maximum load power requires the maximum length of useable load line. This is obtained when $V_0 = 0.5\ V_{CC}$ and the quiescent point is thus chosen so that it bisects the load line.

In the idealised case shown in the diagram,

$$V_{max} = V_{CC} ,$$

$$\therefore P_{L(max)} = \frac{(V_{CC} - V_{min})^2}{8R_L} ,$$

which tends towards

$$P_{L(max)} = \frac{V_{CC}^2}{8R_L} .$$

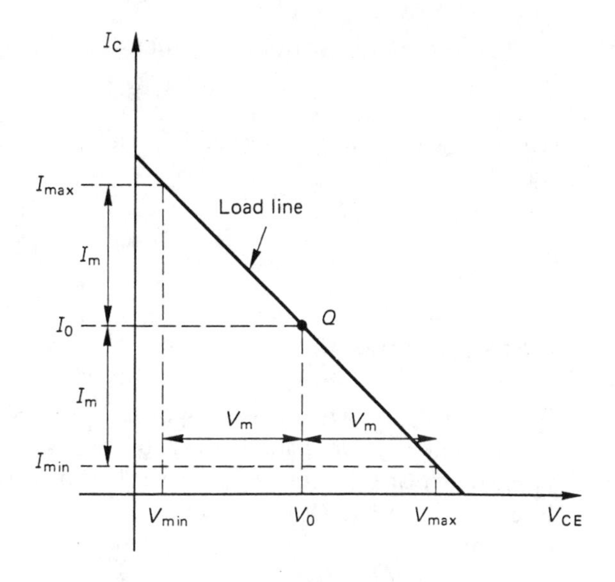

Figure 10.3

Also, the power taken from the supply P_{dc} is given by

$$P_{dc} = V_{CC}\, I_0 = \frac{V_{CC}^2}{2R_L},$$

$$\therefore\ \eta_{max} = \frac{P_{L(max)}}{P_{dc}} = 25 \text{ per cent.}$$

The power loss within the amplifier is due mainly to the dissipation at the collector of the transistor and is called the collector dissipation P_C. In the case of the direct coupled load the d.c. load power also contributes significantly to the power loss.

$$\therefore\ P_C = P_{dc} - P_L - P_{L(dc)},$$

where P_L = signal output power,
and $P_{L(dc)}$ = d.c. power in load.

Now maximum P_C occurs when $P_L = 0$ (i.e. with no signal),

$$\therefore\ P_{C(max)} = \frac{V_{CC}^2}{2R_L},$$

For a transformer-coupled load the maximum possible voltage and current variations are $2V_{CC}$ and $2I_0$ and the optimum load line is with $R_L' = V_{CC}/I_0$.

Now, since $$V_{max} = 2V_{CC}$$

then $$P_{L(max)} = \frac{V_{CC}^2}{2R_L'}$$

and $$P_{dc} = V_{CC}\, I_0 = \frac{V_{CC}^2}{R_L'}.$$

$$\therefore\ \eta_{max} = \frac{P_{L(max)}}{P_{dc}} = 50 \text{ per cent;}$$

i.e. the maximum theoretical efficiency for a transformer-coupled load is twice that for a direct-coupled load.

The collector dissipation is given by

$$P_C = P_{dc} - P_L$$

(the d.c. load power is negligible). This is maximum when $P_L = 0$ (i.e. with no signal),

$$\therefore\ P_{C(max)} = \frac{V_{CC}^2}{R_L'} .$$

(e) Choice of Operating Point

Figure 10.4 shows the constant-power hyperbola for a transistor. To make use of the maximum power-handling capability P_{CM} of the transistor, the d.c. operating point Q is placed on the P_{CM} curve. It can be shown that the point of tangency bisects the load line. This is the necessary condition for maximum conversion efficiency.

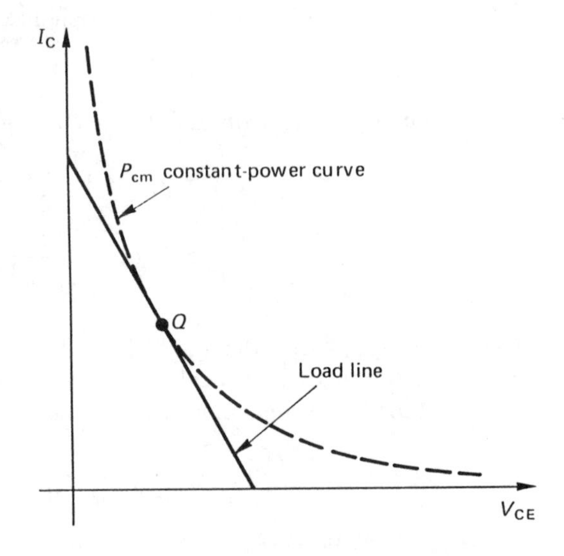

Figure 10.4

(f) Second Harmonic Distortion

The dynamic transfer characteristic relating collector current with base current for a transistor is not a straight line and may be expressed more accurately by the relationship

$$i_c = I_C + G_1 i_b + G_2 i_b^2 .$$

For a sinusoidal input of the form

$$i_b = I_B \cos \omega t,$$

the collector current is of the form

$$i_c = I_C + B_0 + B_1 \cos \omega t + B_2 \cos 2\omega t.$$

The second harmonic distortion is given by $D_2 = \left| \frac{B_2}{B_1} \right|$ (see Example 10.3).

10.3 Push-Pull Amplifiers

(a) Class-A Push–Pull Amplifiers

Much of the distortion introduced by the non-linearity of the single-ended power amplifier may be eliminated by using a push–pull circuit such as that shown in Fig. 10.5. When the signal on Q_1 is positive, the signal on Q_2 is negative and vice versa. The amplification of each half of the voltage waveform is thus shared between Q_1 and Q_2.

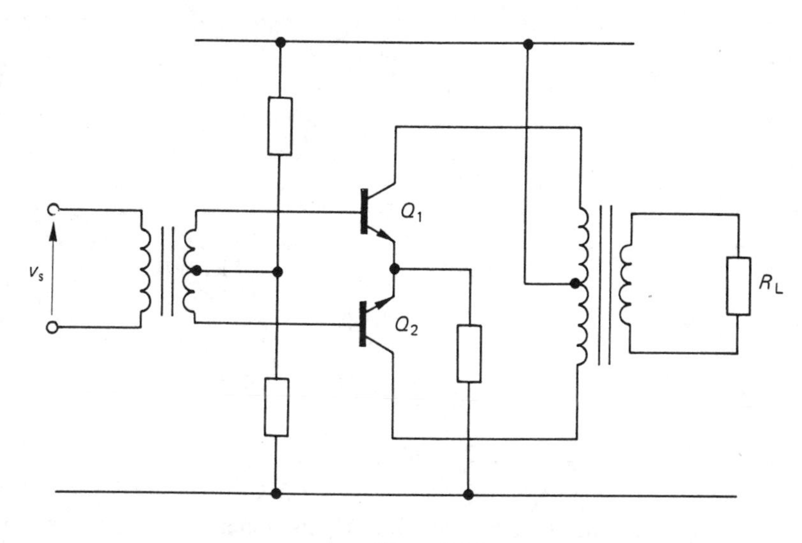

Figure 10.5

The circuit has the following advantages:

(a) Since the transistors are biased towards cut-off, then when the input signal is zero, there is only a small quiescent current and therefore low power dissipation in the transistors.
(b) Distortion is reduced since it may be shown that even harmonics cancel out.

(b) Class-B Push–Pull Amplifiers

In class-B amplification, the transistor is biased approximately at cut-off. The push–pull amplifier is very appropriate for operation in class B. The advantages of class-B operation are:

(a) A greater power output is possible for a given transistor.
(b) The efficiency is greater.
(c) There is negligible power loss when the input signal is zero.

The disadvantages are:

(a) Greater harmonic distortion than in class A.
(b) Self-bias cannot be used.
(c) Supply voltages require good regulation.

Owing to the non-linearity of the mutual characteristic of the transistor, cross-over distortion occurs in class B. This is a form of distortion at low current that for a sinusoidal input signal exhibits the output shown in Fig. 10.6. This distortion may be reduced by biasing the transistors slightly on, so that they take a small collector current under quiescent conditions.

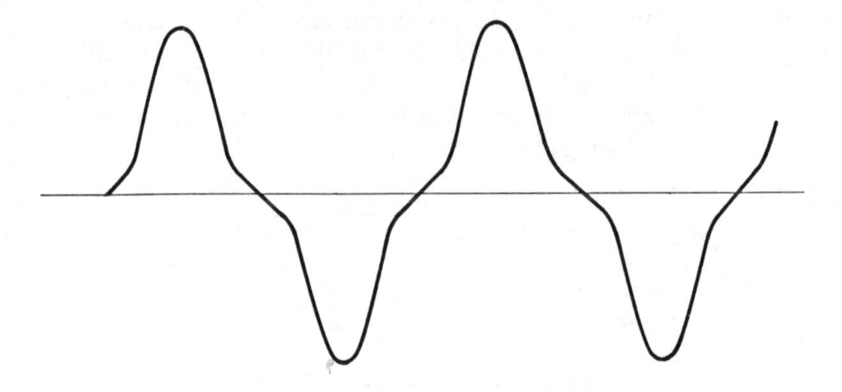

Figure 10.6

(c) Complementary Pairs

A transformerless circuit is now generally used to save on the cost of an output transformer. This circuit uses a matched pair of transistors connected as a complementary emitter follower pair as shown in Fig. 10.7. It is known as the complementary-symmetry class-B circuit. In this circuit, the *npn* transistor conducts during positive half-cycles of the input waveform, while the *pnp* transistor conducts during negative half-cycles, thus giving a sinusoidal load current for a sinusoidal input signal. In practice, the transistors are slightly forward-biased to reduce distortion.

The emitter-coupled transistors have the advantages of:

(a) providing negative feedback, and so reducing the distortion,
(b) good matching due to the low output impedance of the emitter follower.

A number of varieties of the class-**B** complementary-symmetry amplifier are in common use.

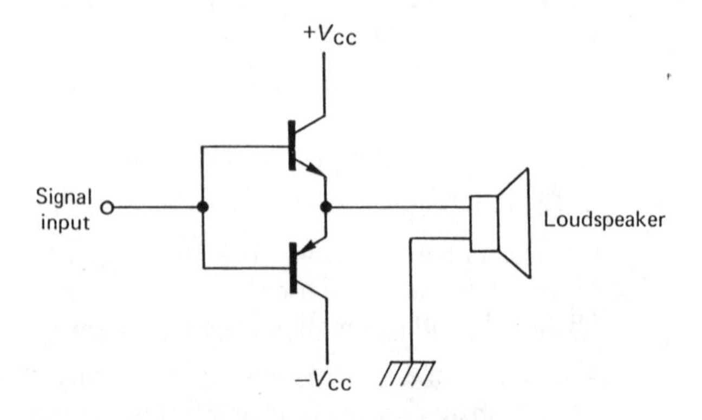

Figure 10.7

(d) Power Conversion Efficiency of Push-Pull Amplifiers

The output characteristic and load line of Fig. 10.8 shows the current and voltage waveforms for one transistor in the push-pull circuit. For the other transistor, the waveforms would be 180° out of phase.

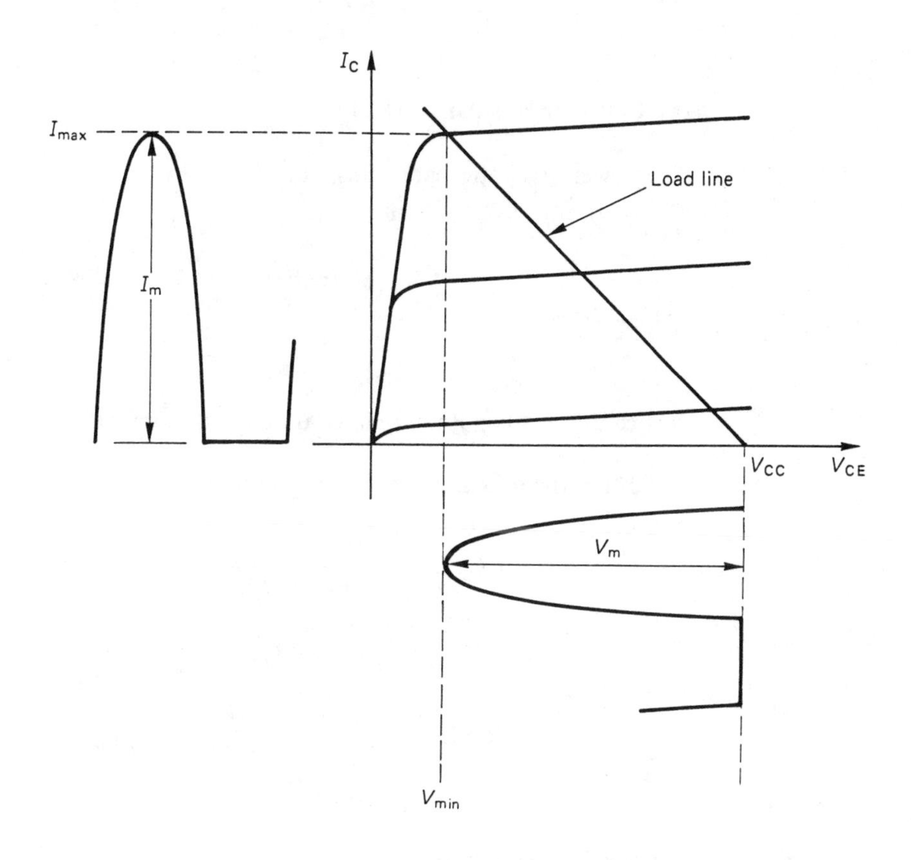

Figure 10.8

With two transistors in push-pull, the power output is given by

$$P_L = \frac{I_m V_m}{2} = \frac{V_m^2}{2R_L} = \frac{(V_{CC} - V_{min})^2}{2R_L}.$$

The average current is that of a half-wave rectified waveform.

$$\therefore\ I_{av} = \frac{I_m}{\pi} = \frac{V_m}{\pi R_L}.$$

The supply power is given by

$$P_{dc} = \frac{2V_m V_{CC}}{\pi R_L}$$

(assuming supplies of $\pm V_{CC}$).

Thus collector efficiency η is given by:

$$\eta = \frac{P_L}{P_{dc}} = \frac{V_m^2}{2R_L}\ \frac{\pi R_L}{2V_m V_{CC}} = \frac{\pi V_m}{4 V_{CC}}.$$

Thus efficiency is proportional to V_m and maximum efficiency occurs at maximum signal level (i.e. at $V_m = V_{CC}$).

$$\therefore \eta_{max} = \tfrac{1}{4}\pi = 78.5 \text{ per cent}$$

and

$$P_{L(max)} = \frac{I_m V_{CC}}{2} = \frac{V_{CC}^2}{2R_L}.$$

(e) Collector Dissipation (P_C)

The power dissipation per transistor is given by

$$P_C = 0.5\,(P_{dc} - P_L)$$

$$= 0.5\left(2\frac{V_m V_{CC}}{\pi R_L} - \frac{V_m^2}{2R_L}\right)$$

$$= \frac{V_m V_{CC}}{\pi R_L} - \frac{V_m^2}{4R_L}.$$

Notice that in class B the quiescent collector dissipation is zero.

To find the maximum power dissipation, equate $\frac{dP_C}{dV_m}$ to zero.

$$\frac{dP_C}{dV_m} = \frac{V_{CC}}{\pi R_L} - \frac{2V_m}{4R_L} = 0,$$

$$\therefore V_m = \frac{2V_{CC}}{\pi},$$

$$\therefore P_{C(max)} = \frac{2V_{CC}^2}{\pi^2 R_L} - \frac{4V_{CC}^2}{4\pi^2 R_L} = \frac{V_{CC}^2}{\pi^2 R_L} \text{ per transistor.}$$

10.4 MOS Power Transistors

A type of MOS transistor called a VMOS (vertical MOS) finds useful application in power output stages. They have the following advantages:

(a) high input impedance such that they require only a small drive power;
(b) high speed as is typical with FET devices;
(c) good linearity at high current levels, unlike bipolar power transistors.

They have the disadvantages that:

(a) they cannot provide the high current and voltage ratings that are available with bipolar transistors;
(b) the saturation voltage is much higher than is the case with bipolar transistors.

10.5 Thermal Dissipation

The maximum average power, $P_{D(max)}$, that a transistor can dissipate depends on the transistor construction. This maximum power is limited by the temperature that the collector–base junction can withstand (150 to 225 °C for silicon).

The junction temperature may rise because of ambient temperature or because of self-heating.

As a result of junction power dissipation, the junction temperature rises, and this in turn increases I_C, with subsequent increase in power dissipation. This effect is referred to as thermal runaway and can permanently damage the transistor.

The temperature rise of the collector junction is given by

$$T = \theta P_D,$$

where $T = T_j - T_A$,
T_j = junction temperature,
T_A = ambient temperature,
θ = thermal resistance,
P_D = power dissipated at collector junction.

The thermal resistance depends on the size of the transistor, on whether or not a heatsink is used, and on whether or not forced air cooling is used.

The maximum collector power at which the transistor may safely operate is normally specified at a case temperature of 25 °C. At ambient temperatures above this, the power rating must be derated as shown in Fig. 10.9, so that at higher case temperatures the power dissipation must be reduced.

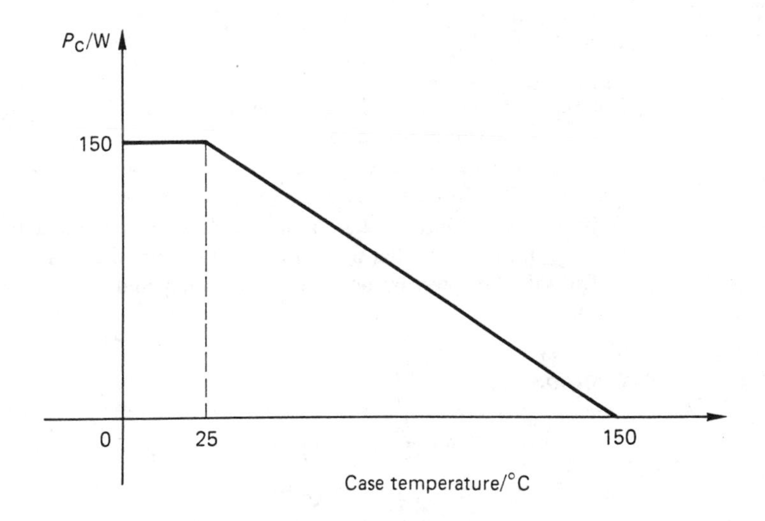

Figure 10.9

10.6 Worked Examples

Example 10.1

For a particular amplifier, the optimum load impedance is 180 Ω. Calculate the turns ratio required to match an 8 Ω load to this transistor.

If the amplifier takes a mean collector current of 2 A from a 15 V supply, and delivers an a.c. load power of 2.5 W to the transformer-coupled load, calculate the efficiency and the collector dissipation (neglecting the losses).

Solution 10.1

For a transformer of turns ratio $n : 1$ the effective resistance seen with regard to the primary, with a load R_L connected to the secondary is

$$R' = n^2 R_L.$$

Now $R_L = 8\ \Omega$ and required $R' = 180\ \Omega$,

$$\therefore n^2 = \frac{180}{8} = 22.5$$

$$\therefore n = 4.7.$$

The d.c. power supplied is

$$P_{dc} = V_{CC} \times I = 15 \times 2 = 30\ \text{W},$$

$$\eta = \frac{\text{a.c. power delivered to load}}{\text{d.c. power supplied}} = \frac{P_L}{P_{dc}}$$

$$= \frac{2.5}{30} \times 100 = 8.3\ \text{per cent.}$$

The collector dissipation is

$$P_c = P_{dc} - P_L = 30 - 2.5 = 27.5\ \text{W}.$$

Example 10.2

Determine the maximum theoretical efficiency of a class-A power amplifier with (a) a resistive load, (b) a transformer-coupled load.

A single-ended class-A amplifier is to drive a 15 Ω load via a transformer. The transistor has a maximum dissipation of 10 W and the characteristics are ideal, apart from a saturation region defined by a line through the origin at a slope corresponding to 5 Ω. Determine the turns ratio for the transformer so that the maximum undistorted output is obtained with a 50 V supply.

Estimate the power output and the efficiency under these conditions. (CEI Part 2)

Solution 10.2

Maximum efficiency is shown in the text to be

(a) 25 per cent for a resistive load,
(b) 50 per cent for a transformer-coupled load.

The load line is shown on the axes of V_{CE} against I_C in the diagram.

The transistor dissipation is

$$P_C = P_{dc} - P_L,$$

which is maximum when the load power $P_L = 0$.

The power rating of the transistor is thus equal to

$$P_{dc} = V_{CC}\, I_0 = 10\ \text{W},$$

$$\therefore I_0 = \frac{10}{50} = 0.2\ \text{A}.$$

Now $R'_L = n^2 R_L$, where R'_L is the effective resistance on the primary side.

Also, $$V_{min} = 5\ \Omega \times I_{max} = 5 \times 2\, I_0,$$

$$= 2\ \text{V}.$$

$$\therefore R'_L = \frac{50 - V_{min}}{I_0} = \frac{50 - 2}{0.2} = 240\ \Omega,$$

$$\therefore n^2 = \frac{240}{15}, \qquad n = 4.$$

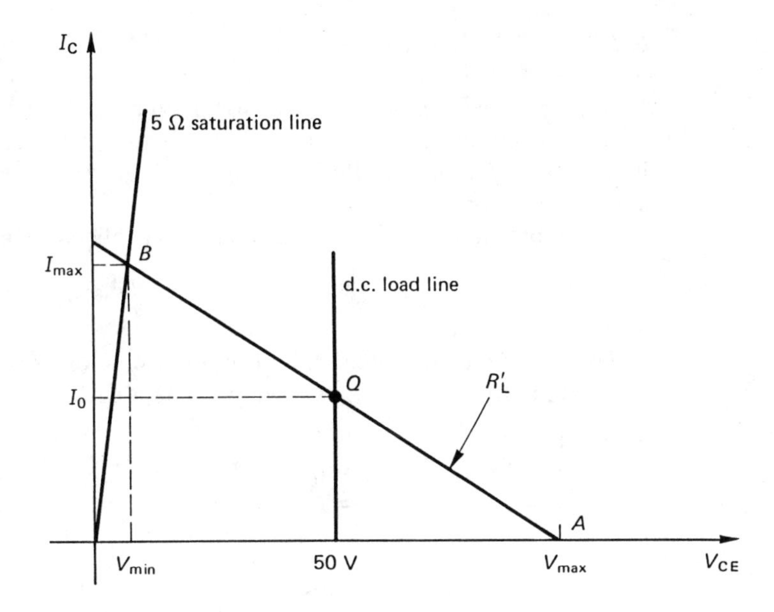

For maximum power, $AQ = BQ$.

$$\therefore \frac{V_{max} - 50}{I_0} = R'_L = \frac{50 - V_{min}}{I_0},$$

$$\therefore V_{max} = 240 \times 0.2 + 50 = 98 \text{ V},$$

$$V_{min} = 50 - 240 \times 0.2 = 2 \text{ V}$$

$\therefore$ Maximum a.c. output power is given by

$$P_{L(max)} = \frac{(V_{max} - V_{min})^2}{8R'_L}$$

$$= \frac{(98 - 2)^2}{8 \times 240} = \underline{4.8 \text{ W}}.$$

$$\text{Efficiency } \eta = \frac{P_L}{P_{dc}} = \frac{P_L}{V_{CC} I_0} = \frac{4.8}{50 \times 0.2}$$

$$= \underline{48 \text{ per cent.}}$$

Example 10.3

Explain why a single-ended, class-A, large-signal amplifier often exhibits second-harmonic distortion. Show that the percentage distortion is given by the expression

$$D = \frac{I_{max} + I_{min} - 2I_C}{2\,(I_{max} - I_{min})} \times 100 \text{ per cent},$$

where I_{max} and I_{min} are the maximum and minimum values of collector current resulting from a sinewave input voltage, and I_C is the quiescent collector current.

A single-ended class-A power amplifier is coupled to an 8 Ω load, using a transformer with a turns ratio of 5 : 1. With a 50 V supply the transistor is biased to have a quiescent collector current of 250 mA. When a sinusoidal signal is applied to the base, the collector voltage varies between a minimum of 5 V and maximum of 90 V. Estimate the efficiency, power output and second-harmonic distortion of this stage.

If the input signal had contained frequencies f_1 and f_2, what frequencies would have been present in the output signal? (CEI Part 2)

Solution 10.3

In a large-signal amplifier, distortion is due to the fact that the static output characteristics of the transistor are not equidistant, straight lines for constant increments of input excitation. The type of distortion is called non-linear or amplitude distortion.

The expression for the instantaneous total collector current i_c is given by

$$i_c = I_C + B_0 + B_1 \cos \omega t + B_2 \cos 2\omega t,$$

where the Bs are constants.

This equation shows that the output current contains a second harmonic term. A typical output characteristic and load line is shown in the diagram,

when $\omega t = 0$, $\quad i_c = I_{max}$,
$\omega t = \frac{1}{2}\pi$, $\quad i_c = I_C$,
$\omega t = \pi$, $\quad i_c = I_{min}$.

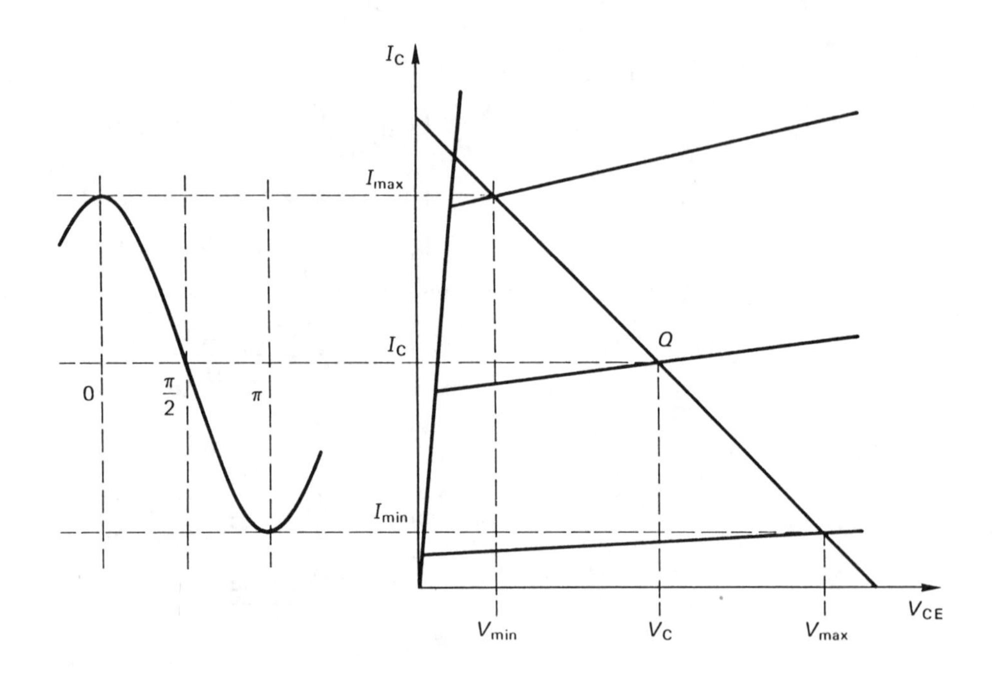

Substituting into the equation for i_c,

$$I_{max} = I_C + B_0 + B_1 + B_2,$$

$$I_C = I_C + B_0 - B_2,$$

$$I_{min} = I_C + B_0 - B_1 + B_2,$$

which leads to

$$B_0 = B_2$$

$$B_1 = \frac{I_{max} - I_{min}}{2}$$

$$B_2 = \frac{I_{max} - I_{min} - 2I_C}{4}.$$

∴ Percentage second-harmonic distortion is given by

$$D = \left| \frac{B_2}{B_1} \right| \times 100$$

$$= \frac{I_{max} + I_{min} - 2I_C}{2\,(I_{max} - I_{min})} \times 100 \text{ per cent.}$$

From the values given,

$$I_0 = 250 \text{ mA},$$

$$R'_L = n^2 R_L = 5^2 \times 8 = 200\ \Omega.$$

$$\therefore I_{min} = 250 \text{ mA} - \frac{(90 - 50)\text{ V}}{200\ \Omega} \times 10^3 \text{ mA}$$

$$= 50 \text{ mA},$$

$$I_{max} = 250 \text{ mA} + \frac{(50 - 5)\text{ V} \times 10^3}{200\ \Omega} \text{ mA}$$

$$= 475 \text{ mA}.$$

∴ Signal power delivered to load is given by

$$P_L = \frac{(V_{max} - V_{min})\,(I_{max} - I_{min})}{8}$$

$$= \frac{(90 - 5)\,(475 - 50)}{8} \text{ mW}$$

$$= \underline{4.516 \text{ W}}.$$

$$\text{Efficiency } \eta = \frac{P_L}{P_{dc}} = \frac{P_L}{V_{CC} I_0}$$

$$= \frac{4.516}{50 \times 0.25} = \underline{36.1 \text{ per cent}}.$$

$$\text{Distortion } D = \frac{0.475 + 0.05 - 2 \times 0.25}{2\,(0.475 - 0.05)}$$

$$= \underline{2.9 \text{ per cent}}.$$

Example 10.4

A transistor has a $V_{CC(max)} = 40$ V, $I_{C(max)} = 1$ A, $P_{CM} = 4$ W and is coupled via a transformer to a 10 Ω load. Calculate an appropriate value for V_{CC} and the transformer turns ratio for maximum output power to the load. Calculate the value of the maximum output power. Assume an ideal transformer.

Solution 10.4

The quiescent operating point is at V_{CC}, I_0 for a transformer-coupled load. This is also the point of maximum power.

$$\therefore P_{CM} = V_{CC}\, I_0,$$

$$\therefore V_{CC} = \frac{P_{CM}}{I_0}.$$

Now $$V_{CC} = I_0 R'_L = I_0 n^2 R_L,$$

$$\therefore \frac{P_{CM}}{I_0} = I_0 n^2 R_L.$$

$$I_0 = \frac{1}{n}\sqrt{\frac{P_{CM}}{R_L}} = \frac{1}{n}\sqrt{\frac{4}{10}} = \frac{0.63}{n}.$$

Also, $V_{CC} = \dfrac{P_{CM}}{V_{CC}} n^2 R_L$.

$$\therefore V_{CC} = n\sqrt{P_{CM} R_L} = n\sqrt{4 \times 10} = 6.3n.$$

Now we choose V_{CC} and n such that the $V_{CE(max)}$ and $I_{C(max)}$ ratings are not exceeded.

$$\therefore V_{CC} < \frac{V_{CE(max)}}{2} < \frac{40}{2} = 20 \text{ V},$$

$$I_0 < \frac{I_{C(max)}}{2} < \frac{1}{2} = 0.5 \text{ A}.$$

$$\therefore \frac{0.63}{n} < 0.5 \qquad \text{and} \qquad 6.3n < 20,$$

$$\therefore 1.26 < n < 3.17.$$

To achieve the maximum voltage swing use the highest V_{CC}.

$$\therefore \underline{n = 3.17} \qquad \text{and} \qquad \underline{V_{CC} = 20 \text{ V}}.$$

$$\therefore I_0 = \frac{0.63}{n} = 0.2 \text{ A}, \qquad \text{and } P_{L(max)} = \frac{P_{CM}}{2} = 2 \text{ W}.$$

Example 10.5

Describe, using appropriate diagrams, the operation of a class-B push-pull power output stage, and calculate the maximum efficiency of such a stage. Comment briefly on the possible deficiencies in performance that may be observed when it is used in a high-quality audio amplifier.

A class-B push-pull amplifier using complementary transistors has supplies ± 40 V and a load of 8 Ω. The transistor characteristics are ideal except for a saturation region defined by a line through the origin at a slope corresponding to 5 Ω. For a sinusoidal input, calculate the maximum signal power output to the load, the corresponding dissipation in each transistor and the efficiency. (CEI Part 2)

Solution 10.5

The output stage is shown in diagram (a) and the corresponding load line for *one* transistor in diagram (b).

The maximum signal power output to the load is given by

$$P_{L(max)} = \frac{V_m I_m}{2} = \frac{V_m^2}{2R_L} = \frac{(V_{CC} - V_{min})^2}{2R_L}.$$

V_{min} is at the point where the load line crosses the saturation line.

Referring to diagram (b),

$$I_{max} = \frac{V_{CC} - V_{min}}{8};$$

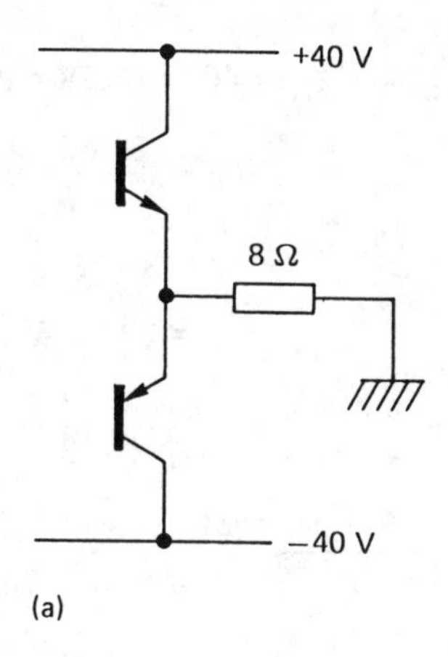

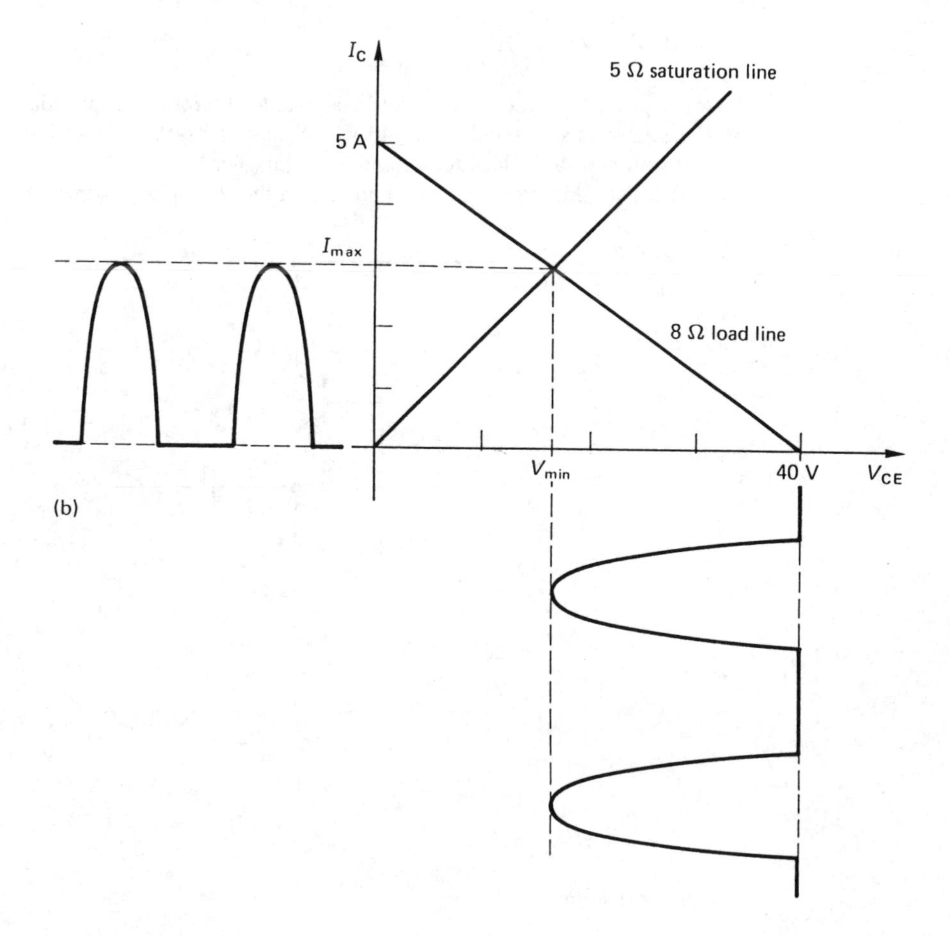

also,
$$I_{max} = \frac{V_{min}}{5},$$
$$\therefore V_{min} = 15.38 \text{ V}$$
$$\therefore P_{L(max)} = \frac{(40 - 15.38)^2}{2 \times 8}$$
$$= \underline{37.88 \text{ W}}.$$

The dissipation in each transistor is given by
$$P_C = \frac{V_m \, V_{CC}}{\pi R_L} - \frac{V_m^2}{4R_L},$$

where $V_m = V_{CC} - V_{min} = 40 - 15.38 = 24.62$ V.

$$\therefore P_C = \frac{24.62 \times 40}{\pi R_L} - \frac{(24.62)^2}{4R_L}$$

$$= 39.18 - 18.94$$

$$= \underline{20.24 \text{ W}}.$$

$$\text{Efficiency } \eta = \frac{\pi}{4} \frac{V_m}{V_{CC}} = \frac{\pi}{4} \times \frac{24.62}{40}$$

$$= \underline{48.3 \text{ per cent}}.$$

Example 10.6

A complementary transistor pair drive a 10 Ω load through a capacitor as shown in the diagram. If the transistors are rated as 1 A, 40 V, 4 W, choose a suitable value for V_{CC} that will give maximum power into the load for a sinusoidal input signal.

Calculate this maximum signal power and the corresponding maximum collector dissipation.

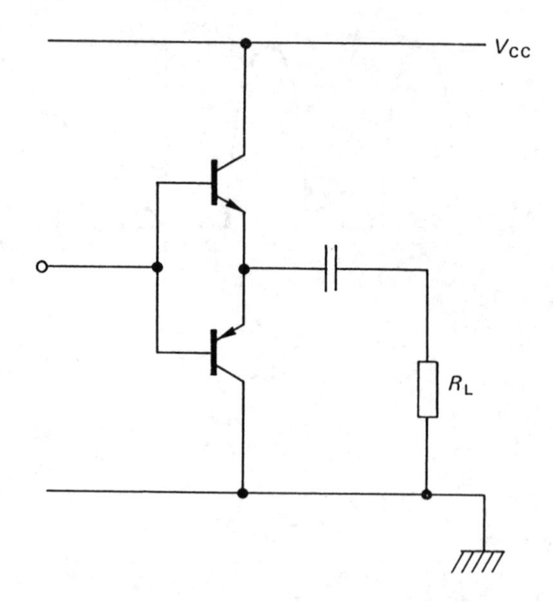

Solution 10.6

In this circuit a single supply V_{CC} is used with the load capacitively coupled.

The analysis in the text considers the case for split supplies. The equations in the text thus require V_{CC} to be replaced by $V_{CC}/2$.

Maximum a.c. power to load:

$$P_{L(max)} = \frac{V_{CC}^2}{8R_L}.$$

$$\text{The peak signal amplitude} = \frac{V_{CC}}{2}.$$

$$\therefore \text{Peak } I_C = \frac{V_{CC}}{2R_L}.$$

The transistor is rated at 1 A,

$$\therefore \frac{V_{CC}}{2R_L} = 1 \text{ A},$$

$$\therefore V_{CC} = 2R_L = \underline{20 \text{ V}}.$$

(The full rated voltage of the transistor cannot be used as then the I_C rating will be exceeded.)

$$P_{L(max)} = \frac{20^2}{8 \times 10} = \underline{5 \text{ W}}.$$

Maximum collector dissipation:

$$P_{C(max)} = \frac{V_{CC}^2}{4\pi^2 R_L} = \frac{20^2}{4\pi^2 \times 10} = \frac{10}{\pi^2}$$

$$\approx \underline{1 \text{ W}}.$$

Example 10.7

A push-pull transformer-coupled amplifier operates in class-B from a 30 V supply. At maximum signal level, the minimum collector-emitter voltage is 1.5 V. The transformer turns ratio is that required to produce a maximum output power of 15 W for a sinusoidal input signal, with a resistive load of 3 Ω. The secondary winding resistance is 0.2 Ω and the primary resistance is negligible.

Determine

(a) the transformer turns ratio;
(b) the overall efficiency at full drive;
(c) the maximum allowable heatsink thermal resistance for each transistor if the amplifier is required to operate in an ambient temperature of 90 °C and relevant transistor parameters are $T_{j(max)} = 175$ °C and $\theta_{jc} = 2$ K/W.

Solution 10.7

(a) For a maximum output power of 15 W, the power supplied to the transformer

$$= \frac{3 + 0.2}{3} \times 15 = 16 \text{ W}.$$

If the peak values of collector voltage and current are V_m and I_m then the power supplied to the transformer

$$= \frac{V_m I_m}{2}.$$

At maximum signal level, the power supplied to the transformer

$$= \frac{(V_{CC} - V_{min})^2}{2R'_L} = P_{L(max)},$$

where $R'_L = n^2 (R_L + R_s)$.

$$\therefore R'_L = \frac{(V_{CC} - V_{min})^2}{2P_{L(max)}} = \frac{(30 - 1.5)^2}{2 \times 16}$$

$$= 25.4 \ \Omega.$$

$$\therefore n = \sqrt{\frac{R'_L}{R_L + R_s}} = \sqrt{\frac{25.4}{3.2}} = \underline{2.8}.$$

(b)

$$\text{Supply power } P_{dc} = \frac{2V_m V_{CC}}{\pi R'_L},$$

$$\text{load power } P_L = \frac{3}{3.2} \times \frac{V_m^2}{2R'_L}.$$

$$\therefore \text{ Efficiency } \eta = \frac{3V_m^2 \pi R'_L}{3.2 \times 2R'_L \times 2V_m V_{CC}}$$

$$= \frac{3}{12.8} \times \frac{V_m \pi}{V_{CC}}.$$

At full drive, V_m has a maximum value of

$$V_{CC} - V_{min} = 30 - 1.5 = 28.5 \text{ V}$$

$$\therefore \eta_{max} = \frac{3\pi}{12.8} \times \frac{28.5}{30}.$$

$$= \underline{70 \text{ per cent.}}$$

(c) Maximum collector dissipation

$$P_{C(max)} = \frac{V_{CC}^2}{\pi^2 R'_L} = \frac{30^2}{\pi^2 \times 25.4}$$

$$= \underline{3.59 \text{ W}}.$$

Let the total thermal resistance be θ_T,

$$\therefore T_j - T_{amb} = \theta_{T(max)} \times P_{C(max)}$$

$$\therefore \theta_{T(max)} = \frac{175 - 90}{3.59}$$

$$= 23.7 \text{ K/W}.$$

$$\therefore \theta_{heatsink(max)} = \theta_T - \theta_{jc} = 23.7 - 2$$

$$= \underline{21.7 \text{ K/W}}.$$

Example 10.8

A transistor has a worst-case power dissipation of 1 W. Determine the maximum transistor junction temperature that would be obtained if the maximum ambient temperature is 40 °C and the transistor is mounted on a separate heatsink of thermal resistance 50 K/W. The transistor is rated at 5 W for case temperatures up to 25 °C and is derated linearly to zero at a case temperature of 175 °C.

Solution 10.8

From the derating data, the transistor is derated from 5 W to zero over a temperature range from 25 °C to 175 °C (i.e. 150 °C).

The slope of the derating curve and hence the thermal resistance between junction and case is thus 30 K/W.

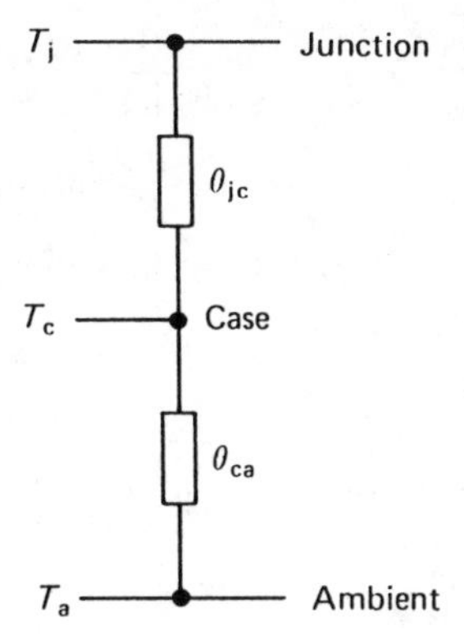

With reference to the resistance analogy of the diagram,

$$T_j - T_c = \theta_{jc} W,$$
$$T_c - T_a = \theta_{ca} W;$$
$$\therefore\ T_j - T_c = 30 \times 1,$$
$$T_c - 40 = 50 \times 1.$$
$$\therefore\quad T_j = 50 + 40 + 30$$
$$= \underline{120\ ^\circ\text{C}}.$$

Example 10.9

A transistor heatsink has a thermal resistance θ (in K/W) given by

$$\theta = 5A^{-0.45},$$

where A is the area in cm^2 of the heatsink.

Calculate the minimum area of the heatsink necessary for the transistor to dissipate 50 W at an ambient temperature of 20 °C if the transistor case has a maximum permitted temperature of 50 °C, the case-to-heatsink thermal resistance is 0.5 K/W and the case-to-air thermal resistance is 5 K/W.

Solution 10.9

The resistive analogy is shown in the diagram.

$$T_c - T_a = \frac{(\theta_{cs} + \theta_{sa})\,\theta_{ca}}{\theta_{cs} + \theta_{sa} + \theta_{ca}} \text{ watts}$$

$$\therefore\ 50 - 20 = \theta \times 50$$

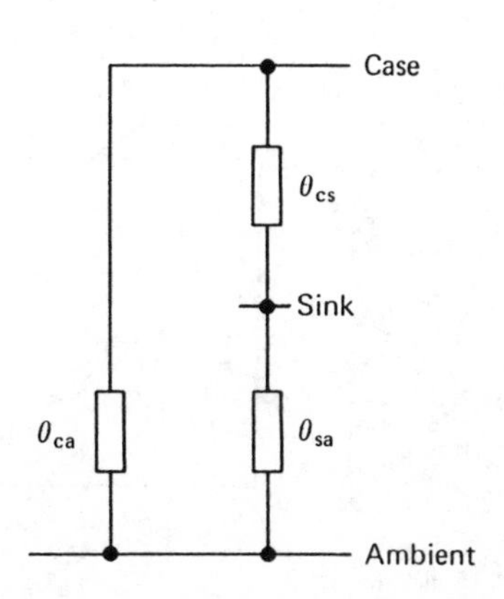

$$\therefore \frac{(0.5 + \theta_{sa})\,5}{0.5 + 5 + \theta_{sa}} = 0.6$$

$$\therefore \theta_{sa} = 0.1818$$

Now $$5A^{-0.45} = \theta_{sa} = 0.1818,$$

$$\therefore A = \underline{1579 \text{ cm}^2}.$$

10.7 Unworked Problems

Problem 10.1

Draw to scale the family of idealised I_C/V_{CE} characteristics for a transistor having $h_{fe} = 200$ and $h_{oe} = 80\ \mu S$, for base currents at 100 μA intervals over the range 0 to 600 μA. Assume a maximum value for V_{CE} of 40 V and a negligibly small saturation voltage.

Draw load lines on the characteristics with a 20 V d.c. supply for (a) a 400 Ω collector load resistance and (b) a collector load consisting of an ideal 1 : 1 transformer with a 400 Ω resistor connected across its secondary. In each case choose an operating point to obtain maximum 'undistorted' swing with minimum quiescent collector current. Compare the transistor dissipations, the powers in the loads with a sinusoidal signal, and the efficiencies for the above two cases.

Draw a circuit for the transformer-coupled amplifier, giving appropriate design values to the bias components to obtain the desired operating point. (EC Part 1)

Problem 10.2

Describe the operation of a class-B push-pull power output stage. What are the advantages of this type of output stage? Calculate the maximum theoretical efficiency of a class-B stage with a sinusoidal input.

A class-B push-pull amplifier using complementary transistors has supplies of ± 30 V and a load of 10 Ω.

The transistor characteristics are ideal except for a saturation region defined by a line through the origin at a slope corresponding to 4 Ω. Calculate the maximum signal power output to the load and the corresponding efficiency for an input signal that is (a) a sinusoid and (b) a square wave. (EC Part 2)

Problem 10.3

Describe the construction and features of field effect and bipolar transistors designed for high-power applications.

The diagram shows the power derating curve for a transistor. If the maximum power dissipation is 120 W, determine the thermal resistance between the junction and the case.

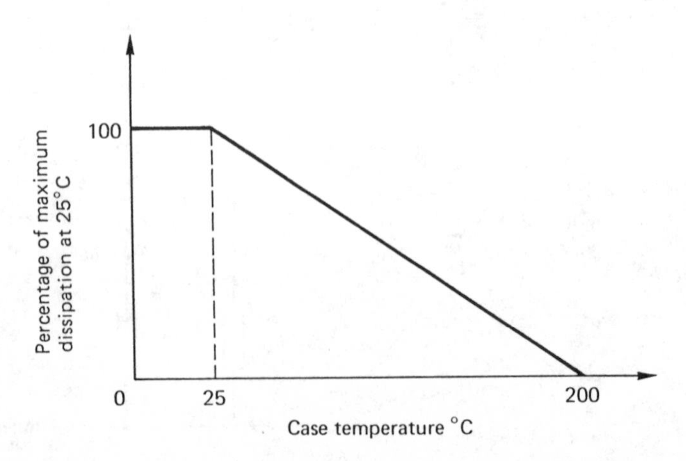

A circuit designer requires this transistor to dissipate 80 W at an ambient temperature of 40 °C. Determine the thermal resistance of the heatsink required if the thermal resistance between the case and the heatsink is 0.4 K/W. Comment on the practicality of this requirement. Describe how the thermal resistance between the transistor and the heatsink may be minimised.

(EC Part 2)

11 Regulated Power Supplies

11.1 Principle of Operation

An ideal regulated power supply provides a constant d.c. voltage source independently of the load current drawn from it, of the variations in the a.c. supply voltage and of the temperature variations.

One type of regulator is the shunt regulator. A simple version using a zener diode is shown in Fig. 11.1. When the voltage across the zener exceeds the breakdown voltage then the current through the zener increases and the voltage across the zener and the load resistor are held approximately constant at the rated voltage of the zener. The design of shunt regulators involves calculating the values and ratings of the resistors and the zener diode for a given input voltage, taking into account the expected variations in load current and supply voltage.

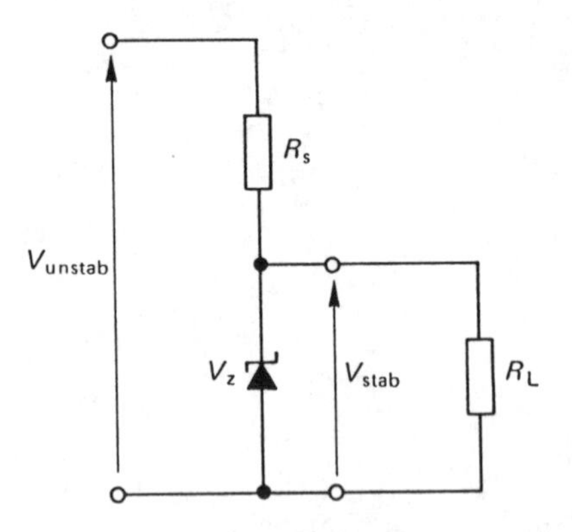

Figure 11.1 Shunt regulator

Another type of regulator is the series regulator, one version being shown in Fig. 11.2. The base voltage of the transistor is maintained at a fixed voltage by the voltage drop across the zener. If the load current increases for some reason then there is an increase in voltage across the transistor causing it to conduct harder, and so tending to maintain the output voltage at a constant value. A closed-loop series regulator is shown in Fig. 11.3. In this case the output voltage V_o is kept approximately constant by the application of voltage series feedback. A proportion of the output voltage βV_o is compared against the reference voltage V_{ref}

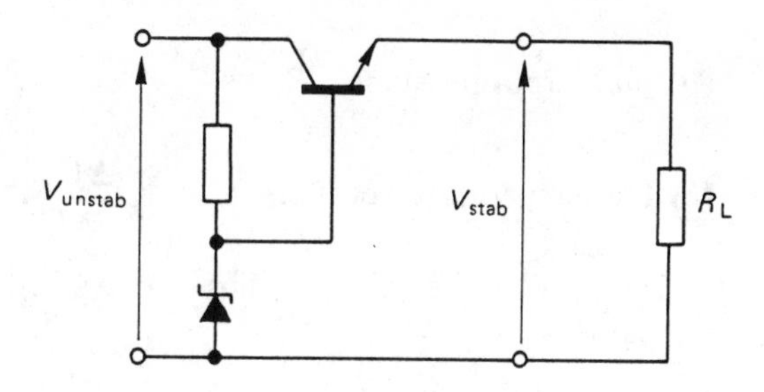

Figure 11.2 Series regulator

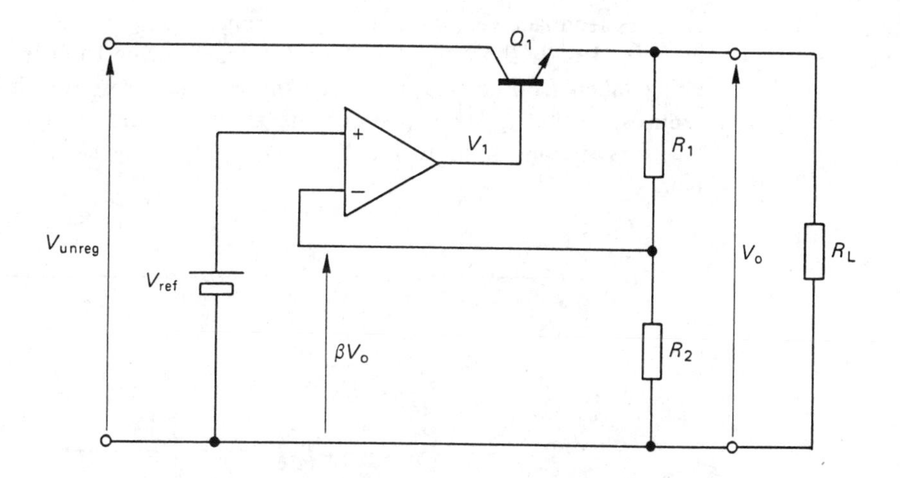

Figure 11.3 Closed-loop series regulator

(normally a zener), and the difference drives the control transistor Q_1. The op-amp output voltage V_1 is given by

$$V_1 = A_v (V_{ref} - \beta V_o) \approx V_o$$

and

$$\beta = \frac{R_2}{R_1 + R_2},$$

$$\therefore \; V_o = V_{ref} \frac{A_v}{1 + \beta A_v}$$

$$\approx V_{ref} \frac{R_1 + R_2}{R_2}.$$

11.2 Power Supply Specification

The output voltage of a regulated power supply is a function of the unregulated input voltage V_s, the load current I_L and the temperature T:

$$V_o = f(V_s, I_L, T).$$

Three coefficients are defined:

(a) the stabilisation ratio (input regulation factor),

$$S_v = \frac{\Delta V_o}{\Delta V_s},$$

(b) the output resistance, $R_o = \dfrac{\Delta V_o}{\Delta I_L}$

(c) the temperature coefficient, $S_T = \dfrac{\Delta V_o}{\Delta T}$.

The smaller the value of these three coefficients, the better the power supply regulation.

11.3 Discrete Component Regulator

A series feedback regulator may be arranged using discrete components as shown in Fig. 11.4. If the output voltage V_o drops owing to an increased load current being taken from the supply, then the voltage across transistor Q_2 b–e junction reduces, so that Q_2 takes less current, allowing more of the current in R_4 to drive Q_2, so switching it more on. If necessary, Q_1 may be replaced by a Darlington pair.

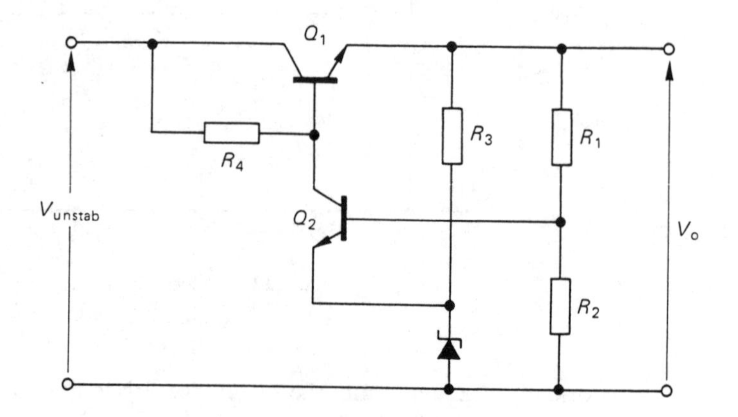

Figure 11.4 Series feedback regulator

11.4 Power Supply Protection

(a) Overcurrent Protection

A power supply must be protected against the possibility of damage due to current overload. Short-circuit overload protection is provided by the current limit circuit of Fig. 11.5. When $I_L r$ exceeds about 0.6 V then transistor Q_2 conducts, so preventing any further increase in the base current drive to Q_1.

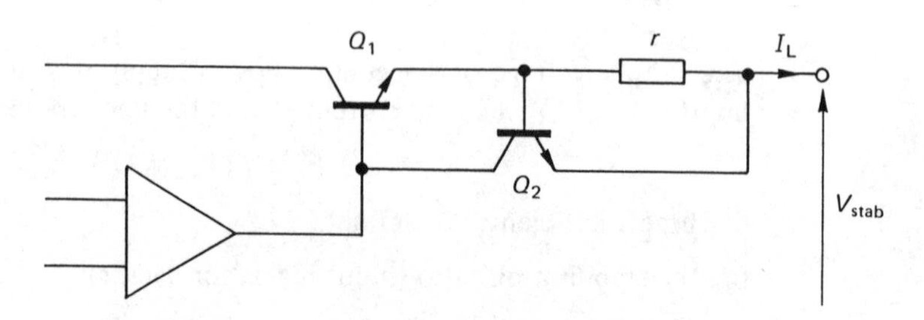

Figure 11.5 Short-circuit overload protection

(b) Overvoltage Protection

Many power supplies provide protection against an overvoltage output due to component failure within the power supply. This is referred to as 'crowbar' protection and one method is shown in Fig. 11.6. When V_o exceeds the 'safe' level of about $V_z + 0.6$ V then the zener conducts and the crowbar thyristor switches on thus effectively shorting the supply output.

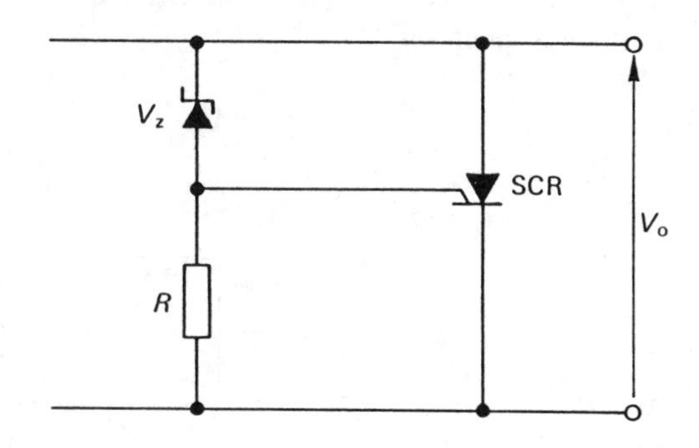

Figure 11.6 Overvoltage protection

(c) Foldback Current Limiting

To prevent high power dissipation in the regulator transistor when the output is short-circuited, foldback current limiting is often used, giving a regulation characteristic as shown in Fig. 11.7.

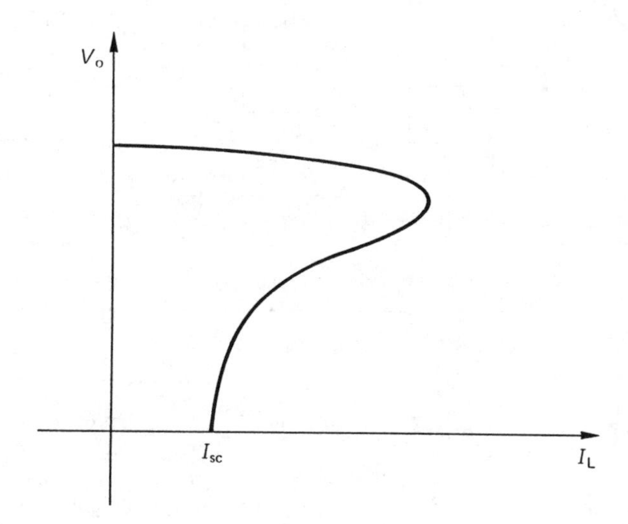

Figure 11.7 Foldback current limiting

11.5 Integrated-circuit Voltage Regulators

Integrated-circuit voltage regulators provide the benefits of low cost, small size and high performance. They are normally used to regulate supply voltages locally on each individual circuit board of a large system. A wide range of types are available as either fixed-voltage (three-terminal) or variable voltage (four-terminal) in which the fourth lead is used as the 'control' terminal. A typical fixed-voltage regulator is the 7805 5 V regulator, which is easily used as shown in Fig. 11.8.

More complex forms such as switching regulators are quite common owing to their flexibility in application and their ability to provide good regulation at low cost. A basic switching regulator schematic is shown in Fig. 11.9. The output of the power switch is a square wave whose mark–space ratio is controlled by the feedback via R_1 and R_2. The square-wave output oscillates between 0 and V_{unreg} and is filtered by the LC combination to provide a steady voltage. The output voltage is given by

$$V_o = V_{ref}\left(\frac{R_1 + R_2}{R_1}\right).$$

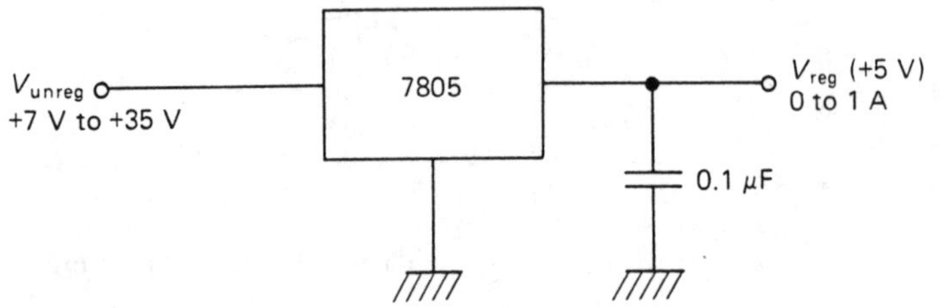

V_{unreg}
+7 V to +35 V
7805
V_{reg} (+5 V)
0 to 1 A
0.1 μF

Figure 11.8 Typical fixed-voltage regulator application

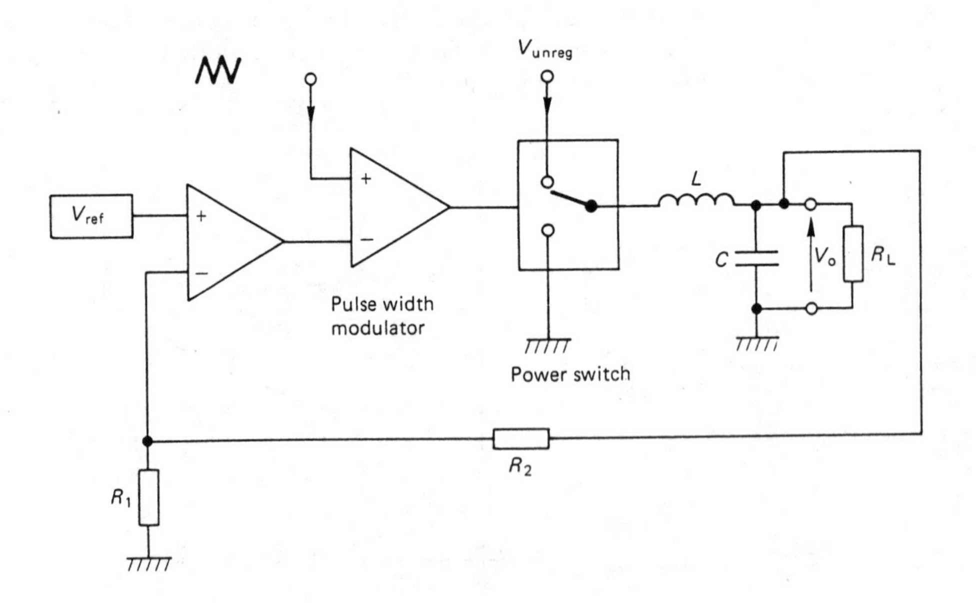

Figure 11.9 Basic switching regulator schematic

11.6 Worked Examples

Example 11.1

Design a zener regulator circuit to provide a 9 V d.c. supply from a 12 V d.c. source. The load current variation is from open circuit to 200 mA. Use a 9.1 V zener diode with a minimum current to sustain breakdown of 10 mA. Calculate the worst-case power dissipation in the series resistor and the zener.

Assuming a maximum dynamic impedance $R_D = 5\ \Omega$, calculate:

(a) the stabilisation ratio,
(b) the output impedance,
(c) the load regulation.

Solution 11.1

In the circuit shown:

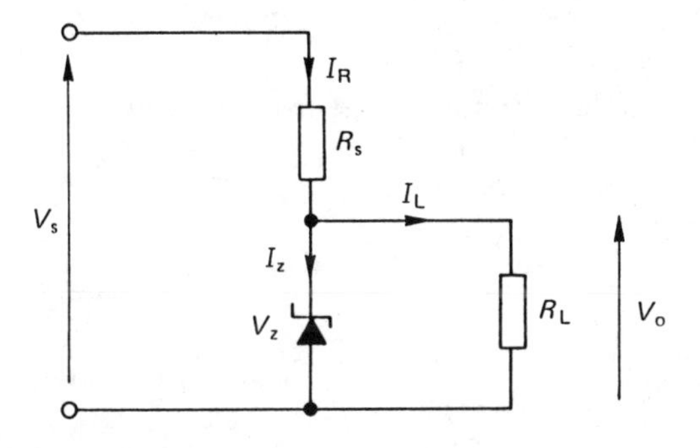

With V_z nominally at 9.1 V and V_s at 12 V,

$$V_R = V_s - V_z = 2.9\text{ V}.$$

The resistor R_s must be small enough to supply the load current of 200 mA and a zener current of, say, 10 mA to ensure that the zener is on the breakdown part of its characteristic.

$$\therefore\ I_{R(max)} = 200 + 10 = 210\text{ mA},$$

$$\therefore\ R_s \leqslant \frac{2.9\text{ V}}{210\text{ mA}} \leqslant 13.8\ \Omega, \qquad \text{say } 10\ \Omega.$$

Worst-case resistor power dissipation is given by

$$P_D = I^2_{R(max)} R_s$$
$$= (210 \times 10^{-3})^2 \times 10\ \Omega$$
$$= 440\text{ mW}.$$

We would probably choose a resistor of $10\ \Omega, \frac{1}{2}$ W.

Worst-case zener current is when $I_L = 0$.

$$\therefore\ I_{z(max)} = I_R = \frac{12 - 9.1}{10}$$
$$= 290\text{ mA}.$$

$\therefore$ Worst-case zener power dissipation is

$$P_z = 290 \times 10^{-3} \times 9.1$$
$$= 2.6\text{ W}.$$

(a) $$\text{Stabilisation ratio} = \frac{\Delta V_o}{\Delta V_s} = \frac{R_z}{R_s + R_z},$$

$$\therefore S_v = \frac{5}{10+5} = \frac{1}{3}.$$

i.e. a 1 V change in V_s would cause a 0.33 V change in V_o

(b) $$\text{Output impedance } Z_o = \frac{R_s R_z}{R_s + R_z} = \frac{10 \times 5}{15} = 3.33\ \Omega.$$

(c) $$\text{Load regulation} = \frac{\Delta V_o}{V_o} \times 100 \text{ per cent}$$

as the current changes from no-load to full-load.

Now $$\Delta V_o = \Delta I_L \times Z_o = 200\text{ mA} \times 3.33\ \Omega = 666\text{ mV}.$$

$$\therefore \text{Load regulation} = \frac{0.67}{9.1} \times 100 = 7.4 \text{ per cent.}$$

Example 11.2

Explain the operation of the linear power supply of diagram (a), stating the function of each transistor.

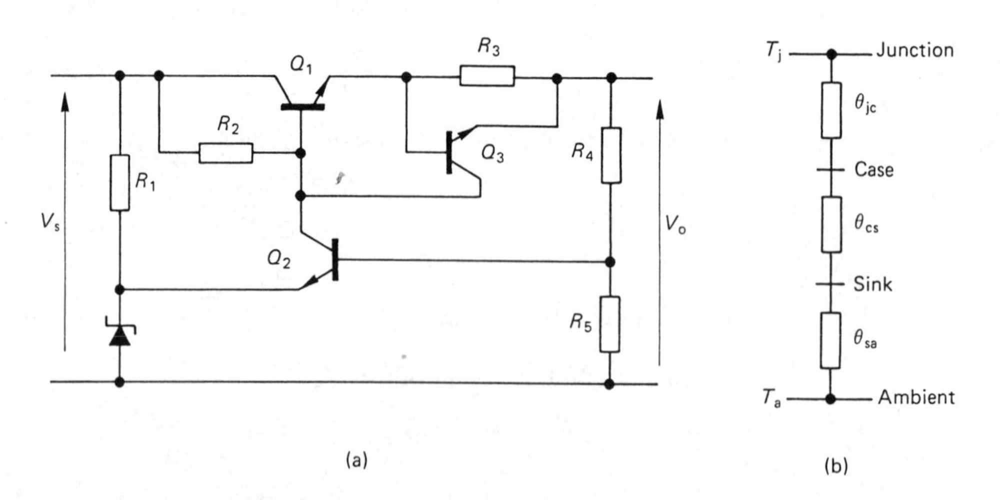

The transistor Q_1 is to be mounted on a heatsink, where the case–sink thermal resistance is 2.3 K/W. The transistor manufacturer specifies that the junction temperature must not exceed 200 K that the maximum power dissipation of the transistor is 11.25 W and that the junction-case thermal resistance is 5.8 K/W. If the ambient temperature around the heatsink is 25 K, determine the maximum thermal resistance of the heatsink.

Solution 11.2

Transistor Q_1 is the series pass element for the power supply. It operates as an emitter follower amplifier, maintaining the emitter voltage approximately 0.6 V less than the base voltage. Thus a fixed voltage applied to the base gives a fixed output voltage.

Transistor Q_2 acts as a comparator and error amplifier.

R_4 and R_5 feed back a proportion of the output voltage to the comparator circuit. The zener diode acts as a reference voltage.

If V_o tends to increase, then Q_2 conducts more heavily, diverting current away from Q_1 base, and thus returning the output to its original value. The circuit is an application of negative feedback and stabilises the output voltage against changes in parameters such as load current and supply voltage.

Transistor Q_3 and resistor R_3 provide a current limit circuit.

If the load current attempts to increase beyond a certain value, then the voltage drop across R_3 causes Q_3 to turn on, thus diverting the drive current from the base of Q_1 and hence preventing damage to Q_1 due to excessive current.

The dissipation analogy is shown in diagram (b).

The total thermal resistance is given by

$$\theta_{TOT} = \frac{T_j - T_a}{P} = \frac{200 - 25}{11.25} .$$

$$\begin{aligned}\theta_{sa} &= \theta_{TOT} - \theta_{jc} - \theta_{cs}\\ &= 15.56 - 5.8 - 2.3\\ &= \underline{7.46 \text{ K/W}.}\end{aligned}$$

Example 11.3

The series-regulated power supply shown in diagram (a) provides a nominal output of 12 V and a supply load current of up to 1 A. The unregulated supply is 25 V ± 5 V with an output resistance of 10 Ω. Calculate

(a) the input regulation factor;
(b) the output resistance;
(c) the worst-case change in output voltage due to input voltage and load current variations.

The following components are used:

(i) a reference zener diode with $V_R = 5.6$ V, $R_z = 12$ Ω at 20 mA;
(ii) a transistor Q_2 with $h_{fe} = 200$ when operated at 10 mA collector current;
(iii) a transistor Q_1 with $h_{fe} = 100$ when operated at 1 A collector current.

Solution 11.3

(a) The input regulation factor is given by

$$S_v = \left.\frac{v_o}{v_s}\right|_{i_L = 0},$$

where v_o, v_s and i_L represent changes.

But $v_s \approx i_3 R_3$ where i_3 is the current change in R_3.

Also, since we are considering only the effect of change in V_s, then the change in V_o is very small and I_L and I_{B1} are almost constant.

$\therefore$ i_3 is also the current change in Q_2's collector.

Referring to the equivalent circuit of diagram (b), the voltage divider $R_1 R_2$ is replaced by its Thévenin equivalent,

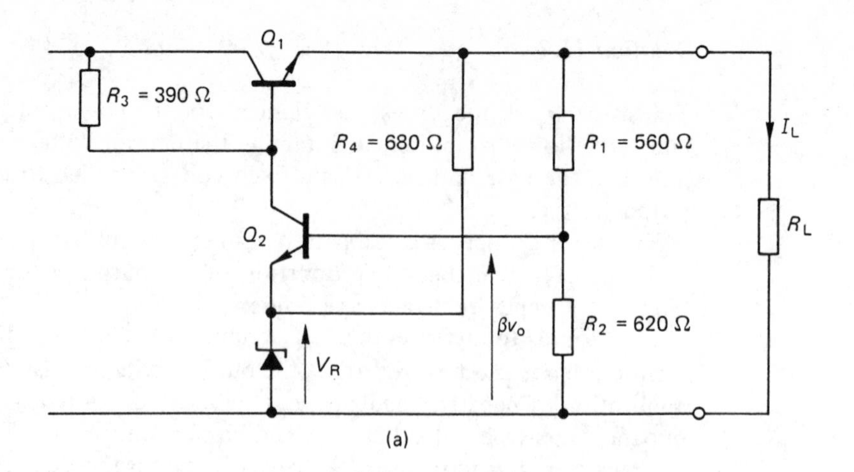

(a)

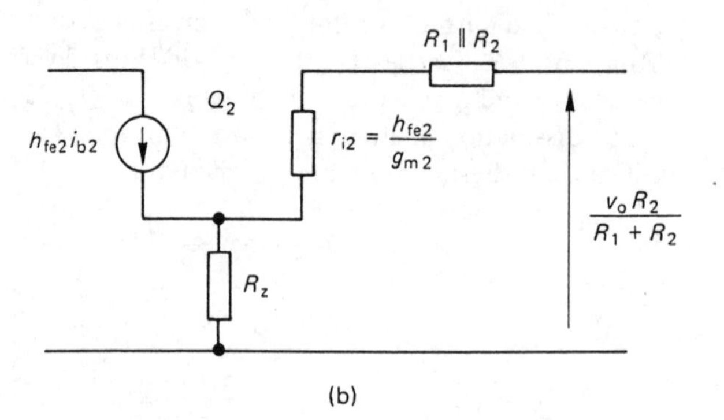

(b)

$$\therefore \quad \frac{v_o R_2}{R_1 + R_2} = i_{b2}\,(R_1 \parallel R_2 + r_{i2}) + (1 + h_{fe2})\, i_{b2}\, R_z .$$

Now $$i_3 = h_{fe2}\; i_{b2} ,$$

$$\therefore \; i_3 = h_{fe2}\left(\frac{R_2}{R_1 + R_2}\right)\frac{v_o}{R_1 \parallel R_2 + r_{i2} + (1 + h_{fe2})\, R_z} ,$$

$$= G_m \; v_o .$$

Now $$S_v = \frac{v_o}{v_s} = \frac{i_3}{G_m}\;\frac{1}{i_3 R_3}$$

$$= \frac{1}{G_m R_3} .$$

Putting $$r_{i2} = \frac{h_{fe2}}{g_{m2}} = \frac{200}{40 \times 10 \times 10^{-3}}$$

$$= 500\ \Omega ,$$

$$\underline{S_v = 0.078 .}$$

(b) The output resistance is calculated using the equivalent circuit of diagram (c) where r_o is the output resistance of the unregulated supply.

Now $$R_o = \frac{v_{oc}}{i_{sc}} ,$$

where $$v_{oc} = v_s + i_3 R_3 - i_{b1} r_{i1}$$

and $$i_{sc} = -\, i_{b1}\;(1 + h_{fe1}) .$$

$$\therefore \; v_{oc} = (i_3 - h_{fe1}\, i_{b1})\, r_o + i_3 R_3 - i_{b1} r_{i1} ,$$

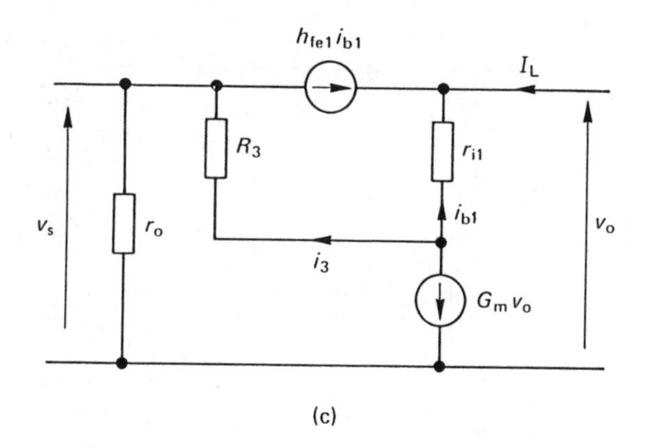

(c)

where $i_3 = -(i_{b1} + G_m\ v_{oc})$,

$$\therefore\ v_{oc} = -h_{fe1}\ i_{b1}\ r_o - (i_b + G_m\ v_{oc})(r_o + R_3) - i_{b1}\ r_{i1},$$

$$\therefore\ v_{oc}\ [1 + G_m\ (r_o + R_3)] = -(1 + h_{fe1})\ i_{b1}\ r_o - i_{b1}\ (R_3 + r_{i1}).$$

$$\therefore\ R_o = \frac{v_{oc}}{i_{sc}} = \frac{(1 + h_{fe1})\ r_o + R_3 + r_{i1}}{[1 + G_m\ (r_o + R_3)]\ (1 + h_{fe1})}.$$

Using $$r_{i1} = \frac{h_{fe1}}{g_{m1}} = \frac{100}{40} = 2.5\ \Omega,$$

$$R_o \approx \underline{0.98\ \Omega}.$$

(c) The worst-case change is given by

$$\Delta V_o = S_v\ \Delta V_s + R_o\ \Delta I_L$$
$$= 0.078 \times 10 + 0.98 \times 1$$
$$= \underline{1.76\ \text{V}}.$$

Example 11.4

The integrated circuit regulator μA 723 is used together with an outboard pass transistor as shown in diagram (a) to provide a regulated voltage supply. Foldback current limiting is provided using Q_3, together with the associated resistors R_3, R_4 and R_5. Calculate

(a) the regulated output voltage;
(b) the full-load current at the regulated output voltage and the short-circuit foldback current;
(c) the worst-case power dissipation for transistor Q_1, assuming that the input voltage is between 25 V and 35 V.

The component values are:

R_1 = 10 kΩ, R_2 = 15 kΩ, R_3 = 2 Ω,
R_4 = 2.7 kΩ, R_5 = 10 kΩ; V_z = 7.15 V.

Solution 11.4

(a) $$V_z \approx V_{reg}\ \frac{R_2}{R_1 + R_2}$$

$$\therefore\ V_{reg} = \frac{10 + 15}{15} \times 7.15 = \underline{12\ \text{V}}.$$

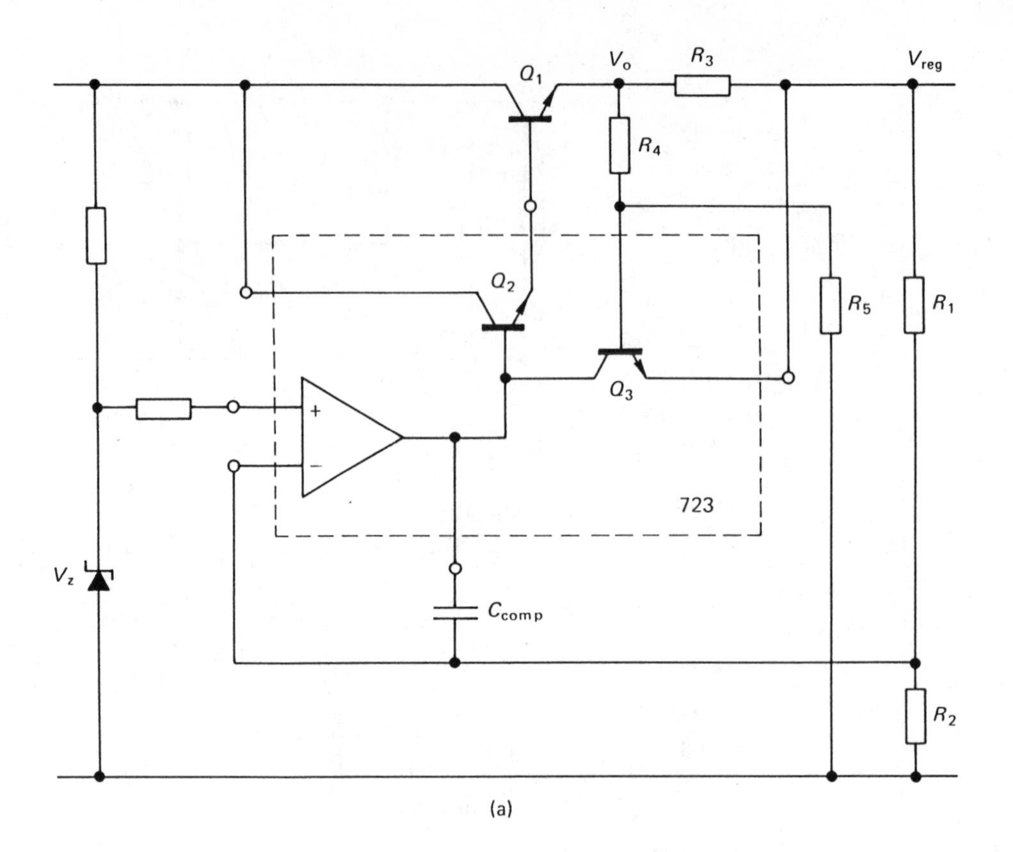

(a)

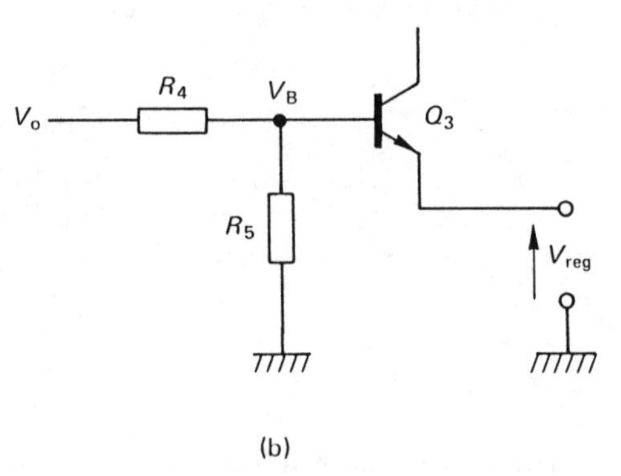

(b)

(b) When V_{reg} = 12 V, we require the base voltage of Q_3 to be approximately 12.5 V. Referring to diagram (b),

with $$V_{reg} = 12\text{ V},$$

$$\frac{12.5\text{ V}}{R_5} \approx \frac{V_o}{R_4 + R_5},$$

$$\therefore V_o = \frac{2.7 + 10}{10} \times 12.5\text{ V}$$

$$= 15.88\text{ V},$$

$$\therefore I_{FL} = \frac{15.88 - 12}{2}$$

$$= \underline{1.94\text{ A}}.$$

When $\quad V_{reg} = 0$ V $\quad$ then $\quad V_B \approx 0.5$ V.

$$\therefore \frac{0.5}{R_5} = \frac{V_o}{R_4 + R_5},$$

$$\therefore V_o = 0.64 \text{ V},$$

$$\therefore I_{sc} = \frac{0.64}{2} = \underline{0.3 \text{ A}}.$$

(c) The worst-case power dissipation in Q_1 is when the input voltage is at its maximum of 35 V, and the load current is also maximum at 1.9 A.

Assuming the output voltage to be 12 V, then the power dissipated in Q_1 under these conditions is

$$P = (35 - 12) \times 1.9$$

$$= \underline{43.7 \text{ W}}.$$

Example 11.5

The LM340T-8, as shown in diagram (a), is a three-terminal voltage regulator *IC* with internal current limiting and over-temperature protection. The major parameters are as follows:

regulated output voltage	8 V;
current limiting operates at	1.5 A;
thermal cut-off operates at die temperature	175 °C;
thermal resistance to case	6 K/W;
regulator quiescent current	10 mA.

Determine the maximum output current obtainable from the regulator when the input voltage is (a) 12 V and (b) 20 V, and the regulator is mounted on a heat sink of thermal resistance 10 K/W in an ambient temperature of 50 °C.

Show how the output voltage may be increased by the addition of two external resistors, and suggest suitable resistor values to give an output of 10 V. Why is the performance of the modified regulator likely to be inferior to that of the basic regulator?

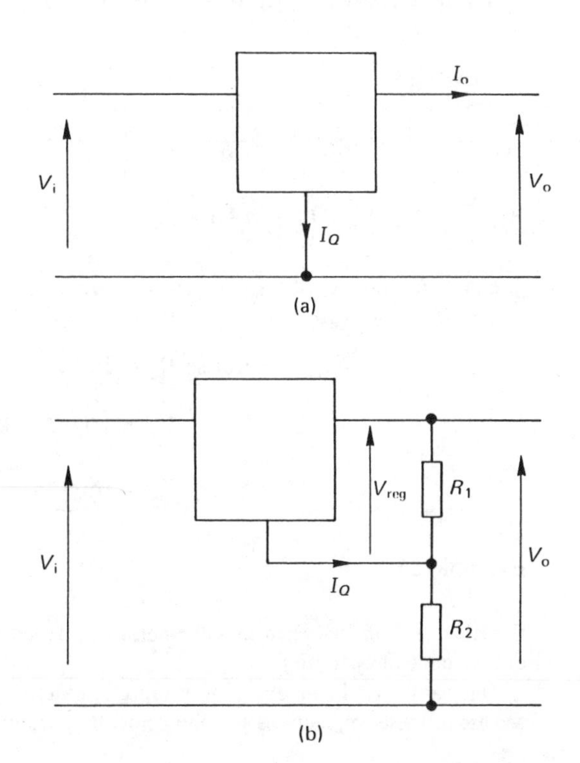

Solution 11.5

The limitation on the value of the output current I_o may be due to the limit circuit, limiting to 1.5 A, or due to the thermal cut-off. To establish which of these causes the limiting in this case, consider the dissipation necessary to give a junction temperature T_j = 175 °C with an ambient temperature of 50 °C.

$$\therefore \; P = \frac{\Delta T}{\theta_{tot}},$$

where θ_{tot} = thermal resistance of case plus heatsink,

$$\therefore \; P = \frac{175 - 50}{6 + 10} = 7.8 \text{ W}.$$

Also,

$$P = (V_i - V_o) I_o + V_i I_Q.$$

For v_i = 12 V,

$$P = (12 - 8) I_o + 12 \times 0.01$$

$$= 4 I_o + 0.12.$$

But P = 7.8 W

$$\therefore \; I_o = 1.9 \text{ A}.$$

The output current will thus be limited to 1.5 A with V_i = 12 V.

For V_i = 12 V,

$$P = (20 - 8) I_O + 20 \times 0.01$$

$$= 12 I_O + 0.2 = 7.8.$$

$$\therefore \; I_O = 0.63 \text{ A}.$$

The output current will thus be limited to 0.63 A with V_i = 20 V.

To increase the output voltage, resistors R_1 and R_2 are added as shown in diagram (b).

Now $V_o = V_{reg} + V_{R2}$

$$= V_{reg} + R_2 \left(\frac{V_{reg}}{R_1} + I_Q \right)$$

$$= V_{reg} \left(1 + \frac{R_2}{R_1} \right) + I_Q R_2.$$

Since I_Q varies with V_i, I_o and temperature, then the term $I_Q R_2$ should be kept small. We thus choose R_1 such that $V_{reg}/R_1 \gg I_Q$

Choose R_1 = 200 Ω.

$$\therefore \; V_o = 10 = 8 + R_2 \left(\frac{8}{200} + 0.01 \right),$$

$$\therefore \; R_2 = 40 \; \Omega.$$

Example 11.6

Explain the term 'switched mode' when it is applied to power supplies and list the advantages of this mode of operation.

On many switched-mode power supplies external sense connections are included. Explain the use of these connections and show how they would be used in practice.

Crowbar over-voltage trip is a common feature of switched-mode power supplies. Describe the function and circumstances under which it is required.

Explain the term 'hold-up time'. What modifications could be made to a switched-mode power supply to increase the hold-up time, and what precautions would be required?

Solution 11.6

'Switched mode' relates directly to the regulating device within the power supply. In this mode of operation the regulating device, usually a transistor, is either switched hard on or hard off.

Driving the device in this way minimises the heat developed by the device, and so increases the efficiency of the supply. Because the efficiency is so high there is no direct need for a step-down transformer, and so the line voltage can be immediately rectified, partially smoothed, and then fed into the regulating device.

External sense connections can be used to compensate for any voltage drop caused by resistance in the distribution wiring. This ensures that the voltage at the input to a circuit board is at the correct level.

Crowbar over-voltage trip is a unit that is connected to the output of a power supply. It consists of a triac connected across the output terminals of the supply, and a trigger circuit, which is set so as to turn on the triac should the output voltage exceed a specified maximum. When the crowbar trips, the triac is turned on and clamps the output to about 0.7 V. It provides an effective method of over voltage protection.

'Hold-up time' is the period between the turning off of the power supply line voltage and the point at which the output voltage falls out of specification. It is usually quoted with the unit supplying a full load.

11.7 Unworked Problems

Problem 11.1

Discuss briefly the principle of operation and characteristics of semiconductor diodes for (a) high-frequency small-signal rectification, (b) power rectification, (c) detection of light, and (d) voltage stabilisation.

The zener diode D shown has a breakdown voltage of 6.0 V and a constant slope resistance of 0.1 Ω. The battery has a terminal voltage E which varies between 9.3 V and 7.3 V as it discharges. Estimate the range of E over which the voltage across the load R is stabilised when R = 680 Ω. Sketch approximately to scale the graph of load voltage against battery voltage. Determine the minimum value of R that will give stabilisation of the whole range of battery voltage. (EC Part 1)

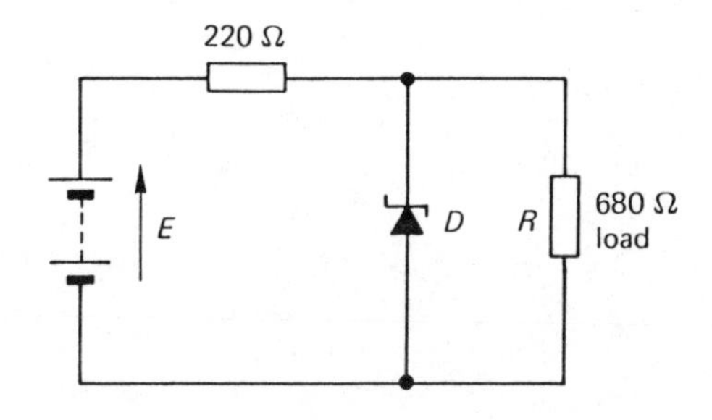

Problem 11.2

State the main disadvantages of power supplies using series-pass regulators and explain how the switched-mode power supply (or switching regulator) overcomes these disadvantages.

Explain the operation of the flyback d.c.-to-d.c. converter shown and calculate appropriate values for R_4 and R_5 to give a nominal output voltage of 10 V for a d.c. supply voltage of 15 V. Sketch the waveforms at v_1 and v_o.

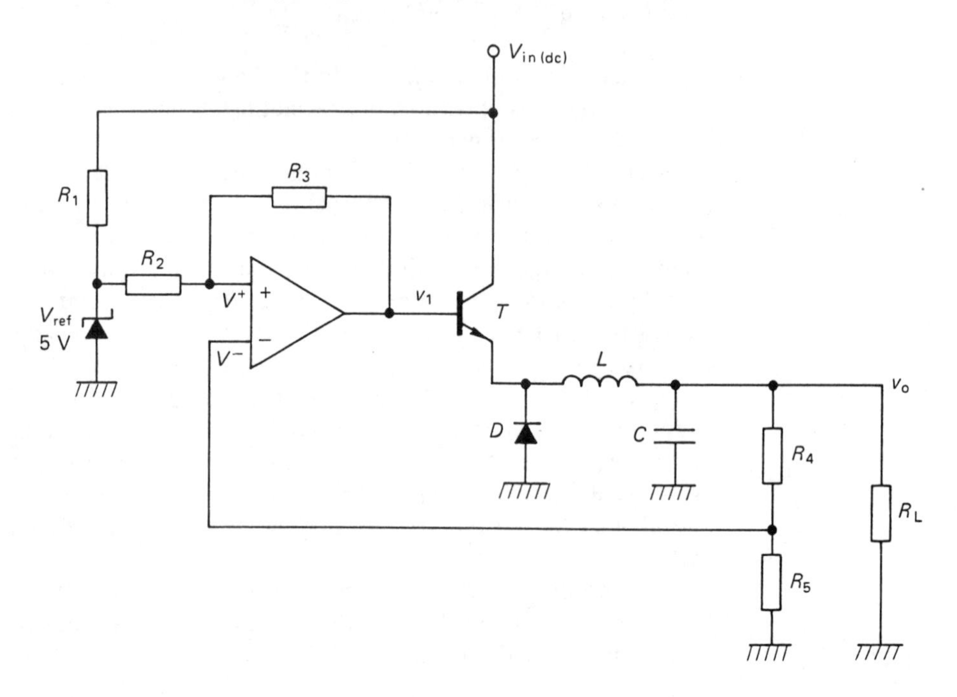

Problem 11.3

Outline the reasons for using a switched-mode pre-regulator to feed a variable-output linear regulator.

A switched-mode regulator of 85 per cent efficiency feeds a linear regulator, whose output can be varied up to 30 V, a voltage not more than 5 V greater than the output of the linear regulator. If the maximum output current is 5 A and the linear-regulator pass transistor has a thermal resistance junction-to-case of 1.5 K/W and a maximum junction temperature of 200 °C, find the thermal resistance of the common heatsink, for an ambient temperature of up to 35 °C. (Neglect the power dissipation in the control circuitry.)

12 Thyristors

12.1 Principle of Operation

The term 'thyristor' refers to a group of semiconductor devices that have a very rapid switch-on characteristic. They are all basically four-layer devices, being made up as a *p n p n* sandwich, and their 'snap-on' switching action is referred to as '*p n p n* regenerative feedback'. They can have two, three or four terminals, depending on the type of device. Some devices pass current only one way (unidirectional), while others pass current both ways (bidirectional).

The SCR (silicon-controlled rectifier) is just one of these devices but is generally referred to simply as the thyristor.

The thyristor is essentially a power diode that can be switched into the conducting state by a low-power switching signal applied to a control input terminal called the 'gate'. It acts like a 'latch', since once switched on it does not require the control current to be maintained to keep it switched on. To switch the device off it is necessary to reduce the anode current to below its holding level. (This is normally done by reducing the anode voltage to zero.)

The main advantages of thyristors are:

(a) low 'on' resistance;
(b) high 'off' resistance;
(c) rapid switching;
(d) low control-power;
(e) long life.

They find particular applications in speed control of rotating electrical machines, in heating control, in car ignition systems and in lamp dimmers.

The symbol for the thyristor and its basic structure are given in Fig. 12.1.

The operation of the thyristor and its characteristics are given in Example 12.1.

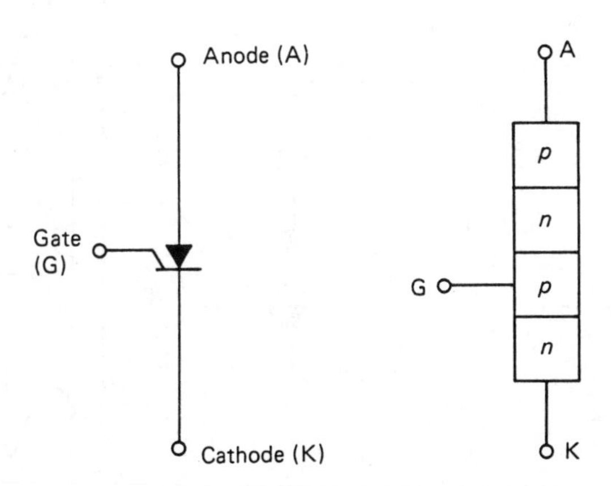

Figure 12.1 Symbol and basic thyristor structure

12.2 Thyristor Control Methods

Thyristors lend themselves most readily to applications in a.c. circuits, since the voltage waveform passes through zero twice per cycle, and thus enables ease of switch-off.

Two basic methods are generally used for the control of thyristors in a.c. circuits.

(a) Burst Triggering Control

The thyristor is switched on and off for whole numbers of consecutive half-cycles, as shown in Fig. 12.2. Mechanical or thermal inertia is used to smooth the effects on the load.

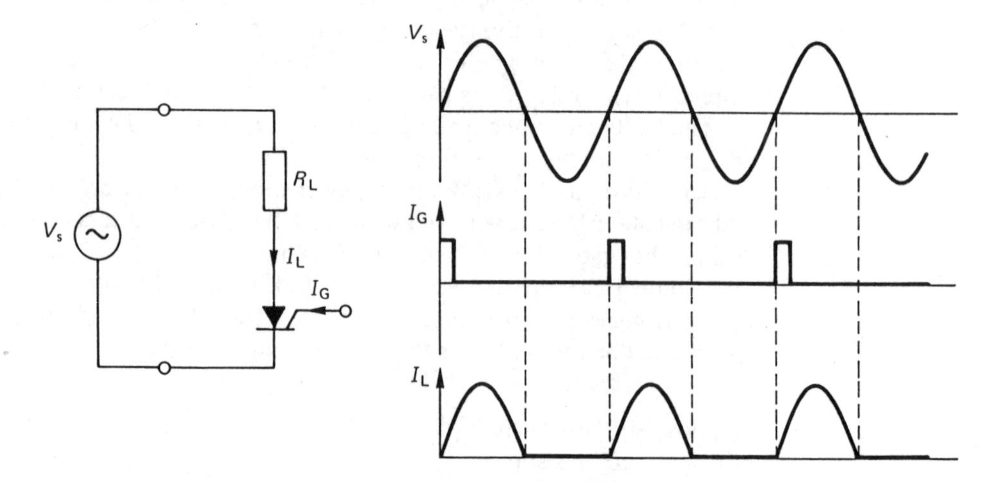

Figure 12.2 Burst triggering control

(b) Phase Control

The device is made to conduct for a fraction of each half-cycle as shown in Fig. 12.3. The advantage of this method of control is that periods of conduction occur

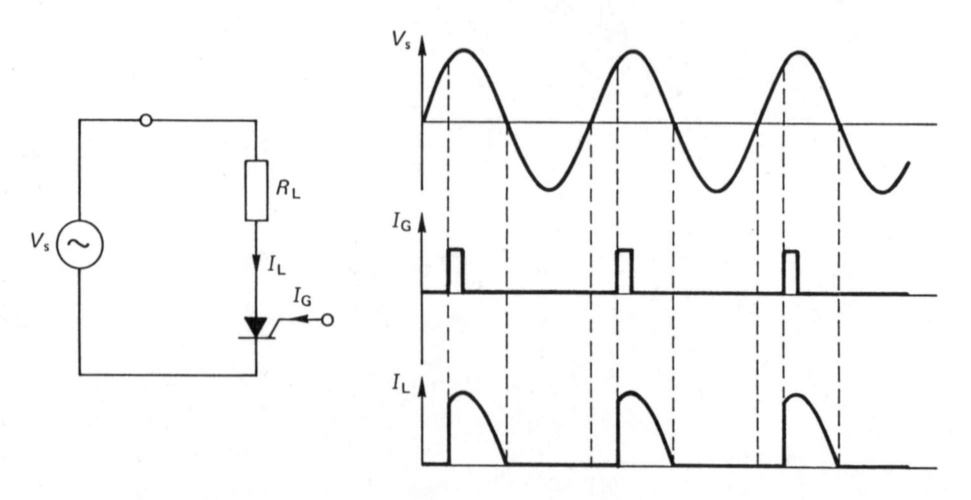

Figure 12.3 Phase control

during every positive half-cycle, thus giving a smoother supply of controllable power. The disadvantage is that unless adequate precautions are taken, mains-conducted and r.f. interference occur.

The angle at which turn-on occurs is called the trigger angle, α.

The angle through which the thyristor conducts is called the conduction angle θ.

12.3 Thyristor Ratings

The ratings of a device define limiting values for voltage, current, power dissipation and temperature. Thyristors are generally rated on an absolute-maximum system, stating values which must not be exceeded. The symbols used are the usual ones of voltage and current (V and I), together with subscripts which specify the condition to which the rating applies.

Generally, the first subscript defines whether the condition is the 'on' state (T) or the 'off' state (D), in the forward direction (F) or the reverse direction (R). The second subscript defines whether the value is the working value (W), the repetitive value (R), or the non-repetitive (surge) value (S).

There are numerous ratings used with regard to thyristors, and it is generally advisable to make full use of manufacturers' data sheets and applications advice when choosing a device for a particular application.

Some typical examples of more commonly used ratings are as follows.

(a) V_{RWM} The working peak reverse voltage (the maximum instantaneous value of the reverse voltage across the thyristor).
(b) V_T The on-state voltage (the voltage across the thyristor in the 'on' state).
(c) $I_{T(rms)}$ The r.m.s. value of on-state current (the total r.m.s. value of the current when in the 'on' state).
(d) V_{GT} The gate trigger voltage (the gate voltage required to produce the gate trigger current).
(e) I_{GT} The gate trigger current (the maximum gate current required to switch the thyristor from the 'off' state to the 'on' state).
(f) dv/dt The maximum rate of rise of anode–cathode voltage that will not trigger the thyristor.
(g) di/dt This indicates the maximum rate of rise of current when the thyristor is triggered that will not cause unequal distribution resulting in 'hot spots' in the junctions of the device. If the rate of rise of current is greater than the permitted value then it must be limited by additional series inductance in the circuit.

12.4 The Diac, Triac and Silicon-controlled Switch

The diac (bidirectional diode thyristor) is a two-directional trigger device. Break-over occurs at a relatively low voltage in either direction, after which the diac exhibits a negative resistance with the current rising rapidly and the voltage falling to a lower working value, as shown in Fig. 12.4. It is mainly used in thyristor and triac triggering circuits.

The triac (bidirectional triode thyristor) may be regarded as two thyristors connected back-to-back with a common-gate electrode. It can be triggered by either positive or negative gate pulses and is used to control the flow of current in either direction, thus making it useful as a full-wave a.c. control device. The triac symbol is shown in Fig. 12.5.

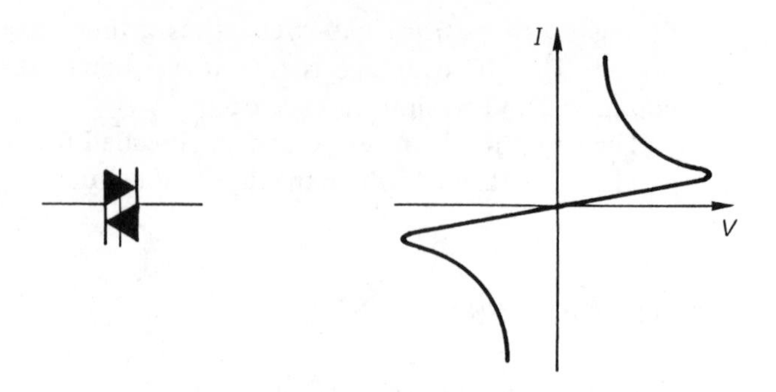

Figure 12.4 The diac symbol and its characteristic

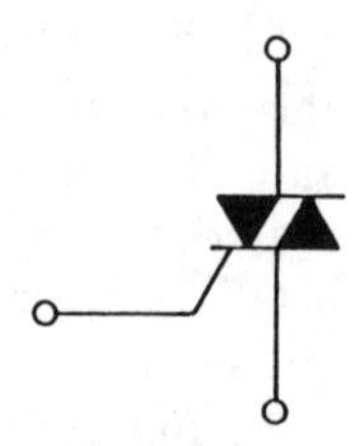

Figure 12.5 The triac

The silicon-controlled switch, SCS, is a thyristor with an additional gate. It may be used as a thyristor but triggered with either positive or negative pulses on one or the other gate. It may also, however, be turned off by applying pulses to the gates.

12.5 Worked Examples

Example 12.1

Explain briefly the physical principles of operation of a silicon-controlled rectifier (SCR) and sketch typical characteristics. Describe a simple method of phase control for an SCR. What are the disadvantages of this type of control? An SCR in series with a 60 Ω heater is connected via a bridge rectifier to a 240 V, 50 Hz supply. The trigger voltage is adjusted so that conduction starts 45° after each voltage zero. Calculate the mean current and the power dissipated in the heater.
(CEI Part 2)

Solution 12.1

The thyristor operation may be understood best by considering it as being made up of two transistors as shown in diagram (a). The two transistors interconnect to form a regenerative feedback pair. With normal bias applied to the thyristor, only leakage current flows, the transistor common-base current gains h_{fb} are low, and the positive feedback around the loop is less than unity.

To switch to the low-impedance 'on' state requires the loop gain to be raised to unity. In this condition the regenerative feedback action drives both devices into saturation.

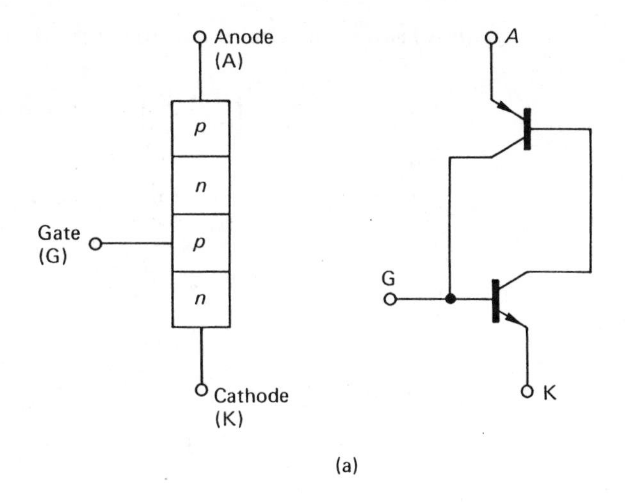

(a)

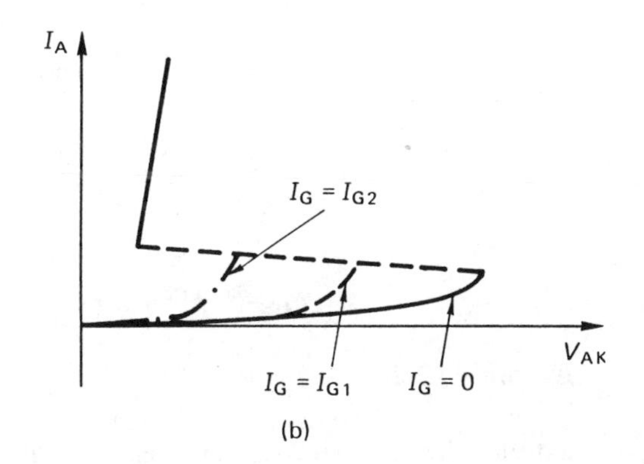

(b)

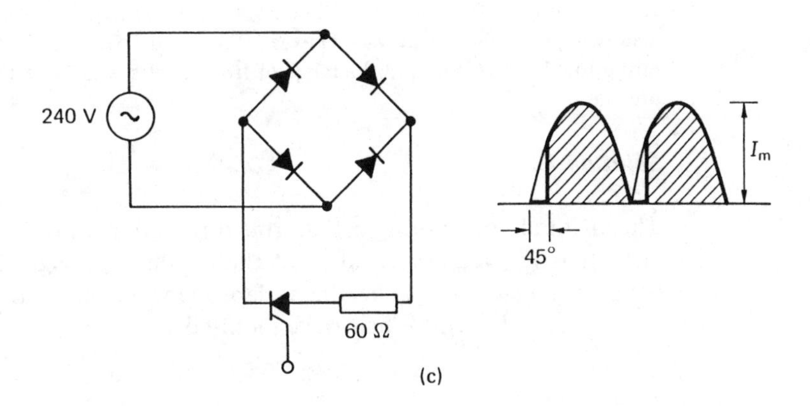

(c)

One way of increasing the loop gain to unity is by injecting a pulse of current into the gate.

The thyristor can now be switched off only by reducing the anode voltage to the point at which the current falls below the holding level.

A typical thyristor characteristic is shown in diagram (b) for various levels of gate current.

Phase control is explained in the text.

Diagram (c) shows the circuit and the current in the load

$$I_{av} = \frac{1}{\pi}\int_{\alpha}^{\pi} I_m \sin\theta \, d\theta$$

$$= \frac{I_m}{\pi}[1 + \cos\alpha]$$

$$= \frac{240\sqrt{2}}{60\pi}[1 + \cos 45°]$$

$$= \underline{3.07 \text{ A}}.$$

The power dissipation

$$P_D = I_{rms}^2 R_L,$$

where

$$I_{rms} = \sqrt{\frac{1}{\pi}\int_{\alpha}^{\pi} I_m^2 \sin^2\theta \, d\theta}$$

$$= I_m \sqrt{\frac{1}{2\pi}(\pi - \alpha + \tfrac{1}{2}\sin 2\alpha)}$$

$$= \frac{240\sqrt{2}}{60\pi}\sqrt{\frac{1}{2\pi}(\pi - \frac{\pi}{4} + \tfrac{1}{2}\sin 90)}$$

$$= 1.21 \text{ A}.$$

$$\therefore P_D = 1.21^2 \times 60 = \underline{88.4 \text{ W}}.$$

Example 12.2

Explain why the r.m.s. current rating for a thyristor, at a given case temperature, may be quoted as a simple numerical value, but the mean current rating must be quoted as a function of conduction angle.

A thyristor is used as a half-wave controlled rectifier in series with a variable-load resistor across a 240 V, 50 Hz supply. If a constant mean current of 10 A is to be maintained in the load, whose resistance may vary between 0.5 Ω and 5 Ω, determine the required r.m.s. current rating for the thyristor, and the range of firing angles which the thyristor trigger circuit must provide.

Solution 12.2

The methods of quoting ratings are different because the heating effect of the current is $I_{rms}^2 R_{on}$ irrespective of the conduction angle. However, for the mean (or average) current, the heating is dependent on conduction angle.

The control circuit is half-wave rectified with

$$\text{firing angle} = \alpha$$

$$\text{conduction angle} = \pi - \alpha.$$

∴ Mean-load current

$$I_{av} = \frac{1}{2\pi}\int_{\alpha}^{\pi} I_m \sin\theta \, d\theta$$

$$= \frac{I_m}{2\pi}(1 + \cos\alpha)$$

(assuming that the 'on' voltage of the thyristor is approximately zero). Using

$$R = 0.5\ \Omega,$$

$$I_m = \frac{240\sqrt{2}}{0.5}.$$

$$\therefore \cos\alpha = \frac{2\pi I_{av}}{I_m} - 1$$

$$= \frac{2\pi \times 10 \times 0.5}{240\sqrt{2}} - 1,$$

$$\therefore \underline{\alpha = 155.2°.}$$

Using $R = 5\ \Omega$,

$$I_m = \frac{250\sqrt{2}}{5}, \text{ leading to}$$

$$\underline{\alpha = 94.3°.}$$

The range of firing angles is thus

$$94.3° < \alpha < 155.2°.$$

The required r.m.s. current rating is found from

$$I_{rms} = \sqrt{\frac{1}{2\pi}\int_{\alpha}^{\pi} I_m^2 \sin^2\theta \, d\theta}$$

$$= \frac{I_m}{2}\sqrt{\frac{\pi - \alpha + \frac{1}{2}\sin 2\alpha}{\pi}}.$$

The worst case is with $\alpha = 155.2°$.

$$\therefore I_{rms(max)} = \underline{43.8\text{ A.}}$$

Example 12.3

Explain why thyristor manufacturers specify di/dt and dv/dt values for their devices, and how it may be ensured that these parameters are not exceeded. With the aid of a sketch, show how the speed of a d.c. motor may be controlled from an a.c. supply using SCRs in a half-wave controlled bridge, and explain the purpose of the flywheel diode.

Solution 12.3

The di/dt rating of a thyristor indicates the maximum rate of rise of current, when triggered, that will not cause unequal current distribution resulting in hot spots in the device junctions.

If the rate of rise of current is greater than that permitted, it must be limited by the inclusion of additional series inductance.

The dv/dt rating is the maximum rate of rise of anode–cathode voltage that will not trigger the thyristor. That this triggering is possible is due to internal parasitic capacitances across the junctions, inherent in the device construction. The rate of rise of voltage may be controlled by the use of an RC series 'snubber' network between anode and cathode.

The control of a d.c. machine may be achieved by control of the current through the armature circuit or the field circuit or both.

The speed of a d.c. motor is given by

$$N = \frac{V - I_a R_a}{kI_F}.$$

Using a single-phase a.c. supply, the half-controlled bridge of the diagram will enable a variable d.c. voltage from zero to almost supply value to be achieved.

The thyristors may be in either the parallel or the series arms.

It is essential to ensure that the thyristor is commutated (switched off) at the end of each half-cycle, or else a short-circuit path will be provided through the diodes. With inductive loads, commutation is not guaranteed and it may not be possible to turn off the load current.

The flywheel diode D shown across the load effectively overcomes the problem. The load current transfers to the flywheel diode at the end of each half-cycle of the supply, thus allowing the thyristor to turn off.

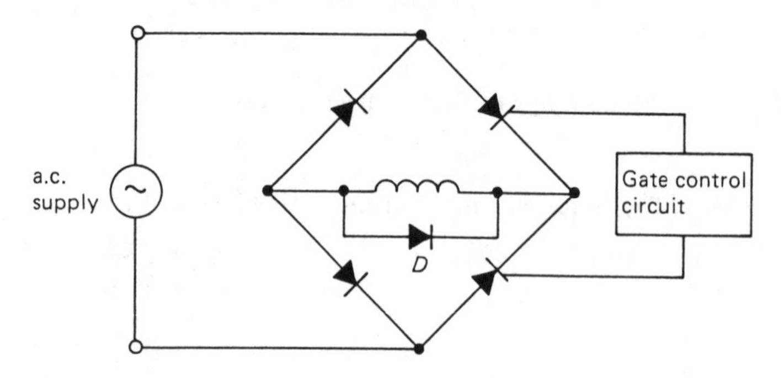

Field or armature winding

Example 12.4

With the aid of a suitable diagram, explain the principal features and operation of a thyristor.

Sketch a circuit diagram of a thyristor chopper circuit incorporating a freewheel diode, suitable for controlling the speed of a d.c. shunt motor.

The thyristor chopper circuit referred to above is supplied from a 72 V battery and is used to control the speed of a d.c. shunt motor which has an armature resistance of 0.2 Ω; the motor field is fully excited. When the chopper is short-circuited, the motor runs at 460 rev/min on half full-load with the armature current of 15 A. If the load is increased to full load and the chopper is brought into use, what will be the chopper mark–space ratio for the motor to run at 320 rev/min?

Solution 12.4

The thyristor features and operation are described in the text.

One possible thyristor chopper circuit for controlling a d.c. motor is shown in the diagram. When T_2 is fired, C is allowed to charge up to the supply voltage. When T_1 is fired, the motor connects to the supply and also causes the charge on C to reverse via L. The oscillation between L and C lasts for one half-cycle. Diode D_1 prevents reverse current.

With the chopper short-circuited,

$$\text{speed } N_1 = 460 \text{ rev/min},$$

$$I_{a1} = 15 \text{ A}.$$

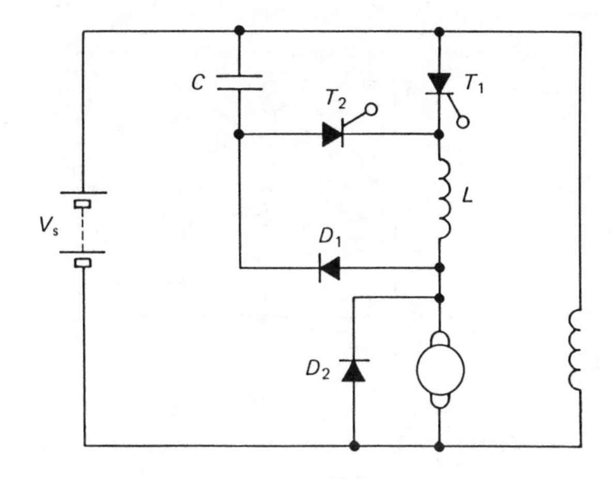

The back e.m.f. is given by

$$E_1 = V - I_{a1} R_a$$
$$= 72 - 15 \times 0.2 = 69 \text{ V}.$$

With the chopper in circuit, speed N_2 relates to the back e.m.f. E_2 by the equation

$$\frac{E_2}{E_1} = \frac{N_2}{N_1}$$

$$\therefore E_2 = 69 \times \frac{320}{460} = 48 \text{ V}.$$

The mean voltage across the armature is given by

$$V_{mean} = E_2 + I_{a2} R_a$$
$$= 48 + 30 \times 0.2$$
$$= 54 \text{ V}.$$

Taking the on period as t_1 and the off period as t_2,

$$\frac{V_{mean}}{V_s} = \frac{t_1}{t_1 + t_2} = \frac{54}{72}.$$

$\therefore$ The mark–space ratio is given by

$$\frac{t_1}{t_2} = \frac{54}{18} = \underline{3}.$$

Example 12.5

In the circuit of diagram (a) the current in the load is controlled by using a thyristor triggered from a timing circuit via a diac.

Explain the operation of the circuit and calculate the value of R to provide a mean current of 0.27 A into the load from a supply voltage of 240 V. The device parameters are as follows:

$R_L = 100\ \Omega$; $C = 0.1\ \mu F$; $V = 72$ V.
V_{BO} (diac breakover voltage) = 30 V;
V_H (diac holding voltage) = 10 V;
I_H (diac holding current) = 100 μA.

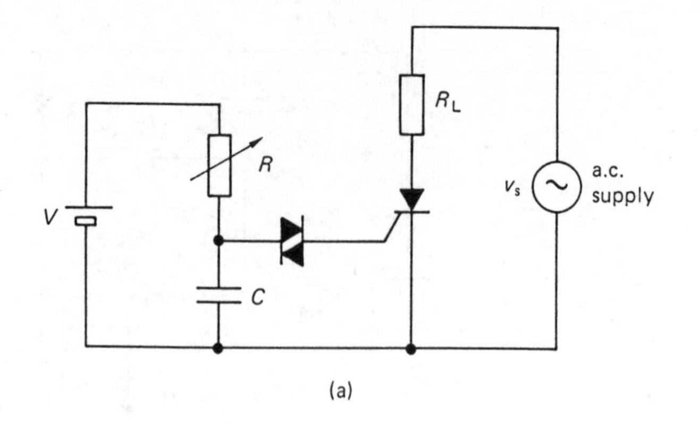

(a)

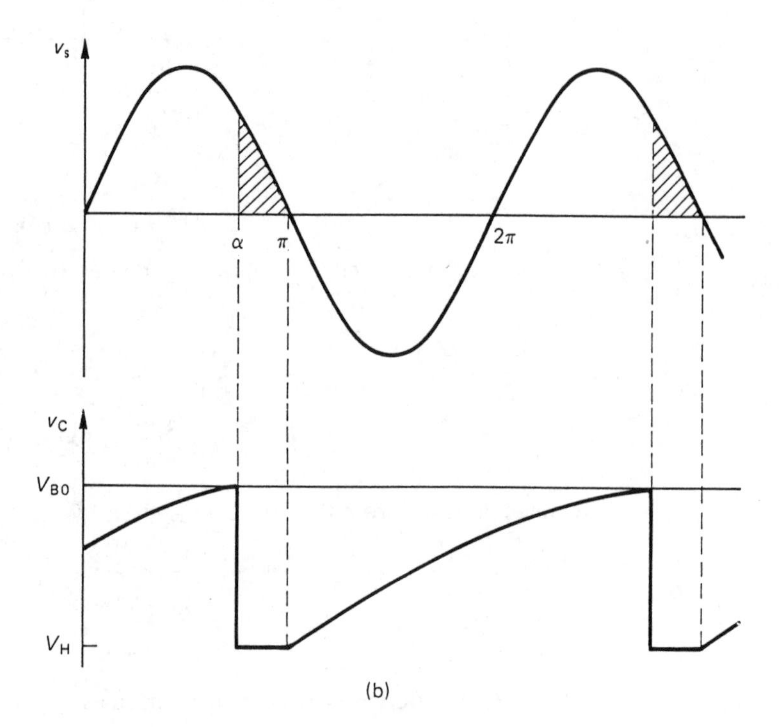

(b)

Solution 12.5

A diac is a silicon bi-directional trigger diode. It is non-conducting until the voltage across it exceeds the breakover voltage. It then conducts with a holding voltage V_H until the current through it falls below the holding current I_H.

The capacitor C changes through the resistor R until the diac reaches its breakover voltage V_{BO}. At this point the diac switches on and triggers the thyristor.

The waveforms are shown in diagram (b).

$$I_{av} = \frac{1}{2\pi} \int_{\alpha}^{\pi} I_m \sin\theta \, d\theta$$

$$= \frac{I_m}{2\pi} [1 + \cos\alpha]$$

$$= \frac{240\sqrt{2}}{100 \times 2\pi} [1 + \cos\alpha]$$

$$= 0.27 \text{ A.}$$

$$\therefore \underline{\alpha = 120°}.$$

The voltage across the capacitor during the charging period is

$$v_C = V_H + (V - V_H)(1 - e^{-t/CR}).$$

The capacitor charging period is

$$T = \frac{300}{360} \times 20 \text{ ms}$$

$$= 16.67 \text{ ms}.$$

After time T, $v_c = V_{BO}$,

$$\therefore V_{BO} = V_H + (V - V_H)(1 - e^{-T/CR}),$$

$$\therefore 30 = 10 + (72 - 10)(1 - e^{-T/CR}),$$

$$\therefore T/CR = 0.389,$$

$$\therefore R = \underline{428 \text{ k}\Omega}.$$

For the diac to remain on during the conduction period the diac current must exceed the holding current I_H.

$$\therefore \frac{V - V_H}{R} > I_H.$$

Now

$$\frac{72 - 10}{428 \times 10^3} \approx 145 \ \mu\text{A},$$

which *does* exceed the required 100 μA holding current.

12.6 Unworked Problems

Problem 12.1

Explain briefly the physical principles of operation of a silicon-controlled rectifier (SCR) and sketch typical characteristics. Compare the features of

(a) phase shift;
(b) pulse; and
(c) burst methods of controlling the power delivered to a resistive load in series with an SCR.

Draw circuits showing for *two* of the following:

(i) The use of an SCR in a crowbar circuit giving over-voltage protection to load.
(ii) A simple phase-shift controller applied to an SCR in series with a 50 Ω resistor across a 240 V, 50 Hz mains supply. Estimate the firing angle to give a dissipation of 100 W in the resistor.
(iii) An SCR inverter to produce 240 V, 50 Hz at power up to 100 W from a 24 V battery.

(CEI Part 2)

Problem 12.2

Describe the physical operation of a thyristor. Briefly discuss (a) methods used to turn on and turn off thyristors, (b) how two thyristors may be operated in parallel.

It is proposed to control a 240 V, 3 kW domestic heater by use of a single series-connected thyristor, the resultant circuit being supplied from the 240 V a.c. mains. Sketch the variation in power output to a base of firing angle, deriving the relevant equation. Draw a circuit diagram for a network to control the firing angle of the thyristor. Comment on the shortcomings of the single-thyristor system and suggest a better one.

(CEI Part 2)

Problem 12.3

What factors limit the maximum power that can be controlled by a thyristor?

A thyristor has a maximum permitted junction temperature of 180 °C. The thermal resistances are 1.8 K/W between junction and case, 0.2 K/W between case and heatsink and 4 K/W between case and air. The thermal resistance θ (in K/W) of the heatsink is given by the relationship

$$\theta = 5A^{-0.45},$$

where A is the area in cm^2 of the heatsink.

Calculate the minimum area of heatsink for a thyristor dissipation of 75 W and an ambient temperature of 20 °C. What percentage reduction of the thyristor dissipation will be necessary if the ambient temperature rises to 35 °C? What will be the corresponding temperature of the case?

(CEI Part 2)

13 Combinational Logic Circuits

Digital devices operate on a binary number system, making it possible to use Boolean algebra as a method of analysis and design of digital circuits. The expressions may be implemented using logic gates in various combinations (small-scale integration, SSI). More complicated logic functions are available in one package (medium-scale integration, MSI). These include decoders, multiplexers, demultiplexers, comparators and adders.

13.1 Boolean Logic Operations

There are two operators associated with Boolean algebra:

Operator	Function
$+$	OR
$\cdot$	AND

Boolean variables (A, B, C, etc.) may take values of either 0 or 1 and are connected by these operators, for example:

$$F = A + B \cdot \overline{C}$$

means F is equal to A or (B and not C), which may be represented by the switch network shown in Fig. 13.1.

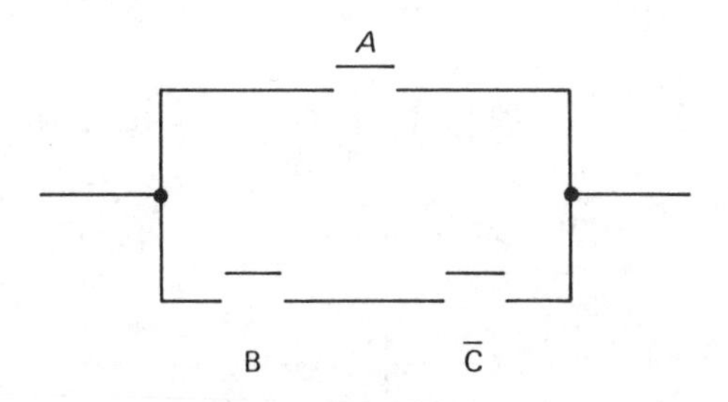

Figure 13.1 Switch network

A table that relates the function value to the logic variables is called a truth table (see Table 13.1).

Table 13.1 Truth tables for various functions

AND

A	B	$A \cdot B$
0	0	0
0	1	0
1	0	0
1	1	1

NAND

A	B	$\overline{A \cdot B}$
0	0	1
0	1	1
1	0	1
1	1	0

OR

A	B	$A + B$
0	0	0
0	1	1
1	0	1
1	1	1

NOR

A	B	$\overline{A + B}$
0	0	1
0	1	0
1	0	0
1	1	0

Excl-OR

A	B	$A \oplus B$
0	0	0
0	1	1
1	0	1
1	1	0

Equivalence

A	B	$\overline{A \oplus B}$
0	0	1
0	1	0
1	0	0
1	1	1

Inverter

A	$\overline{A}$
0	1
1	0

The logic symbols by which the various functions are represented are given in Fig. 13.2.

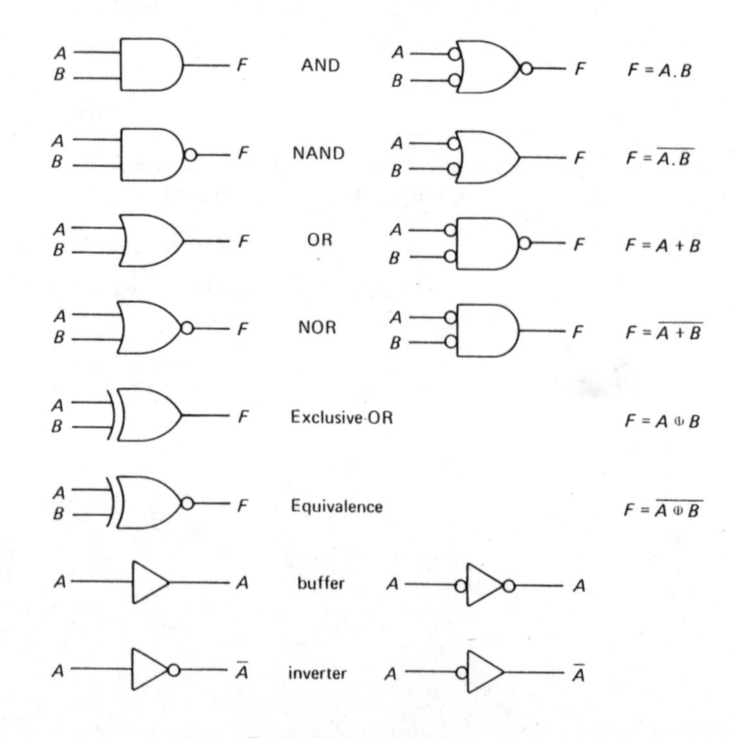

Figure 13.2 Logic symbols

13.2 Boolean Theorems and De Morgan's Rule

(a) Boolean Theorems

A number of theorems are given to assist in simplification of Boolean logic expressions:

$$
\begin{array}{ll}
x \cdot 0 = 0 & x\,(y + z) = xy + xz \\
x \cdot 1 = x & x + xy = x \\
x \cdot x = x & x + \overline{x}y = x + y \\
x \cdot \overline{x} = 0 & xy + x\overline{y} = x \\
x + 0 = x & x\,(x + y) = x \\
x + 1 = 1 & x\,(\overline{x} + y) = xy \\
x + x = x & (x + y)\,(x + \overline{y}) = x \\
x + \overline{x} = 1. &
\end{array}
$$

(b) De Morgan's Theorem

$$\overline{x + y} = \overline{x} \cdot \overline{y}$$

$$\overline{x \cdot y} = \overline{x} + \overline{y}.$$

13.3 Classification of Switching Functions

Switching functions may be classified either as sum of products (SOP) or as product of sums (POS).

Table 13.2

A	B	F
0	0	0
0	1	1
1	0	1
1	1	0

In the truth table of Table 13.2 the function F may be expressed in SOP form as:

$$F = \overline{A} \cdot B + A \cdot \overline{B}.$$

The function may also be obtained by extracting from the table those combinations that do not produce an output:

$$\overline{F} = \overline{A} \cdot \overline{B} + A \cdot B.$$

By applying De Morgan's rule we obtain

$$F = \overline{\overline{A} \cdot \overline{B}} \cdot \overline{A \cdot B} = (\overline{A} + \overline{B}) \cdot (A + B),$$

which is referred to as the POS form.

13.4 Minterms and Maxterms

The SOP form is also called the normal minterm form, and the POS is called the normal maxterm form when referring to switching circuits. The minterms are all of the possible combinations of the variables joined by the AND operation. The maxterms are the combinations joined by the OR operation. Considering the truth table of table 13.2, the minterm representation is

$$F = \Sigma\,(m1, m2),$$

or more simply

$$F = \Sigma\,(1, 2)$$

where Σ means 'summation of'.

The maxterm representation for the truth table is

$$F = \Pi\,(M0, M3),$$

or more simply,

$$F = \Pi\,(0, 3),$$

where Π means 'product of'.

13.5 Minimisation Using K-maps

Minimisation is used to derive the most efficient implementation of a logic function. A K-map (abbreviated from Karnaugh map) uses a graphical technique based on the Venn diagram. Every possible combination of the binary input variables is represented on the map by a square. The squares in the matrix are generally coded using the reflected binary notation for columns and rows, which ensures that there is a change in only one variable between adjacent horizontal or vertical squares. By grouping together the appropriate squares with ones in them it may be seen which terms can be combined to give an effective minimisation. On a K-map the groups are 'looped' together. The looped terms that appear on the map and in the final expression are called the prime implicants. K-maps are convenient for between two and six variables, above which the method becomes too cumbersome. The cells containing zeros may also be combined, except that, in reading the result, the inverse of the variables must be used and combined in POS form.

13.6 Minimisation Using Tabular Methods

For problems with a large number of switching variables the K-map becomes too complicated and a tabular method may be used. It may be used as a hand computation technique or programmed for a computer. One method is the Quine–McCluskey tabular method (QMcC). Tabular methods will not be considered here, except that a modified version of QMcC is used in the multiple output logic problem of Example 13.6.

13.7 Alternatives to Minimisation

Other methods are often used to implement logic functions making use of multiplexers, decoders, read-only memories (ROMs) or programmable logic arrays (PLAs).

13.8 Decoders, Multiplexers and Demultiplexers

A great deal of digital design is concerned with data routing applications. Types of device useful in these applications are the multiplexer (or data selector) in which logic levels, applied to the data select inputs, select one of the inputs and routes it through to the output. The multiplexer may be considered as a multiway switch as shown in Fig. 13.3.

Demultiplexers have a dual role. They may be used to activate one of the 2^n outputs by the state of the n bit control signal at the input as shown in Fig. 13.4. In this application they are referred to as decoders. Alternatively, they may be used to route a data signal to one of the outputs at any given time, in which case they are referred to as demultiplexers.

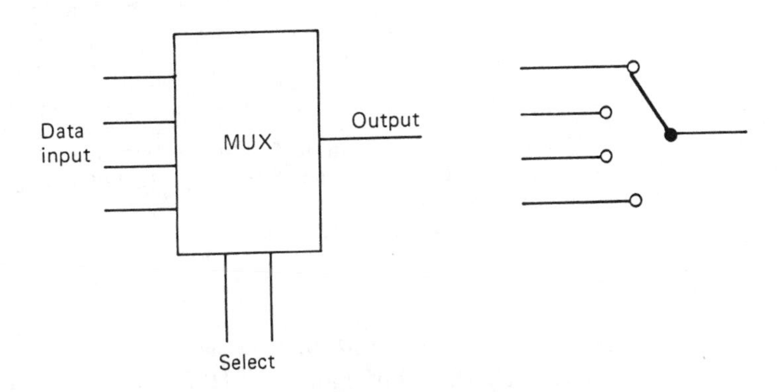

Figure 13.3 The multiplexer and its analogy

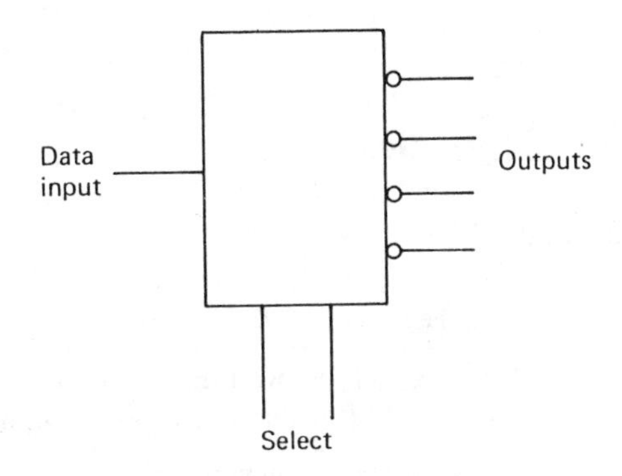

Figure 13.4 The decoder

13.9 Logic Family Characteristics

The main logic families in use today are CMOS, TTL and ECL. Within each family are subgroups that are more or less compatible with each other. As regards speed of operation, the slowest family is CMOS, which will operate at clock rates up to about 10 MHz, but which is normally used at slower speeds. TTL has a wide range of speed options ranging from a few megahertz for standard TTL to about 200 MHz for advanced Schottky devices. ECL is the fastest, with clock rates up to several hundred megahertz. The power supply requirements are between 5 V and 15 V for CMOS, 5 V for TTL and −5.2 V for ECL. Typical logic gates for the CMOS and TTL families are shown in Fig. 13.5. CMOS power consumption is

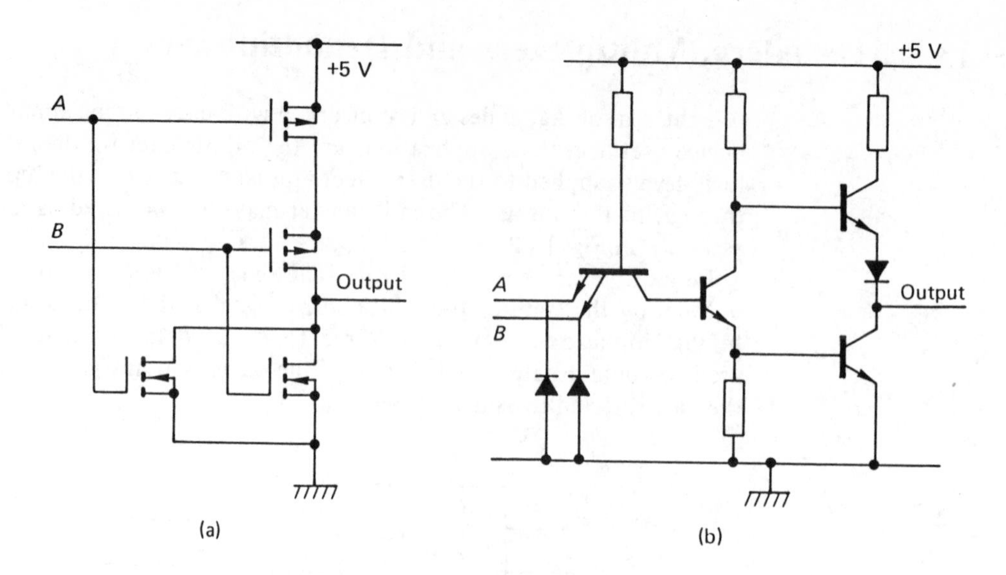

Figure 13.5 Typical logic gates
(a) CMOS NOR gate (b) TTL NAND gate

generally low, whereas TTL may be as high as 10 mW. CMOS fan-out is up to 50, whereas with TTL it is maximum of 10 for the standard family of devices.

13.10 Worked Examples

Example 13.1

Show, giving the relevant circuits, how

(a) an AND function of two logical variables A and B can be realised using only NOR gates,
(b) an exclusive-OR function can be realised using only NAND gates.

Use Boolean algebra to show that if

$$Q = \overline{\overline{A}\,\overline{B}CD + \overline{A}BC\overline{D} + A\overline{B}\,\overline{C}D + AB\overline{C}\,\overline{D}}\,,$$

then $Q = AC + \overline{A}\,\overline{C} + BD + \overline{B}\,\overline{D}$.
Draw the Karnaugh map for Q.

(EC Part 1)

Solution 13.1

(a), (b) The two functions are implemented in diagrams (a) and (b).
Notice that in diagram (b),

$$\begin{aligned}\overline{\overline{A}\cdot\overline{AB}} &= \overline{A\cdot(\overline{A}+\overline{B})}\\ &= \overline{A\overline{B}}.\end{aligned}$$

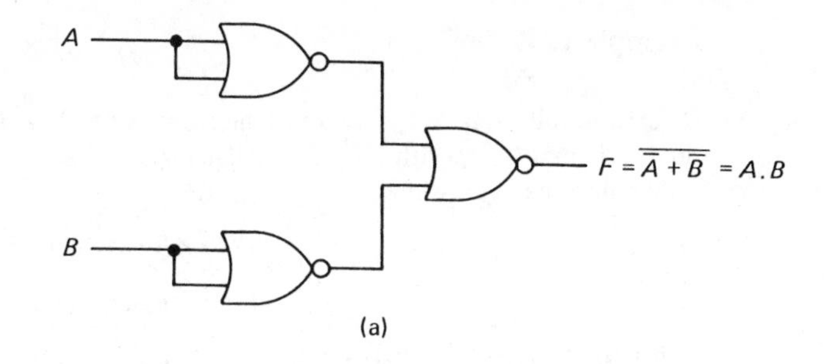

(a)

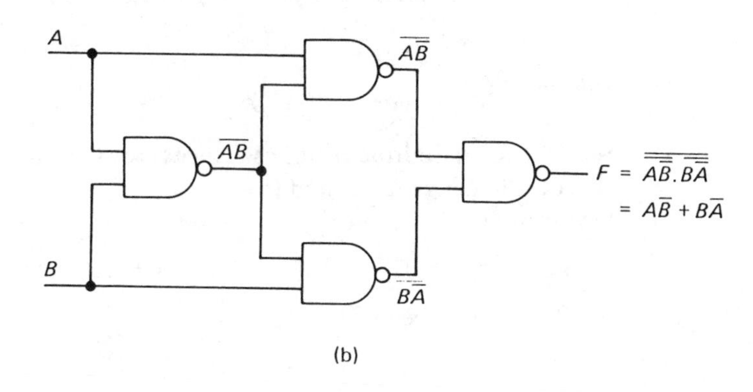

(b)

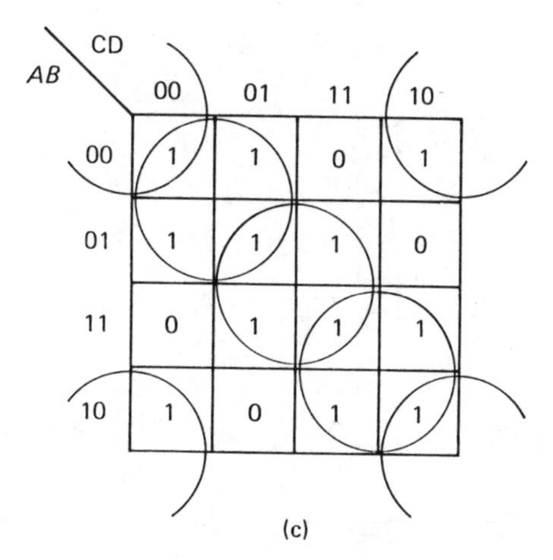

(c)

$$
\begin{aligned}
Q &= \overline{\bar{A}\bar{B}CD + \bar{A}BC\bar{D} + A\bar{B}\bar{C}D + AB\bar{C}\bar{D}} \\
&= \overline{\bar{A}C(\bar{B}D + B\bar{D}) + A\bar{C}(\bar{B}D + B\bar{D})} \\
&= \overline{(\bar{A}C + A\bar{C})(\bar{B}D + B\bar{D})} \\
&= \overline{\bar{A}C + A\bar{C}} + \overline{\bar{B}D + B\bar{D}} \\
&= \overline{\bar{A}C} \cdot \overline{A\bar{C}} + \overline{\bar{B}D} \cdot \overline{B\bar{D}} \\
&= (A + \bar{C}) \cdot (\bar{A} + C) + (B + \bar{D}) \cdot (\bar{B} + D) \\
&= AC + \bar{A}\bar{C} + BD + \bar{B}\bar{D}.
\end{aligned}
$$

The K-map is shown in diagram (c), with the minimal groups encircled.

Example 13.2

Define the following terms used in logic: associative law, De Morgan's theorem, complementation, commutation, positive logic, negative logic.

Simplify the expressions

(a) $$\bar{A}BC + AB + A\bar{B}C + B\bar{C};$$

(b) $$A\bar{B}\bar{C} + A\bar{B}\bar{C}D + \bar{C}A.$$

Briefly compare MOSFETs with bipolar transistors for the fabrication of logic circuits. Give truth tables for the gates shown in diagrams (a) and (b) and explain their operation.

(CEI Part 2)

Solution 13.2

See text for definition of terms. The expressions may be simplified using a K-map as shown in diagrams (c) and (d).

(a) simplified becomes

$$F = B + AC;$$

(b) simplified becomes

$$F = A\bar{C}.$$

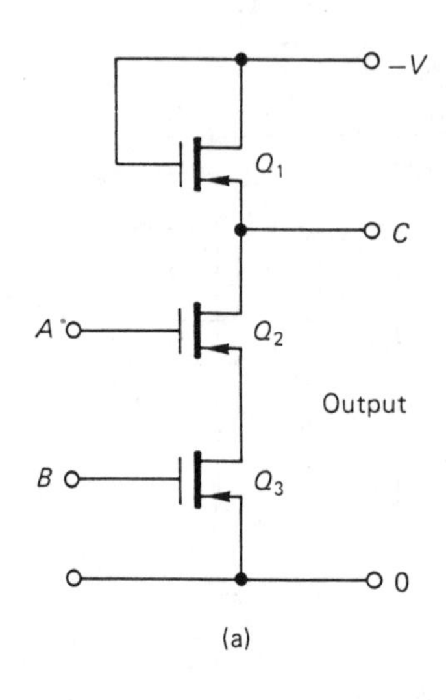

(a)

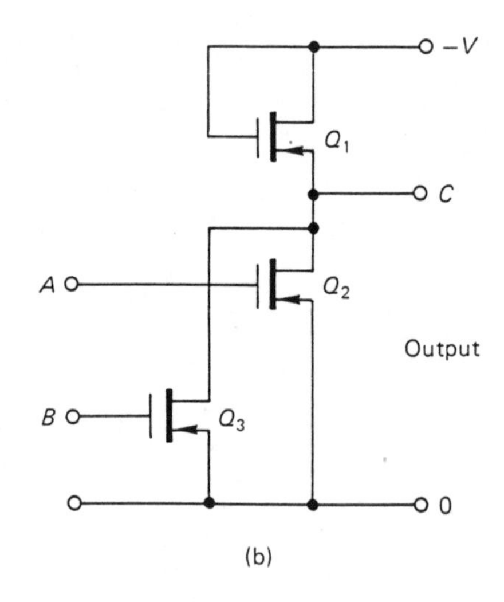

(b)

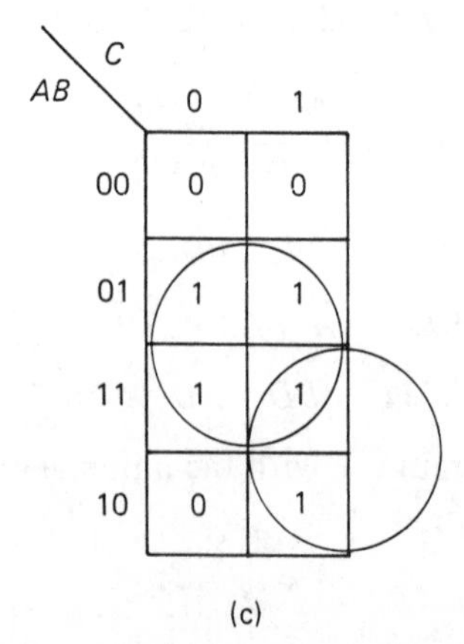

(c)

K-map for $F = \bar{A}BC + AB + A\bar{B}C + B\bar{C}$

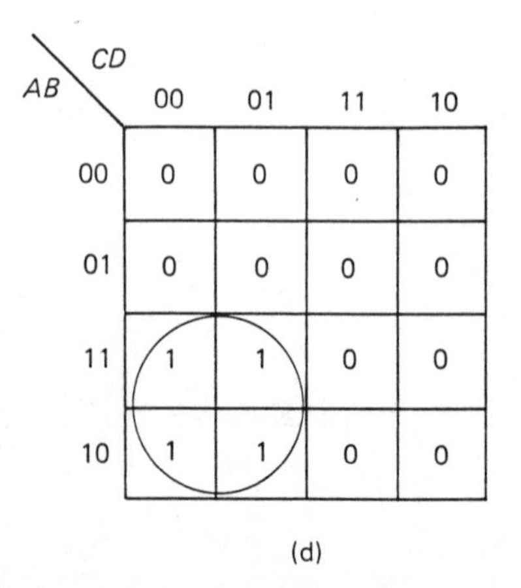

(d)

K-map for $F = A\bar{B}\bar{C} + A\bar{B}\bar{C}D + \bar{C}A$

The advantages of MOSFETs for the fabrication of logic circuits is that they are relatively simple and inexpensive to fabricate, use small physical geometry, use less stages in the masking and diffusion process, consume very little power, can be used at higher voltages and have good noise immunity. Disadvantages over bipolar devices are relatively slow operating speed.

Truth tables for diagrams (a) and (b):

(a)

A	B	C
0	0	0
0	1	0
1	0	0
1	1	1

(b)

A	B	C
0	0	0
0	1	1
1	0	1
1	1	1

The circuits thus provide an AND gate and an OR gate. In both cases, Q_1 acts as the 'load', and Q_2 and Q_3 as the active devices.

Example 13.3

Sketch a circuit, with approximate component values, for a two-input positive TTL NAND gate.

Explain the logic function performed by F in the circuit of diagram (a).

Using the data supplied, calculate the range of suitable values for R.

Suggest two ways of driving a CMOS powered by V_{DD} = 12 V from a TTL device.

74LS03 ratings:

V_{IL} = 0.8 V, I_{IL} = − 0.4 mA,
V_{IH} = 2 V, I_{IH} = 20 μA,
V_{OL} = 0.4 V, I_{OL} = 8 mA,
I_{OH} = 100 μA.

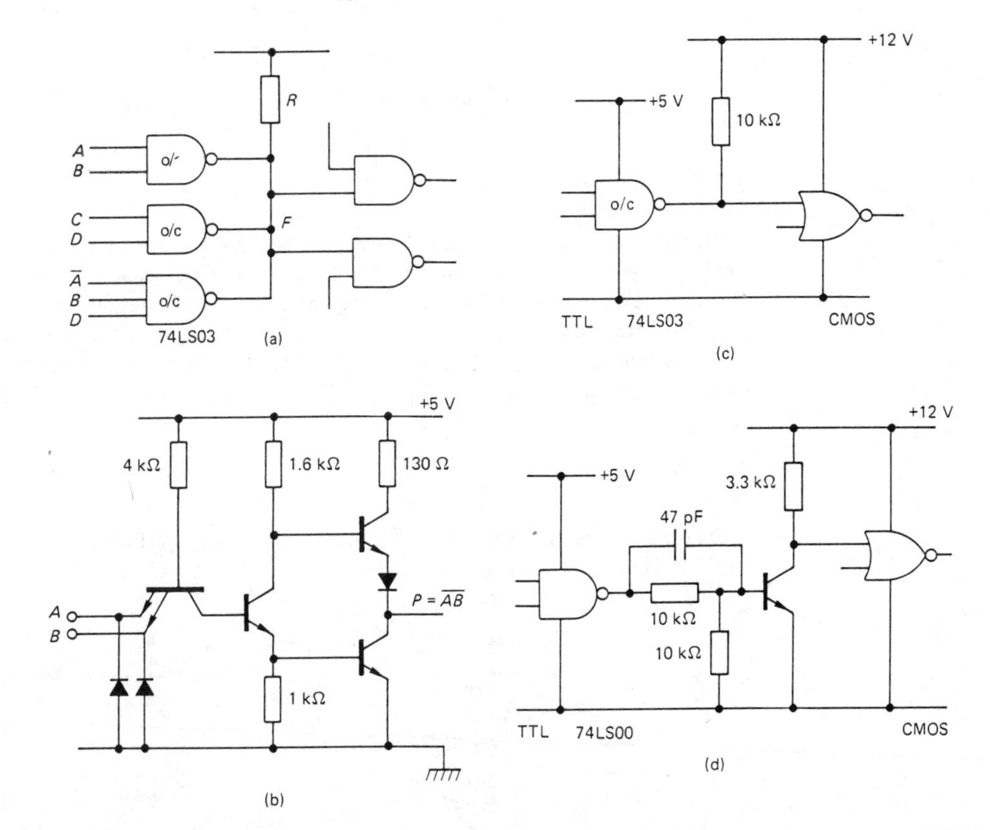

Solution 13.3

The circuit for a two-input TTL NAND gate is shown in diagram (b).

The NAND gates in diagram (a) are open-collector and thus provide a wired-OR function.

Notice that F only goes high when all open-collector outputs are high.

$$\therefore\ F = \overline{AB} \cdot \overline{CD} \cdot \overline{\bar{A}BD},$$

$$\therefore\ F = \overline{AB + CD + \bar{A}BD}.$$

The maximum value of R is when all of the open-collector outputs are high and the output voltage must be > 2 V.

In this condition

$$\begin{aligned} I_R &= 3 \times I_{OH} + 2 \times I_{IH} \\ &= 3 \times 100\ \mu A + 2 \times 20\ \mu A \\ &= 340\ \mu A, \end{aligned}$$

$$\therefore\ R_{(max)} = \frac{V_{CC} - 2}{340\ \mu A} = \frac{5 - 2}{340\ \mu A} = 8.8\ k\Omega.$$

The minimum value of R is when only one of the open-collector outputs is low and the output voltage must be < 0.4 V.

In this condition

$$\begin{aligned} I_R &\approx I_{OL} + 2 \times I_{IL} \\ &= 8\ mA + 2 \times (-0.4\ mA) \\ &= 7.2\ mA. \end{aligned}$$

$$\therefore\ R_{(min)} = \frac{5 - 0.4}{7.2\ mA} = 639\ \Omega.$$

$$\therefore\ \underline{639\ \Omega < R < 8.8\ k\Omega.}$$

Two methods of driving 12 V CMOS from TTL are shown in diagrams (c) and (d).

Example 13.4

Define the terms associative law, distributive law, De Morgan's theorem and minterm as they apply to logic circuits. Give the truth table for a two-input exclusive-OR circuit and show how it may be realised using (a) NOR gates only, and (b) NAND gates only.

A logic function is defined by the minterms $Q(A, B, C) = m(0, 1, 4, 6, 7)$.

Show that a simplified Boolean expression for this function is $Q = AB + \bar{B}(\bar{A} + \bar{C})$. Implement this function using (a) 2-input AND, OR and invert gates and (b) and 8-line to 1-line multiplexer. (CEI Part 2)

Solution 13.4

See text for definitions and Example 13.1 for NAND gate version. NOR gate version is shown in diagram (a).

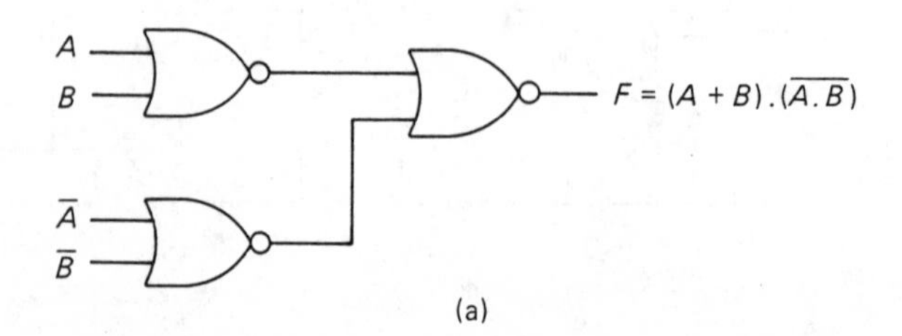

(a)

The truth table for the function Q is shown in the table.

A	B	C	Q
0	0	0	1
0	0	1	1
0	1	0	0
0	1	1	0
1	0	0	1
1	0	1	0
1	1	0	1
1	1	1	1

$$\begin{aligned} \therefore\ Q &= \overline{A}\,\overline{B}\,\overline{C} + \overline{A}\,\overline{B}C + A\overline{B}\,\overline{C} + AB\overline{C} + ABC \\ &= AB\,(C+\overline{C}) + \overline{B}\,(\overline{A}\,\overline{C} + \overline{A}C + A\overline{C}) \\ &= AB + \overline{B}\,(\overline{A}\,(C+\overline{C}) + \overline{C}\,(A+\overline{A})) \\ &= AB + \overline{B}\,(\overline{A} + \overline{C}). \end{aligned}$$

The implementations are shown in diagrams (b) and (c).

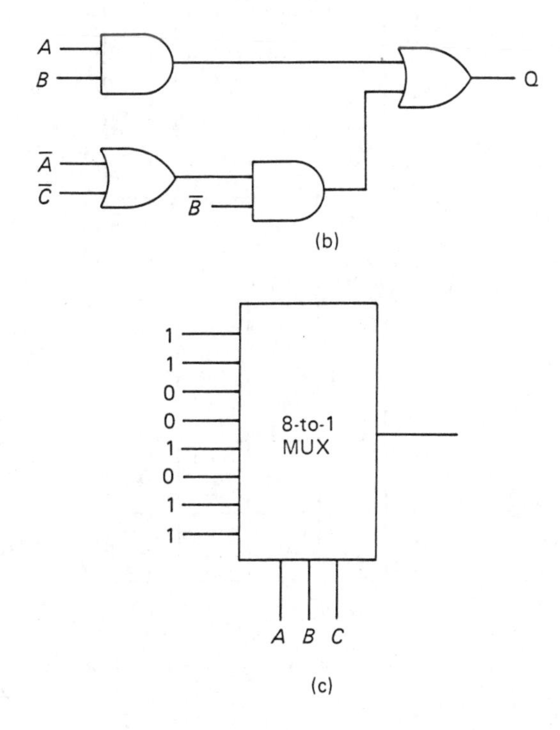

Example 13.5

Discuss the features of the Gray code that make it suitable for use with mechanical displacement transducers. Sketch the pattern on a 3-bit coding disk for use with a rotary transducer. How many bits are required to give an angle resolution of better than 0.1 degree?

Give the truth table and Karnaugh map for a 4-bit Gray code and design a logic network to convert it into binary code, using exclusive-OR gates. (CEI Part 2)

Solution 13.5

Gray code is a 'minimum-change' code, in which only one bit in the code group changes when going from one step to the next.

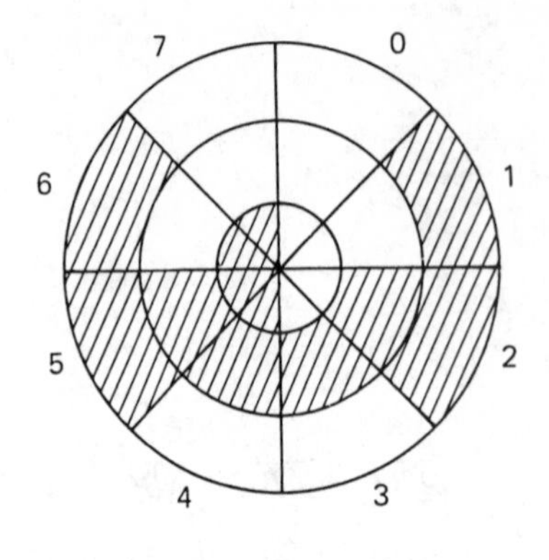

Outer ring is LSB

Code	Angle (degrees)
0	0
1	45
2	90
3	135
4	180
5	225
6	270
7	315

(a)

B_4

G_2G_1 \ G_4G_3	00	01	11	10
00	0 (0)	0 (7)	1 (8)	1 (15)
01	0 (1)	0 (6)	1 (9)	1 (14)
11	0 (2)	0 (5)	1 (10)	1 (13)
10	0 (3)	0 (4)	1 (11)	1 (12)

B_3

G_2G_1 \ G_4G_3	00	01	11	10
00	0	1	0	1
01	0	1	0	1
11	0	1	0	1
10	0	1	0	1

B_2

G_2G_1 \ G_4G_3	00	01	11	10
00	0	1	0	1
01	0	1	0	1
11	1	0	1	0
10	1	0	1	0

B_1

G_2G_1 \ G_4G_3	00	01	11	10
00	0	1	0	1
01	1	0	1	0
11	0	1	0	1
10	1	0	1	0

(b)

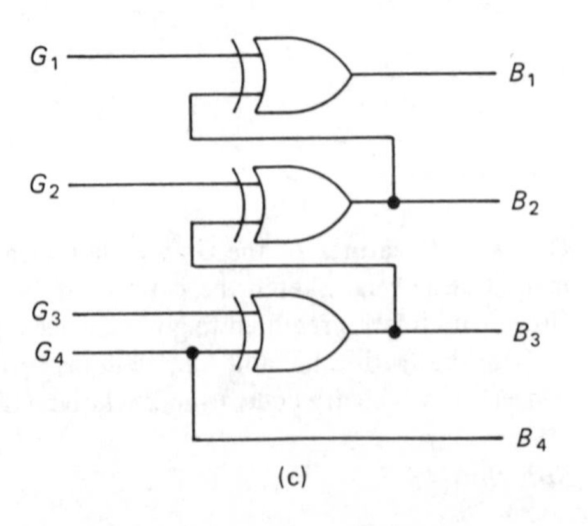

(c)

It is often used in situations where a binary code might produce erroneous results during those transitions in which more than one bit of the code changes.

Decimal	Binary B_4	B_3	B_2	B_1	Gray G_4	G_3	G_2	G_1
0	0	0	0	0	0	0	0	0
1	0	0	0	1	0	0	0	1
2	0	0	1	0	0	0	1	1
3	0	0	1	1	0	0	1	0
4	0	1	0	0	0	1	1	0
5	0	1	0	1	0	1	1	1
6	0	1	1	0	0	1	0	1
7	0	1	1	1	0	1	0	0
8	1	0	0	0	1	1	0	0
9	1	0	0	1	1	1	0	1
10	1	0	1	0	1	1	1	1
11	1	0	1	1	1	1	1	0
12	1	1	0	0	1	0	1	0
13	1	1	0	1	1	0	1	1
14	1	1	1	0	1	0	0	1
15	1	1	1	1	1	0	0	0

The pattern of a 3-bit coded disk is shown in diagram (a). A resolution of 0.1° in 360° would require 12 bits, giving a resolution of 1 in 4096.

The truth table for a 4-bit code is given in the table. The K-maps for a Gray-to-binary converter are given in diagram (b).

From the symmetry of these tables, a simple Gray-to-binary converter is shown in diagram (c) using exclusive-OR gates.

Example 13.6

The following sum of products functions represent three outputs, x, y, z, using three input variables, a, b and c. Obtain the minimum reduction by identifying shared terms.

$$F_x = \Sigma\,(1, 2, 4, 5, 7)$$
$$F_y = \Sigma\,(0, 3, 4, 5, 6)$$
$$F_z = \Sigma\,(1, 2, 3, 4, 5)$$

Show how these functions could be implemented using a ROM.

Solution 13.6

The truth tables are given in the table.

a	b	c	F_x	F_y	F_z
0	0	0	0	1	0
0	0	1	1	0	1
0	1	0	1	0	1
0	1	1	0	1	1
1	0	0	1	1	1
1	0	1	1	1	1
1	1	0	0	1	0
1	1	1	1	0	0

The output functions may be represented by the following K-maps:

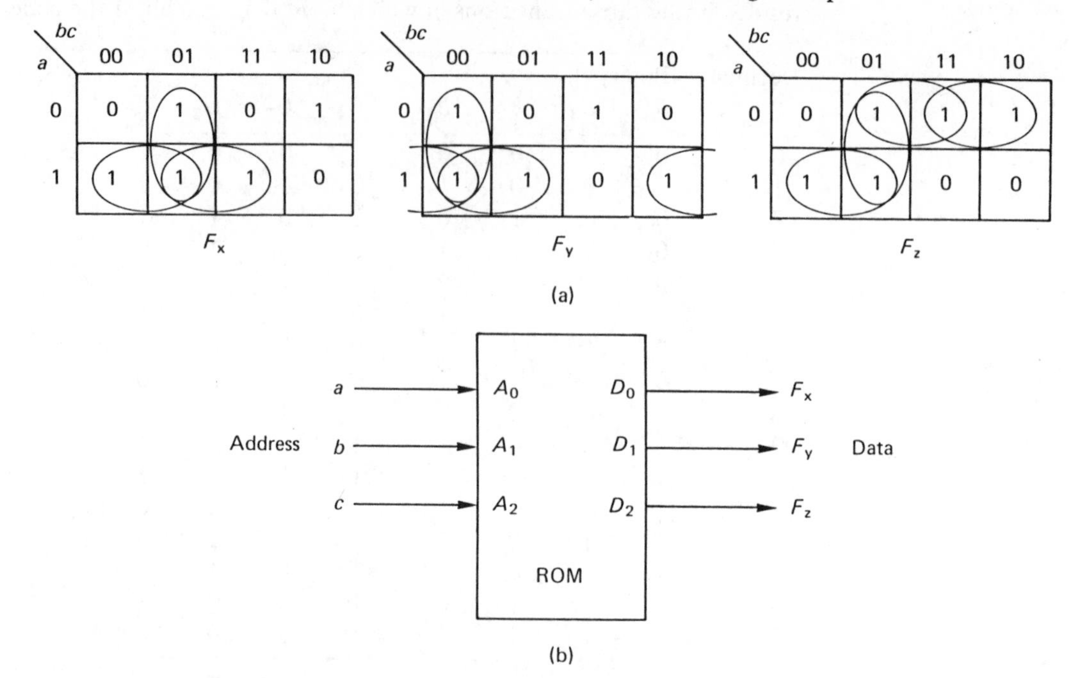

(a)

(b)

These functions could be implemented directly, thus obtaining a minimal solution for each individual function. However, we can minimise collectively using a modified form of the M[c] Cluskey method.

First, identify all implicants by forming *all* possible groups and construct an implicant/minterm chart as shown in the table.

Now apply the rule that any row with only one entry represents an essential implicant and choose the optimal grouping (0, 6 and 7 in this case).

Next examine the rows with two entries and select the optimal groupings (1, 2 and 3 in this case).

Next examine the rows with three entries and select optimal groupings.

The minimal equations are thus given by

$$F_x = \overline{b}c + \overline{a}b\overline{c} + a\overline{b} + ac,$$

$$F_y = \overline{b}\overline{c} + \overline{a}bc + a\overline{b} + a\overline{c},$$

$$F_z = \overline{b}c + \overline{a}b\overline{c} + \overline{a}bc + a\overline{b}.$$

The functions could easily be implemented using a ROM, as shown in diagram (b), with appropriate programming.

Minterm	F_x	F_y	F_z
0		$\overline{a}\overline{b}\overline{c}$ / ($\overline{b}\overline{c}$)	
1	$\overline{a}\overline{b}c$/ ($\overline{b}c$)		$\overline{a}\overline{b}c/\overline{a}c/$ ($\overline{b}c$)
2	($\overline{a}b\overline{c}$)		($\overline{a}b\overline{c}$)/$\overline{a}b$
3		($\overline{a}bc$)	($\overline{a}bc$) /$\overline{a}b/\overline{a}c$
4	$a\overline{b}\overline{c}$/ ($a\overline{b}$)	$a\overline{b}\overline{c}/\overline{b}\overline{c}$/ ($a\overline{b}$) /$a\overline{c}$	$a\overline{b}\overline{c}$/ ($a\overline{b}$)
5	$a\overline{b}c$/ ($a\overline{b}$) /ac	$a\overline{b}c$/ ($a\overline{b}$)	$a\overline{b}c$/ ($a\overline{b}$) /$\overline{b}c$
6		$ab\overline{c}$/ ($a\overline{c}$)	
7	abc/ (ac)		

Example 13.7

Multiplexers may be used to implement logic functions. Compare the costs of implementing the function:

$$F = \Sigma\,(1, 2, 4, 6, 8, 9, 14, 15), \text{ using}$$

(a) a single 8-to-1 multiplexer,
(b) two 4-to-1 multiplexers and appropriate gates;
(c) a single 4-to-1 multiplexer and appropriate gates.

Device	Price
74LS251 8-to-1 MUX	0.97
74LS253 Dual 4-to-1 MUX	0.92
74LS04 Hex inverter	0.42
74LS32 Quad 2-i/p OR gate	0.37
74LS86 Quad 2-i/p Excl-OR gate	0.45

Solution 13.7

The truth table is shown below and the corresponding implementations in diagrams (a), (b) and (c).

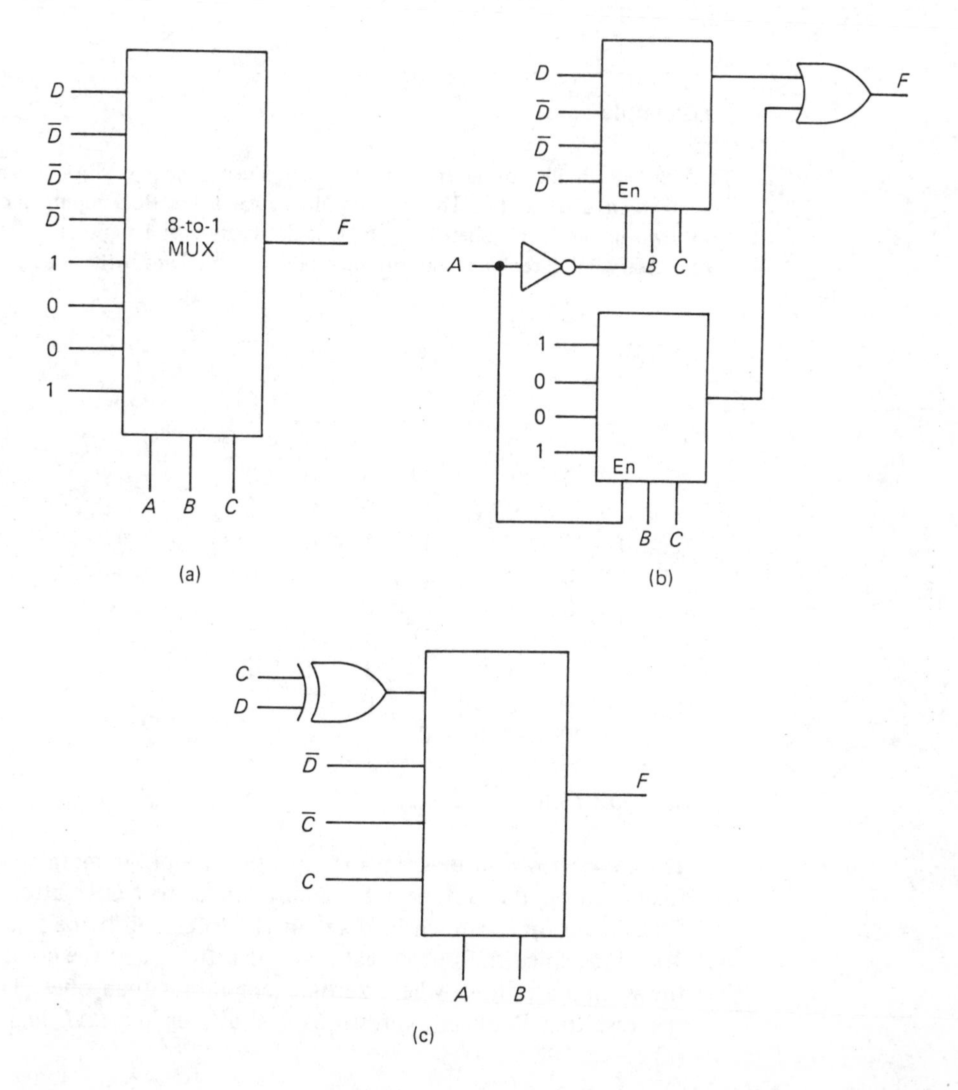

The corresponding costings are:

(a) 1 x 74LS251 + 1 x 74LS04 = £1.39
(b) 1 x 74LS253 + 1 x 74LS04 = £1.71
(c) 1 x 74LS253 + 1 x 74LS04 + 1 x 74LS86 = £1.79

A	*B*	*C*	*D*	*F*
0	0	0	0	0
0	0	0	1	1
0	0	1	0	1
0	0	1	1	0
0	1	0	0	1
0	1	0	1	0
0	1	1	0	1
0	1	1	1	0
1	0	0	0	1
1	0	0	1	1
1	0	1	0	0
1	0	1	1	0
1	1	0	0	0
1	1	0	1	0
1	1	1	0	1
1	1	1	1	1

Example 13.8

A BCD-to-Braille converter is to be designed using the 4 x 4 PAL (Programmable Array Logic) shown in diagram (a). The relationship between the BCD inputs and the Braille outputs is shown in the table. Sketch the PAL logic diagram and show on it, by means of crosses, the necessary link interconnections to be retained to implement the code conversion.

BCD				Braille			
A	*B*	*C*	*D*	*W*	*X*	*Y*	*Z*
0	0	0	0	0	1	1	1
0	0	0	1	1	0	0	0
0	0	1	0	1	0	0	1
0	0	1	1	1	1	0	0
0	1	0	0	1	1	1	0
0	1	0	1	1	0	1	0
0	1	1	0	1	1	0	1
0	1	1	1	1	1	1	1
1	0	0	0	1	0	1	1
1	0	0	1	0	1	0	1

Solution 13.8

The PAL shown in diagram (a) consists of a programmable AND gate array using fusible links, the outputs of which connect to NOR gates. The four K-maps are first drawn up as shown in diagram (b) to establish the logic functions for W, X, Y and Z. Since the output gates are inverting, then the conditions extracted from the K-map are those where zeros occur rather than ones. The implementation of the resulting Boolean expressions is shown in the PAL logic diagram of diagram (c).

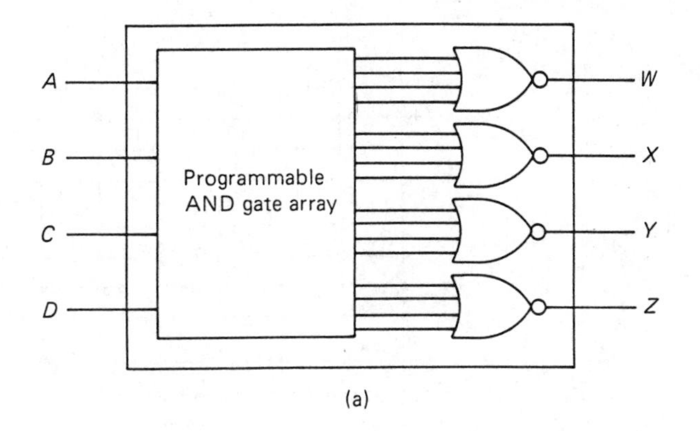

(a)

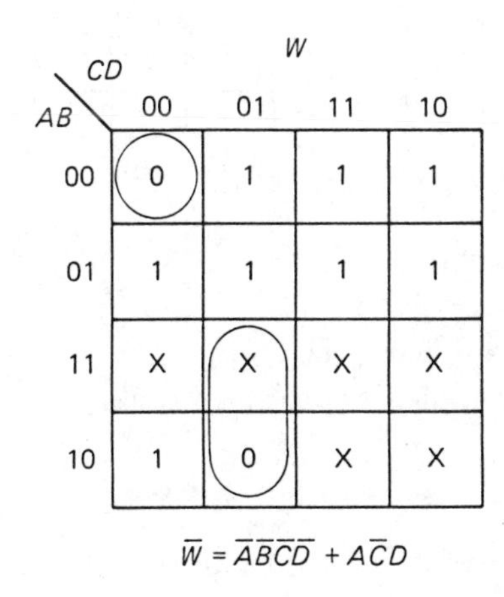

$\overline{W} = \overline{A}\overline{B}\overline{C}\overline{D} + A\overline{C}D$

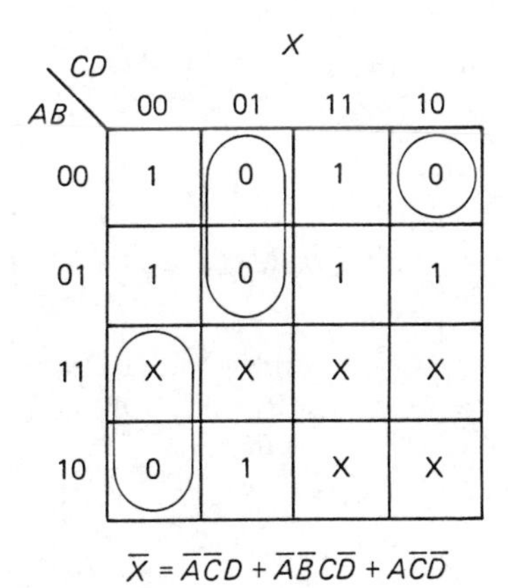

$\overline{X} = \overline{A}\overline{C}D + \overline{A}\overline{B}C\overline{D} + A\overline{C}\overline{D}$

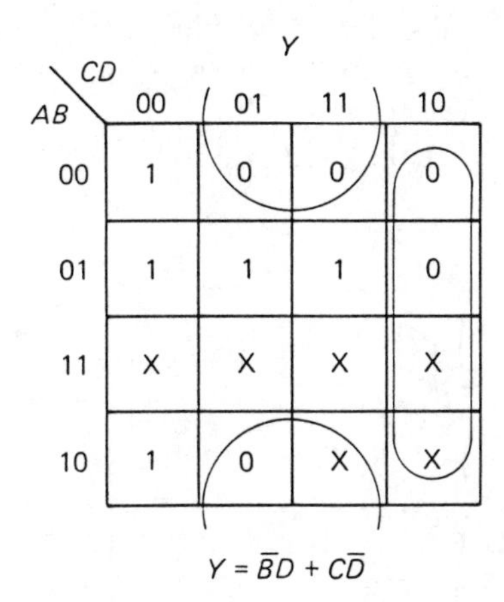

$Y = \overline{B}D + C\overline{D}$

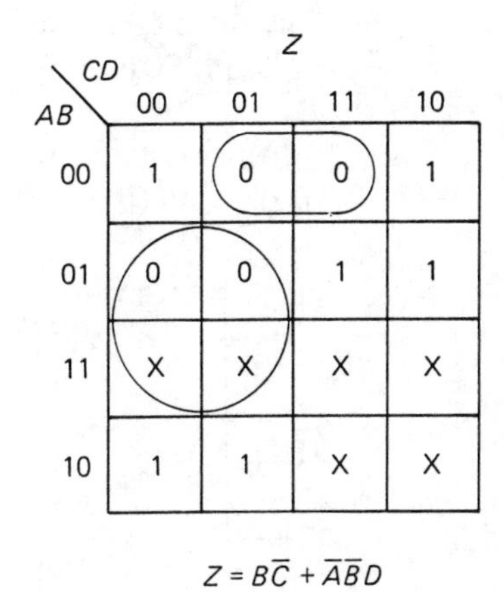

$Z = B\overline{C} + \overline{A}\overline{B}D$

(b)

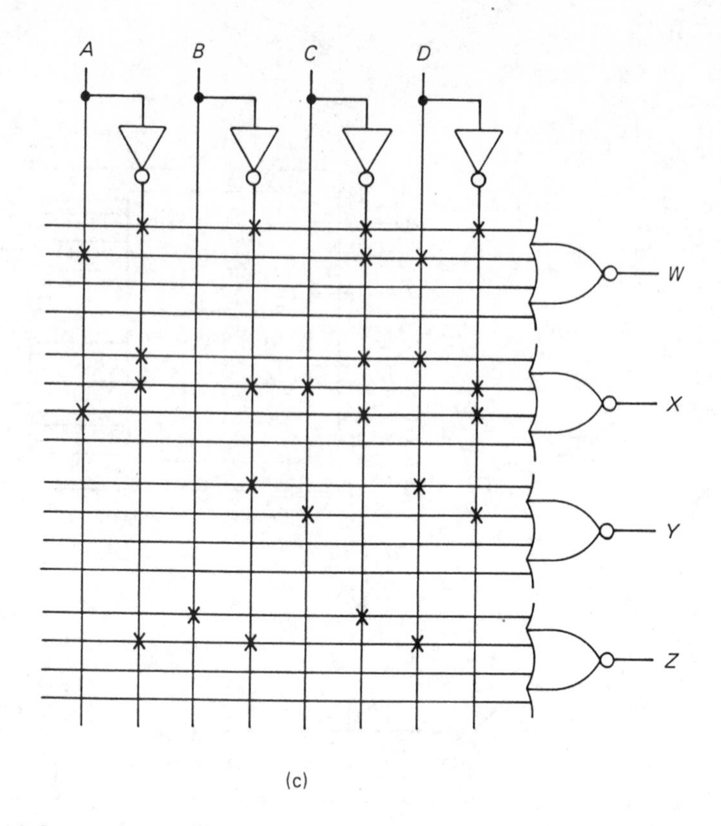

(c)

Example 13.9

Give the truth table for the 74LS138 3-to-8 line decoder.

By using SOP and POS terms, show two methods of using a 4-to-16 decoder to implement the function

$$F = \Sigma\,(0, 2, 3, 5, 7, 8, 9).$$

assuming that you need only consider the first 10 states.

Solution 13.9

The truth table for the 74LS138 3-to-8 line decoder is shown in the table.

Using SOP terms, the circuit of diagram (a) is one method of implementing the function F.

An alternative is to look for the cases where F is zero, which leads to the circuit of diagram (b).

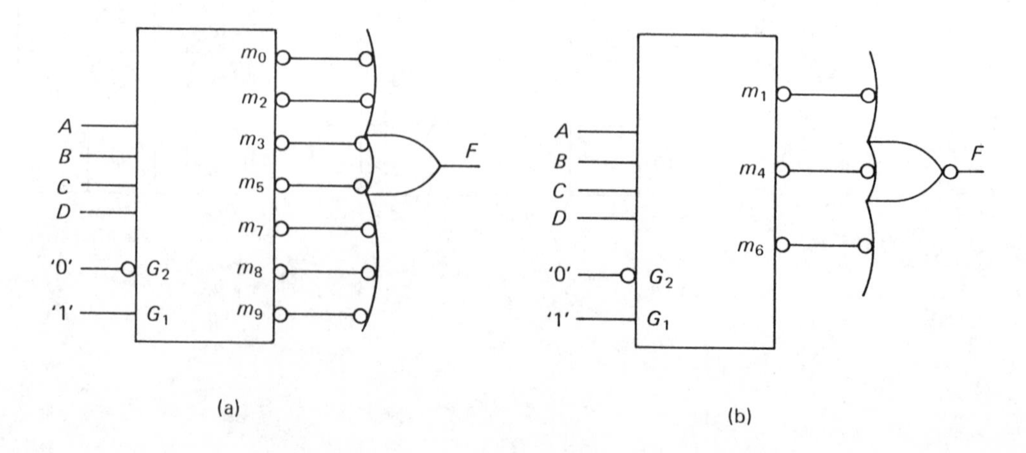

Inputs					Outputs							
Enable		Select										
G_1	G_2	C	B	A	Y_0	Y_1	Y_2	Y_3	Y_4	Y_5	Y_6	Y_7
X	1	X	X	X	1	1	1	1	1	1	1	1
0	X	X	X	X	1	1	1	1	1	1	1	1
1	0	0	0	0	0	1	1	1	1	1	1	1
1	0	0	0	1	1	0	1	1	1	1	1	1
1	0	0	1	0	1	1	0	1	1	1	1	1
1	0	0	1	1	1	1	1	0	1	1	1	1
1	0	1	0	0	1	1	1	1	0	1	1	1
1	0	1	0	1	1	1	1	1	1	0	1	1
1	0	1	1	0	1	1	1	1	1	1	0	1
1	0	1	1	1	1	1	1	1	1	1	1	0

X = don't care.

Example 13.10

Show that the circuit of the diagram implements the function $F = A \oplus B$ and hence show how you would construct, using NAND gates only,

(a) a half-adder;
(b) a full-adder;
(c) a 2-bit adder (with no carry-in).

If the gates used have a propagation delay of T, calculate the maximum delay time for the circuit in diagram (b).

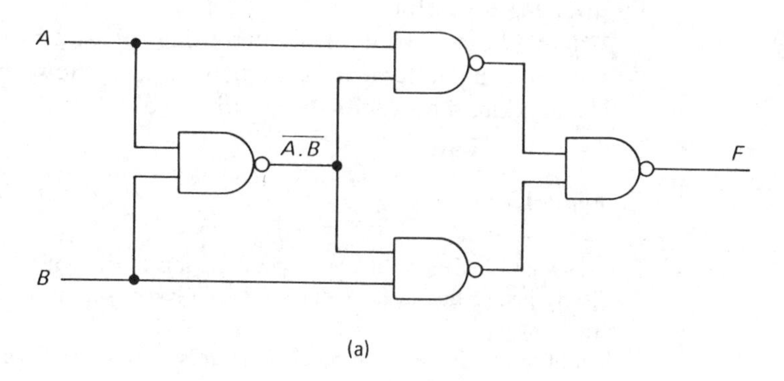

(a)

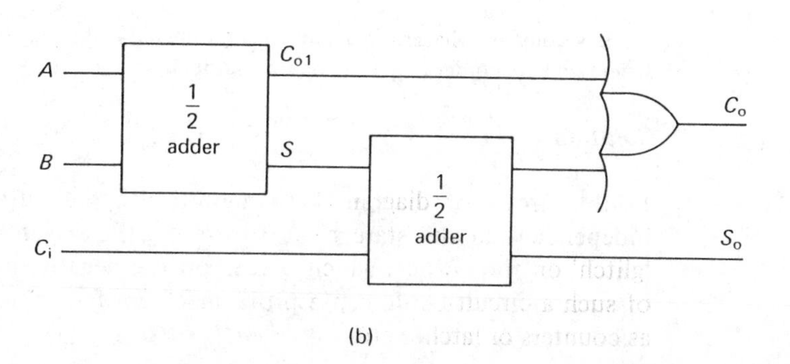

(b)

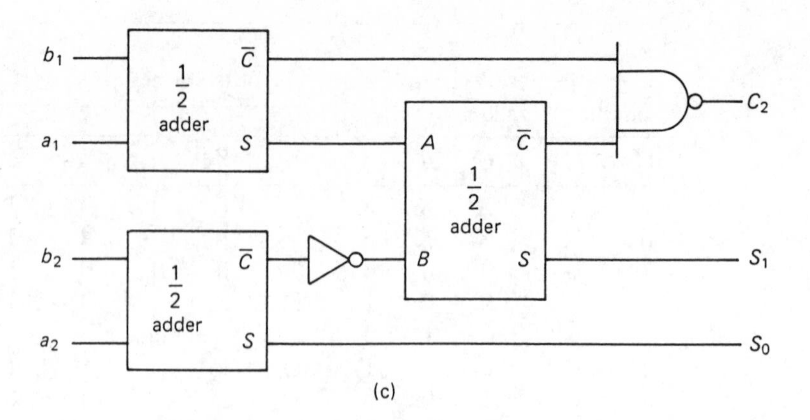

(c)

Solution 13.10

Referring to diagram (a):

$$F = \overline{\overline{A \cdot \overline{A \cdot B}} \cdot \overline{B \cdot \overline{A \cdot B}}}$$

$$= A \cdot \overline{AB} + B \cdot \overline{AB}$$

$$= (A + B) \cdot \overline{AB} = A \oplus B$$

A half-adder requires the generation of Sum and Carry, as shown in the table.

A	B	Sum	Carry
0	0	0	0
0	1	1	0
1	0	1	0
1	1	0	1

The circuit of diagram (a) generates the Sum. The Carry can easily be generated by inverting the output $\overline{A \cdot B}$.

A full-adder may be implemented using half-adders as shown in diagram (b).

A two-bit full-adder may be implemented as shown in diagram (c).

The maximum delay for the circuit in (b) is $6T$.

Example 13.11

Explain what is meant by the term 'combinational logic static hazard'.

Clearly justify the need to eliminate such hazards from certain types of combinational logic circuit.

Identify and eliminate all possible hazards from the following function, assuming all input conditions occur in random order:

$$F = A\overline{B} + \overline{A}\,\overline{D} + B\overline{C}D.$$

How could the hazard-free implementation be simplified if the input conditions occur in strict sequential order (e.g. from an up-counter)?

Solution 13.11

In the circuit of diagram (a), theoretically the output Y should remain high independent of the state of X. However, the waveform of diagram (b) shows a 'glitch' on the output, which occurs on the negative-going edge of X. The output of such a circuit could cause problems when fed to additional logic circuits such as counters or latches etc.

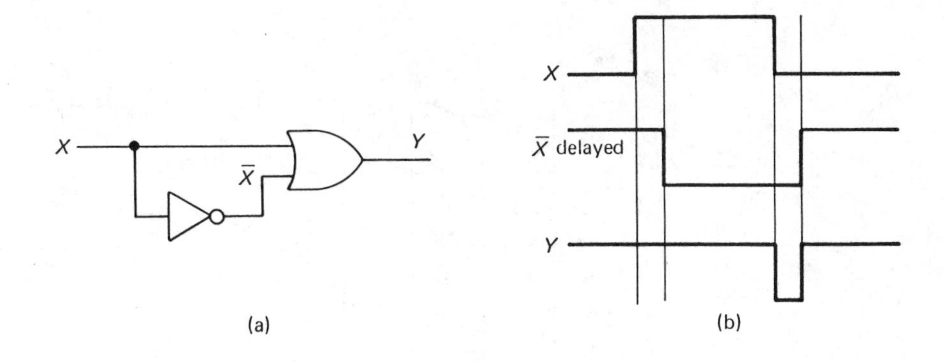

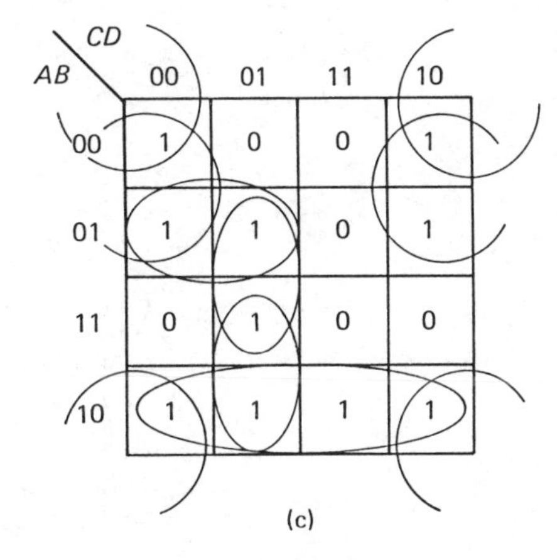

The K-map for the function F is shown in diagram (c).

The hazard-free implementation considers all overlapping implicants giving

$$F = A\overline{B} + \overline{B}\,\overline{D} + \overline{A}\,\overline{D} + \overline{A}B\overline{C} + B\overline{C}D + A\overline{C}D.$$

Example 13.12

Two binary numbers a and b each have two digits. A logic circuit is required to compare the magnitudes of the two numbers, and give outputs L, M and N, such that $L = 1$ if $a < b, M = 1$ if $a = b$ and $N = 1$ if $a > b$.

For the three output variables, give the truth tables and Karnaugh maps, and hence derive Boolean expressions. Describe how the logic circuit must be modified if it is to be interconnected with similar circuits in order to compare numbers having more than two binary digits.

(CEI Part 2)

Solution 13.12

The truth table is given in the table.

The K-maps for L, M and N are shown in the diagram, giving

$$L = \overline{a}_1 b_1 + \overline{a}_1 \overline{a}_0 b_0 + \overline{a}_0 b_1 b_0$$

$$M = \overline{a}_1 \overline{a}_0 \overline{b}_1 \overline{b}_0 + \overline{a}_1 a_0 \overline{b}_1 b_0 + a_1 a_0 b_1 b_0 + a_1 \overline{a}_0 b_1 \overline{b}_0$$

$$N = a_1 \overline{b}_1 + a_0 \overline{b}_1 \overline{b}_0 + a_1 a_0 \overline{b}_0$$

Notice that $M = \overline{L} \cdot \overline{N}$ and this would therefore be a simplification.

If the circuit is to be interconnected with similar circuits to compare larger numbers, then static hazard conditions will need to be eliminated by considering all of the prime implicants in the K-map.

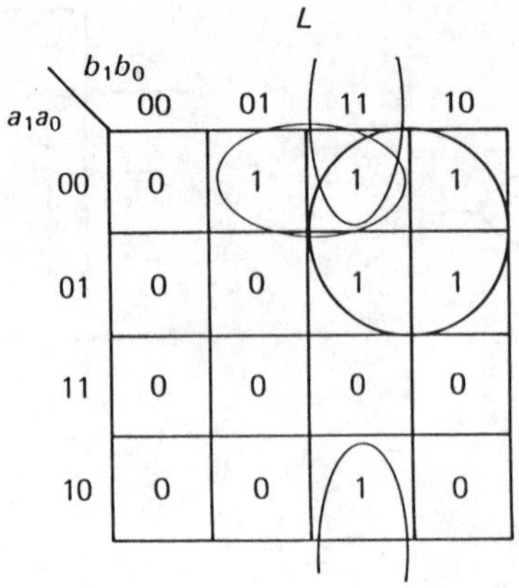

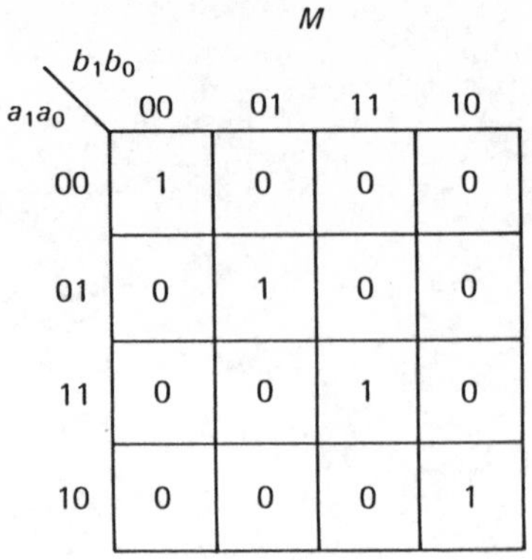

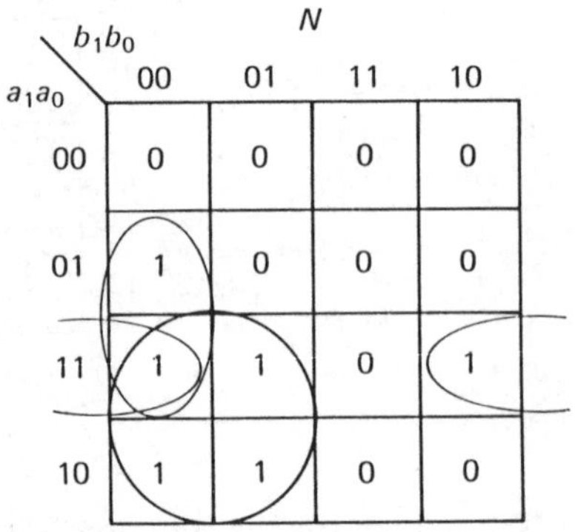

L

a_1a_0 \ b_1b_0	00	01	11	10
00	0	1	1	1
01	0	0	1	1
11	0	0	0	0
10	0	0	1	0

M

a_1a_0 \ b_1b_0	00	01	11	10
00	1	0	0	0
01	0	1	0	0
11	0	0	1	0
10	0	0	0	1

N

a_1a_0 \ b_1b_0	00	01	11	10
00	0	0	0	0
01	1	0	0	0
11	1	1	0	1
10	1	1	0	0

a_1	a_0	b_1	b_0	L	M	N
0	0	0	0	0	1	0
0	0	0	1	1	0	0
0	0	1	0	1	0	0
0	0	1	1	1	0	0
0	1	0	0	0	0	1
0	1	0	1	0	1	0
0	1	1	0	1	0	0
0	1	1	1	1	0	0
1	0	0	0	0	0	1
1	0	0	1	0	0	1
1	0	1	0	0	1	0
1	0	1	1	1	0	0
1	1	0	0	0	0	1
1	1	0	1	0	0	1
1	1	1	0	0	0	1
1	1	1	1	0	1	0

13.11 Unworked Problems

Problem 13.1

State de Morgan's theorems. Show how an OR function of two inputs may be realised using only NAND gates. A combinational logic circuit has three inputs, A, B, C, and a single output Q defined by

$$Q = \bar{A}B\bar{C} + A\bar{B}\bar{C} + \bar{A}\bar{B}C + ABC.$$

Show by means of Boolean algebra that an equivalent expression for Q is

$$Q = \overline{\bar{A}BC + A\bar{B}C + \bar{A}\,\bar{B}\,\bar{C} + AB\bar{C}}.$$

Give a circuit using only NAND gates which will implement the function Q.

Draw the Karnaugh map for Q. Suggest how it could be used to devise a circuit consisting of NAND gates so that any one of three single-pole switches could be used to control a light independently of the other two switches. (EC Part 1)

Problem 13.2

The diagram shows the truth table for a code converter to drive a seven-segment display unit from the numbers 0 to 9 in pure binary. Design a logic circuit for this converter using NAND gates. Describe briefly other methods of realising this converter, which do not rely on simple logic gates. (EC Part 2)

D	C	B	A	a	b	c	d	e	f	g
0	0	0	0	1	1	1	1	1	1	0
0	0	0	1	0	1	1	0	0	0	0
0	0	1	0	1	1	0	1	1	0	1
0	0	1	1	1	1	1	1	0	0	1
0	1	0	0	0	1	1	0	0	1	1
0	1	0	1	1	0	1	1	0	1	1
0	1	1	0	0	0	1	1	1	1	1
0	1	1	1	1	1	1	0	0	0	0
1	0	0	0	1	1	1	1	1	1	1
1	0	0	1	1	1	1	0	0	1	1

Problem 13.3

Design a multiplier circuit that takes two 2-bit binary numbers a_1a_0 and b_1b_0 and produces a binary result $x_3x_2x_1x_0$, which is the arithmetic product of the two input numbers.

Explain how a programmable ROM could be used to provide 4×4 bit binary multiplication.

Problem 13.4

Define the function of an n-to-1 line multiplexer giving the truth table and Boolean expression describing its output. Draw a circuit using simple logic gates of a 4-to-1 line multiplexer.

A 2-bit digital comparator compares the magnitudes of 2-bit binary numbers a and b. It has outputs L, M and N such that $L = 1$ if $a < b$, $M = 1$ if $a = b$ and $N = 1$ if $a > b$. Give the truth table for this comparator and show how it can be realised using 8-to-1 line multiplexers.

(EC Part 2)

14 Sequential Logic Circuits

14.1 Flip-flops

In combinational logic circuits the output levels at any instant are dependent purely on the levels present at the input at that time. In sequential logic circuits the output levels are dependent also on the previous states, and include some form of memory elements. The most widely used memory element is the flip-flop.

(a) The *RS* Flip-flop

A simple *RS* bistable may be constructed using two NOR gates interconnected as shown in Fig. 14.1. As may be seen from the truth table, when both inputs are low, the *Q*-output is stable in whichever state it was at previously. Pulsing the *S*-input high sets the *Q* output low and the $\bar{Q}$ output high. Pulsing the *R* input has just the opposite effect. However, when both inputs go from high to low together then the output state is unpredictable.

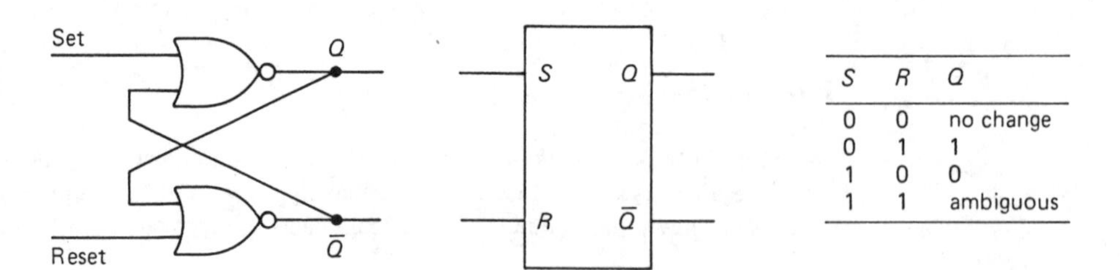

S	R	Q
0	0	no change
0	1	1
1	0	0
1	1	ambiguous

Fig. 14.1 A simple *RS* flip-flop

The simple *RS* flip-flop has two disadvantages for certain applications:

(a) Its operation out of the 1, 1 state is unpredictable.
(b) It is transparent (i.e. when enabled, the output changes whenever the input changes).

(b) The *JK* Flip-flop

The *JK* flip-flop as shown in Fig. 14.2 overcomes both of these difficulties. As well as having *J* and *K* inputs, it has the addition of a clock. As may be seen from

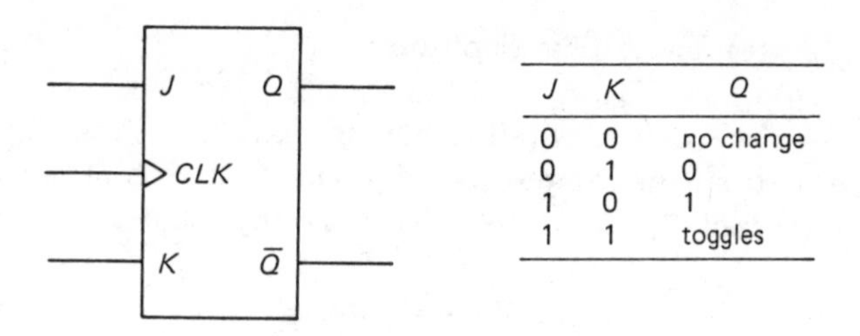

J	K	Q
0	0	no change
0	1	0
1	0	1
1	1	toggles

Fig. 14.2 The *JK* flip-flop

the truth table, the 1, 1 state, rather than being unpredictable, causes the flip-flop to toggle, such that, on receipt of a clock input, the output changes to the opposite state.

JK flip-flops are available with preset and clear inputs. This provides a means of setting the flip-flop into a known condition independent of the clock and the *JK* inputs.

There are two ways in which the clock input may be arranged to control the transitions. These are described in (c) and (d).

(c) Edge-triggered Flip-flops

In this case, one edge is used to define the time at which the output changes. Either the rising or the falling edge may be used, depending on the device. The *JK* inputs must be present for a minimum time (the set-up time) before the clock edge occurs, and remain for a minimum time (the hold time) after the clock edge. The output will be delayed for a period of time after the clock edge (the propagation delay).

(d) Pulse-triggered Flip-flops

In this case the device is pulse-triggered. It is referred to as the master-slave flip-flop. Its operation may be understood by considering the two edge-triggered flip-flops in series as shown in Fig. 14.3, although it is not necessarily made in this way. On the rising edge of the clock, data enters the first flip-flop. On the falling edge, data is transferred to the output. At no time is the input connected through to the output, and feedback oscillations are impossible.

The edge-triggered and master-slave flip-flops operate by clock levels, and are not dependent on the clock edge speed.

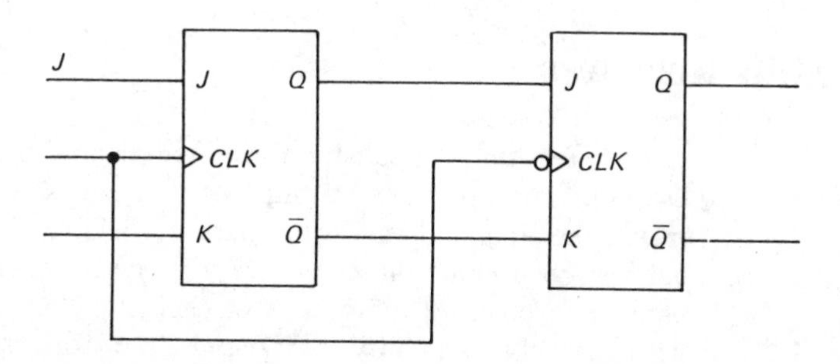

Figure 14.3 *JK* master–slave flip-flop schematic

(e) The *D*-type Flip-flop

The *D*-type flip-flop has one input as shown in Fig. 14.4(a), and data transfers from the *D*-input to the *Q*-output on receipt of a clock pulse. The *JK* flip-flop can easily be converted to a *D*-type by the use of an inverter as shown in Fig. 14.4(b).

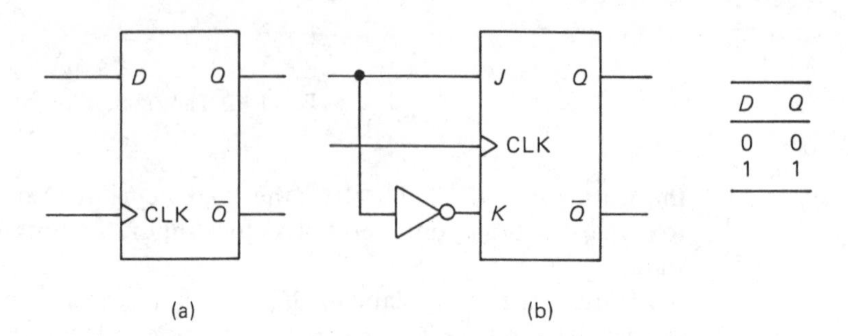

Figure 14.4 The D-type flip-flop

(f) The *D*-type Latch

These are similar to *D*-type flip-flops, except that the *Q*-output follows the *D*-input while the clock is high (i.e. the latch is transparent).

14.2 Counters

Consideration of the clock and output waveforms for a *JK* flip-flop with $J = K = 1$ show that the frequency of the output waveform is half that of the clock waveform. If the output of such a flip-flop is used as the input to another flip-flop, then successive division of frequency will occur, thus providing the basis of a binary counter. If n flip-flops are used then 2^n count states are possible. Count states other than 2^n can be achieved by detecting the required restart point with a logic gate and applying its output to the clear input of the flip-flop. The disadvantages with asynchronous counters are 'glitches' on the output logic states and cumulative propagation delays through the flip-flops. Synchronous counters overcome these difficulties by making all stages of the counter synchronous with the clock. Counters are available in medium-scale integration (MSI) packages with various cycle-lengths.

Also, some MSI synchronous counters are provided with a parallel load facility that can be used to 'short-cycle' a counter by counting to full and then loading a preset value.

14.3 Shift Registers

A shift register is a serially connected arrangement of flip-flops that store data. Data entered into the serial input moves through the shift register under the control of the clock pulses. Output data is available either serially at the serial output, delayed by n clock pulses (where n is the number of stages of the register), or in parallel form using the output of each flip-flop. Shift registers are also available with parallel load facility (PISO) and either forward or reverse shifting.

The shift register can be used as a counter (twisted ring or 'Johnson' counter) by connecting the inverse of the shift register output back into the input.

14.4 Memory Systems

Semiconductor memories can be separated into two types, static and dynamic.
Static memories use a flip-flop for each memory cell; the flip-flops remain in a given state as long as the power supply is maintained or until the data is altered by writing to them.
Dynamic devices store the data as a charge on the interelectrode capacitor, and need continual refreshing.
Both types of semiconductor RAM (random-access memory) are normally provided with tristate outputs and may be multiplexed together to form larger memories using the chip-select control.
Semiconductor memories are available in a variety of different device types including RAM (may be read from or written into) and ROM (read-only memory; non-volatile).

14.5 Worked Examples

Example 14.1

Explain why it is good engineering practice to construct digital systems so that they operate synchronously, and comment on the clock requirements of synchronous systems.
The circuit shown in diagram (a) is devised to accept inputs from a push-button switch and to actuate the relay when seven operations have occurred. Explain why it may not work reliably and draw an alternative circuit (using other TTL devices if necessary) that overcomes defects you have listed.

Solution 14.1

Synchronous operation in sequential systems avoids problems of glitch generation and settling time. Asynchronous systems have indeterminancy during data changes, which synchronous systems avoid by using a synchronous clock, thus giving well-ordered data transfer.

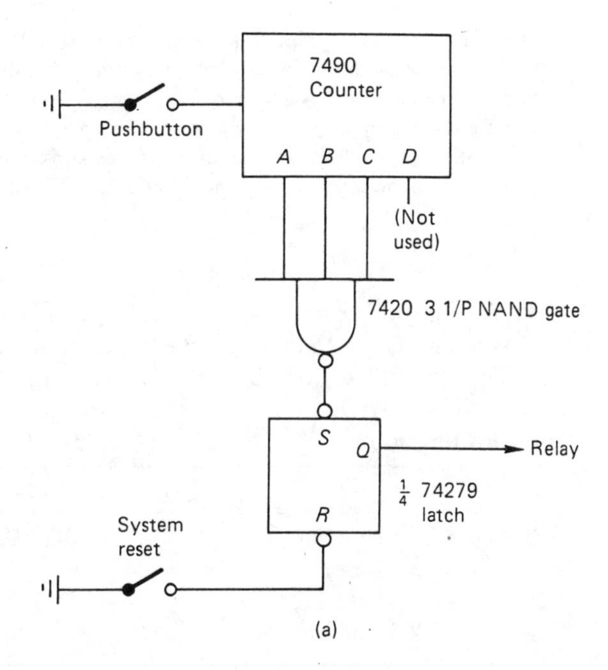

(a)

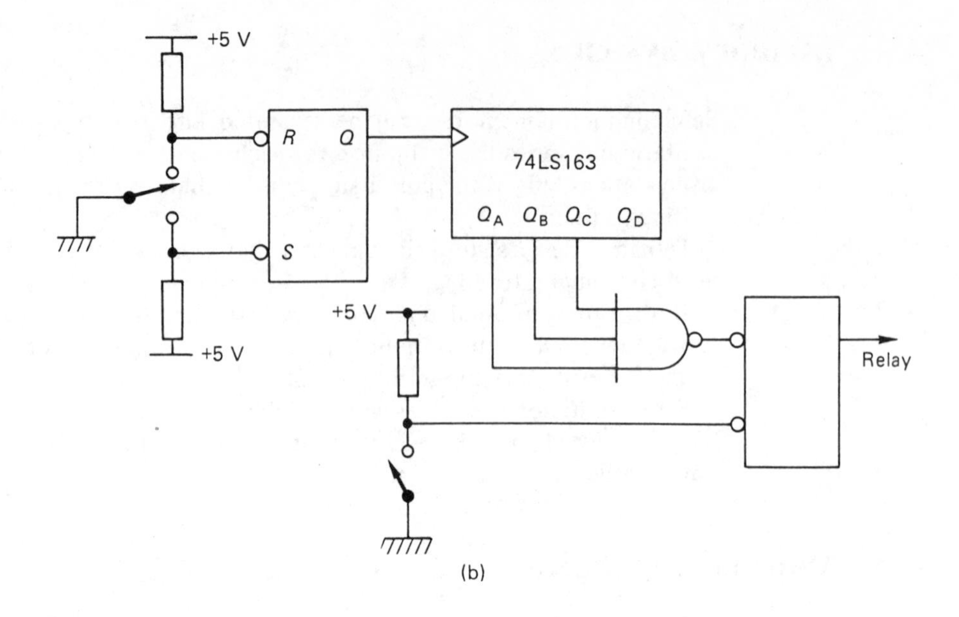

(b)

The clock requirements for synchronous systems are low skew, high fan-out and stable operation. The basic requirement is that all clocked devices within the system receive the required clock edge simultaneously.

The circuit of diagram (a) will suffer from the following:

(a) switch bounce;
(b) decoding spikes through the asynchronous counter;
(c) open-circuit input.

The relay will thus latch-up randomly.

There are many alternatives, although a solution using monostables should be avoided since they introduce asynchronous switching edges.

The solution shown in diagram (b) includes a switch debounce circuit and a synchronous counter.

Example 14.2

Draw circuits showing how a single *RS* flip-flop can be implemented using two NOR gates, and how it can be modified for clocked operation. Give the state tables of clocked *RS* and *JK* flip-flops and explain the advantages of the latter.

Give schematic diagrams and explain the operation of (a) a 4-bit shift register using clocked *RS* flip-flops, (b) a 4-bit ripple or asynchronous up-counter using clocked *JK* flip-flops.

Distinguish between asynchronous and synchronous counters. (EC Part 1)

Solution 14.2

An *RS* flip-flop using cross-coupled NOR gates is shown in diagram (a).

A modification to allow clocked operation is shown in diagram (b).

The state tables for clocked *RS* and *JK* flip-flops are given in tables (a) and (b).

(a) *RS* flip-flop

S	R	Q_{n+1}
0	0	Q_n No change
0	1	0 Reset
1	0	1 Set
1	1	X Not allowed

(b) *JK* flip-flop

J	K	
0	0	Q_n No change
0	1	0 Reset
1	0	1 Set
1	1	$\bar{Q}_n$ Toggles

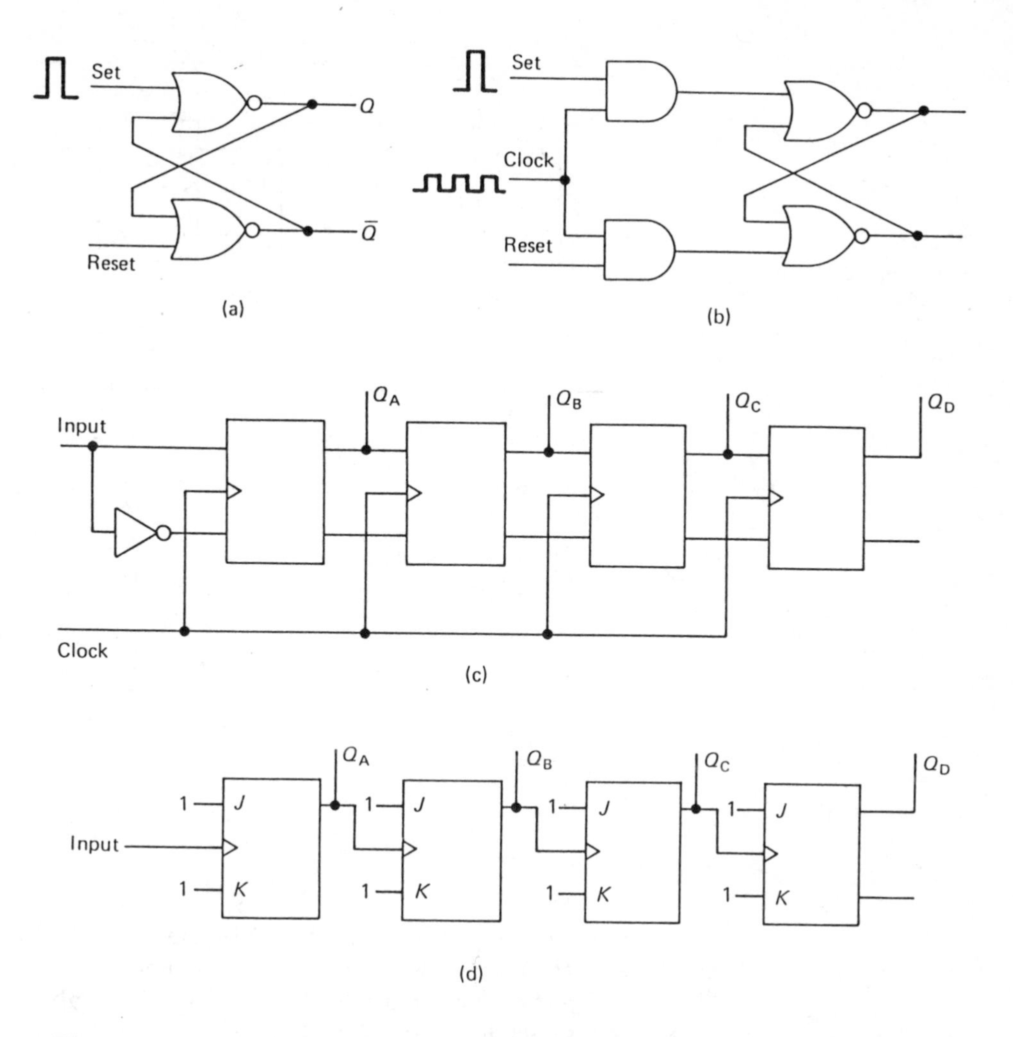

The *RS* and *JK* flip-flops are functionally identical except for the case where *J* and *K* are asserted together; here the *JK* 'toggles'. This state is not allowed in the *RS* since the output state is indeterminate.

(a) A 4-bit shift register using clocked *RS* flip-flops is shown in diagram (c).

(b) A 4-bit ripple-through counter using *JK* flip-flops is shown in diagram (d).

Synchronous counters have all flip-flops clocked synchronously, which ensures that all output edges change state at substantially the same time. The code sequence is determined by the gate combinations connected to the *JK* inputs.

Asynchronous counters do not use a synchronous clock and each stage is clocked by the preceding stage. This causes speed limitations and aggravates 'glitch' problems related to combinational decoding of the output code. It also limits the count sequence to straight binary.

Example 14.3

Show how a simple clocked *RS* flip-flop can be constructed using NAND gates only. Comment on its shortcomings and explain, using a truth table, how these are overcome in a *JK* flip-flop. Using the truth table for the *JK* flip-flop, show that if the initial and final (i.e. after clocking) states of the output are Q_n and Q_{n+1} respectively, then

$$Q_{n+1} = Q_n\bar{K} + \bar{Q}_nJ.$$

The outputs Q_A, Q_B, Q_C and Q_D in the circuit of diagram (a) are initially at logic state 0. Tabulate a full sequence of the states of these outputs after successive clock pulses. What is a practical application of this circuit?

(EC Part 1)

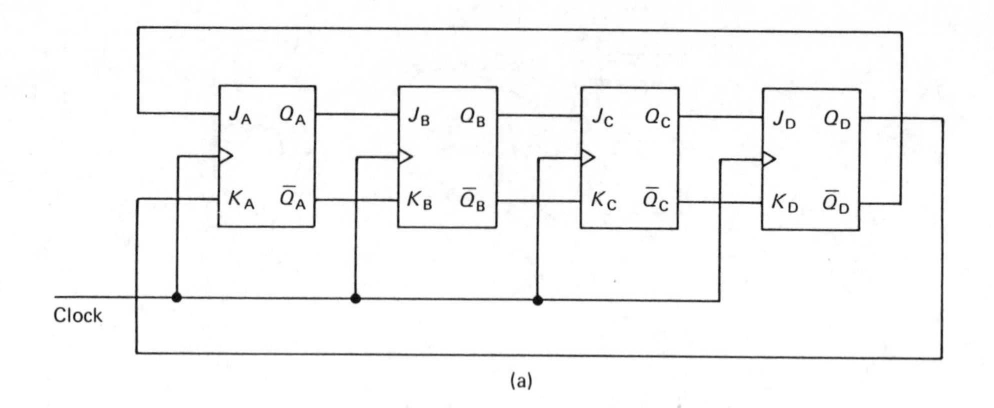

(a)

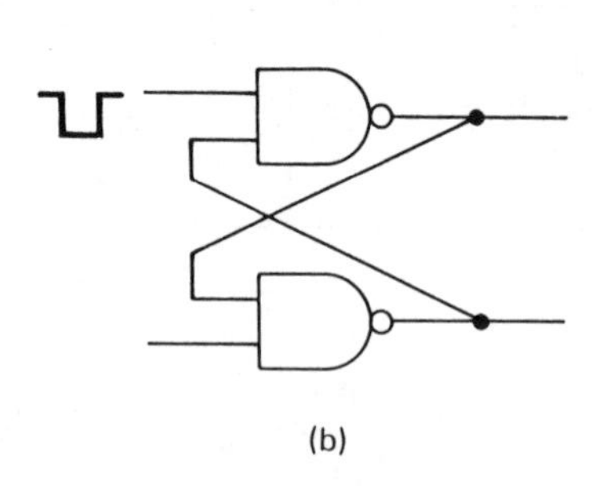

(b)

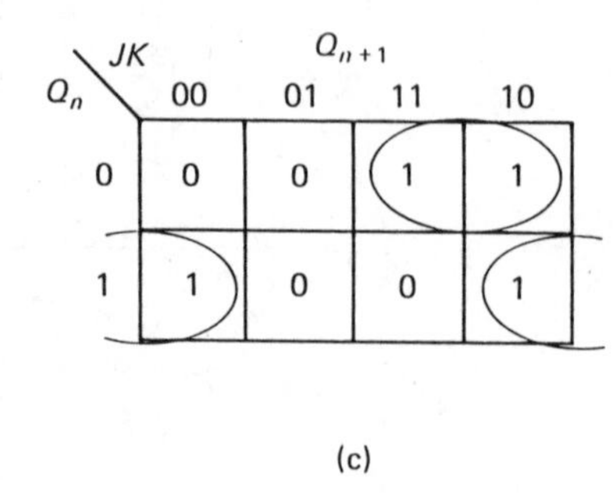

(c)

Solution 14.3

An *RS* flip-flop using cross-coupled NAND gates is shown in diagram (b). Comparison with *JK* flip-flops was considered in Example 14.2.

The full truth table for the *JK* flip-flop is shown in table (a). The corresponding K-map is shown in diagram (c).

Table (a)

J	K	Q_n	Q_{n+1}
0	0	0	0
0	0	1	1
0	1	0	0
0	1	1	0
1	0	0	1
1	0	1	1
1	1	0	1
1	1	1	0

From the K-map the function Q_{n+1} may be seen to be

$$Q_{n+1} = Q_n\overline{K} + \overline{Q}_n J.$$

The sequence of states of the circuit of diagram (a) is shown in table (b). The circuit is a Johnson (or twisted-ring) counter. It has the advantage of being a 'minimum-change' code counter in which only one bit changes at any one time. The count states are $2n$, where n is the number of flip-flops. The one shown is modulo-8. Decoding is particularly simple for this counter, requiring only a 2-input gate for each decoder.

Table (b)

Q_A	Q_B	Q_C	Q_D
0	0	0	0
1	0	0	0
1	1	0	0
1	1	1	0
1	1	1	1
0	1	1	1
0	0	1	1
0	0	0	1
0	0	0	0

Example 14.4

The counter shown in diagram (a) has synchronous clear and asynchronous load inputs. D = most significant bit.

Show how to produce a modulo-7 counter and list the missing codes of your circuit.

Show how to produce a modulo-612 counter using these devices.

Determine the output sequence on A, B, C and D as the shift register of diagram (b) is clocked, assuming an initial state of $ABCD$ = 1101.

Solution 14.4

A modulo-7 counter using the 74LS162 decade counter is shown in diagram (c). The count states are given in table (a). Notice that the CLEAR is synchronous with the clock edge.

Table (a)

Q_D	Q_C	Q_B	Q_A
0	0	0	0
0	0	0	1
0	0	1	0
0	0	1	1
0	1	0	0
0	1	0	1
0	1	1	0
0	0	0	0

The missing codes are 7, 8 and 9. An alternative is shown in diagram (d), with missing codes 0, 1, 2.

A modulo 612 counter is shown in diagram (e).

The output sequence for diagram (b) is shown in table (b), assuming the sequence has initial state 1101.

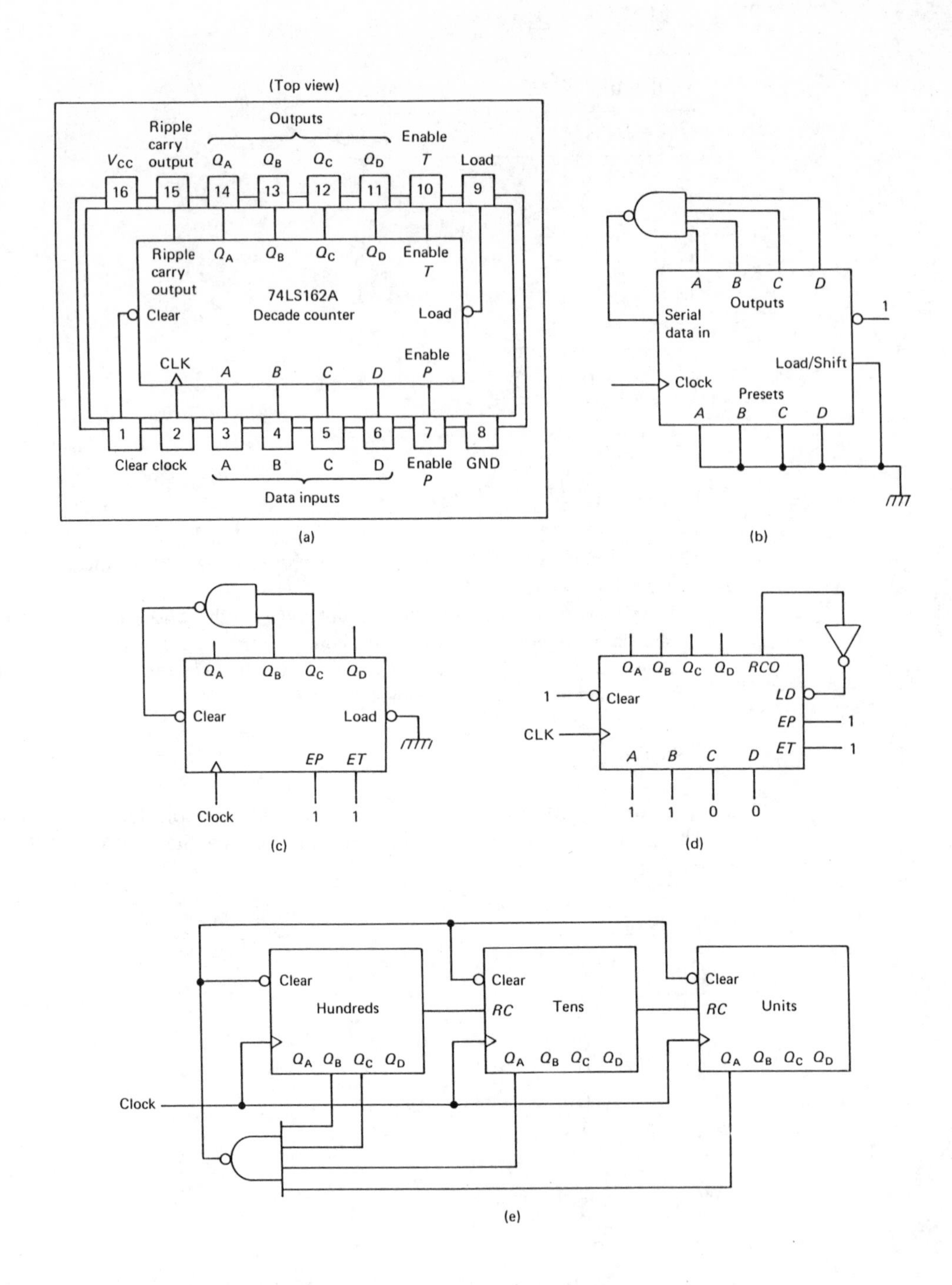

Table (b)

A	B	C	D
1	1	0	1
1	1	1	0
1	1	1	1
0	1	1	1
1	0	1	1
1	1	0	1

Example 14.5

It is required to build a synchronous counter that counts in Gray code up to the equivalent of decimal 5 before returning to zero.

(a) Design and sketch the circuit.
(b) Sketch a complete state diagram for the circuit.

Solution 14.5

The state table for a Gray code counter using three *JK* flip-flops is shown in table (a), where the required *J* and *K* inputs at each state are shown that will cause the outputs to change to the appropriate levels.

The required count sequence for Gray code 0 to 5 is 000, 001, 011, 010, 110, 111. Also included in the table are two states which must be checked to ensure that the counter cannot 'lock up' if these states occur at switch-on.

From this table the K-maps may be drawn up for the *J* and *K* inputs as shown in diagram (a).

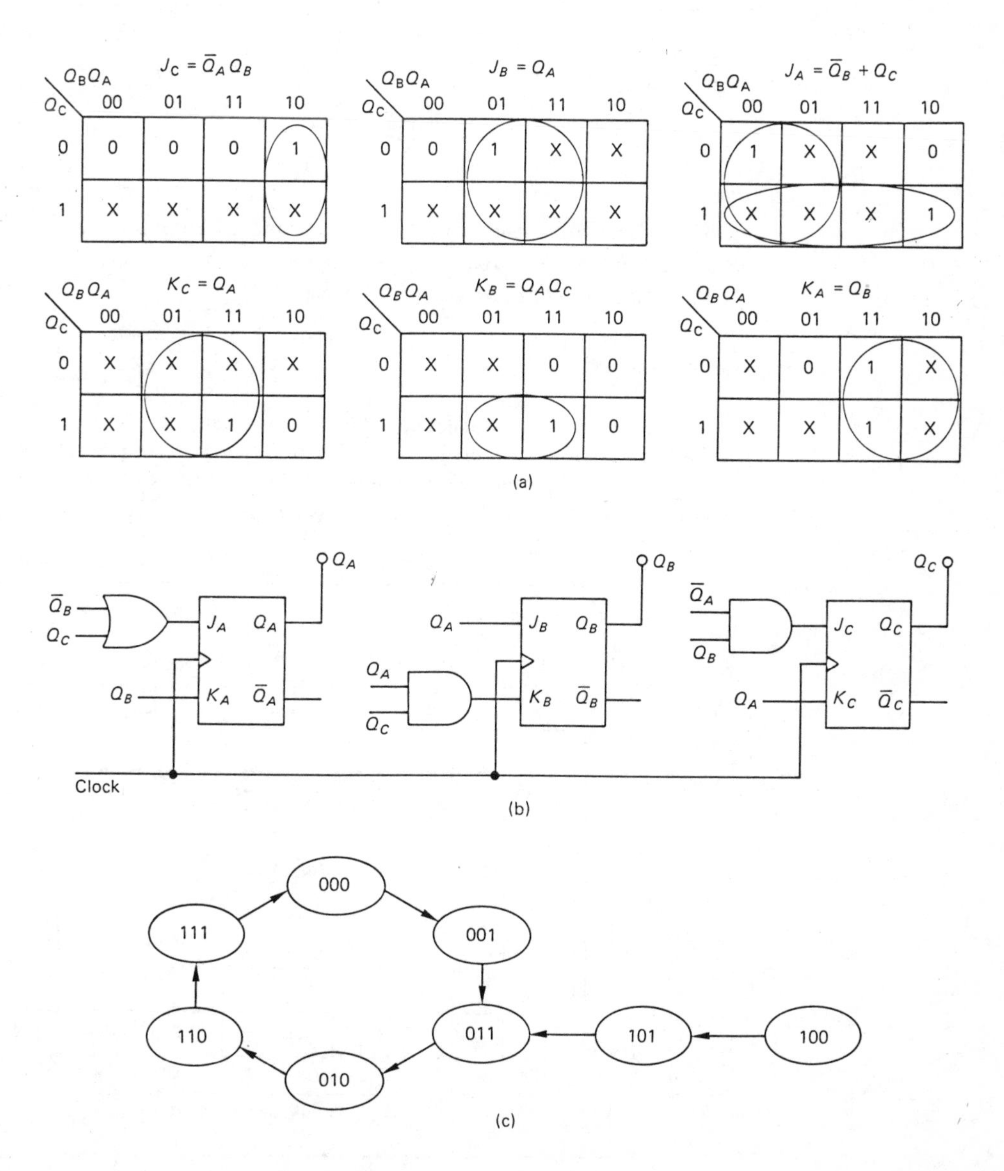

(a)

(b)

(c)

Table (a)

t_n			t_{n+1}								
Q_C	Q_B	Q_A	Q_C	Q_B	Q_A	J_C	K_C	J_B	K_B	J_A	K_A
0	0	0	0	0	1	0	X	0	X	1	X
0	0	1	0	1	1	0	X	1	X	X	0
0	1	0	1	1	0	1	X	X	0	0	X
0	1	1	0	1	0	0	X	X	0	X	1
1	0	0	X	X	X	X	X	X	X	X	X
1	0	1	X	X	X	X	X	X	X	X	X
1	1	0	1	1	1	X	0	X	0	1	X
1	1	1	0	0	0	X	1	X	1	X	1

X = don't care.

The resultant circuit diagram is shown as diagram (b).

The unused states must now be checked as shown in table (b).

Table (b)

t_n									t_{n+1}		
Q_C	Q_B	Q_A	J_C	K_C	J_B	K_B	J_A	K_A	Q_C	Q_B	Q_A
1	0	0	0	0	0	0	1	0	1	0	1
1	0	1	0	1	1	1	1	0	0	1	1

Since both of these states result in one of the states already considered, the counter cannot 'lock up'.

The state diagram is shown in diagram (c).

Example 14.6

Using three D-type flip-flops, design a 3-bit shift register.

By the addition of appropriate 'steering' gates, show how the register may be made reversible.

The 74LS194 4-bit bidirectional universal shift register is shown in diagram (a). Show how two of these registers may be connected together to make a reversible 8-bit shift register.

(Top view)

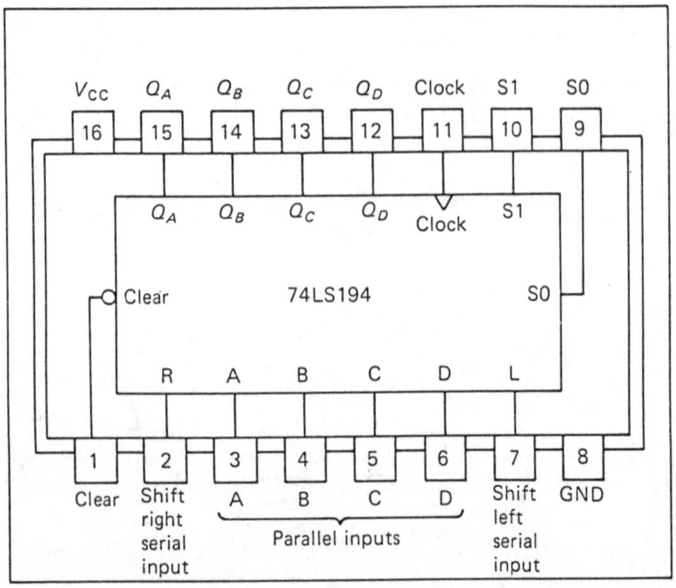

V_{CC} Q_A Q_B Q_C Q_D Clock S1 S0

16 15 14 13 12 11 10 9

Q_A Q_B Q_C Q_D Clock S1

Clear 74LS194 S0

R A B C D L

1 2 3 4 5 6 7 8

Clear Shift right serial input A B C D Shift left serial input GND

Parallel inputs

(a)

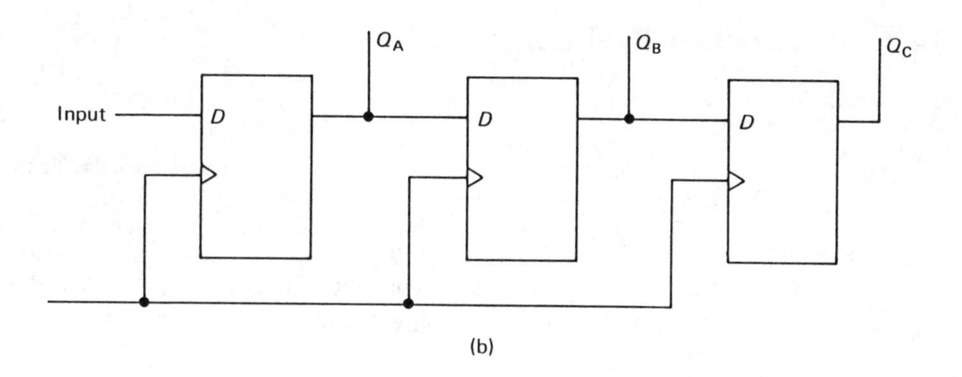

(b)

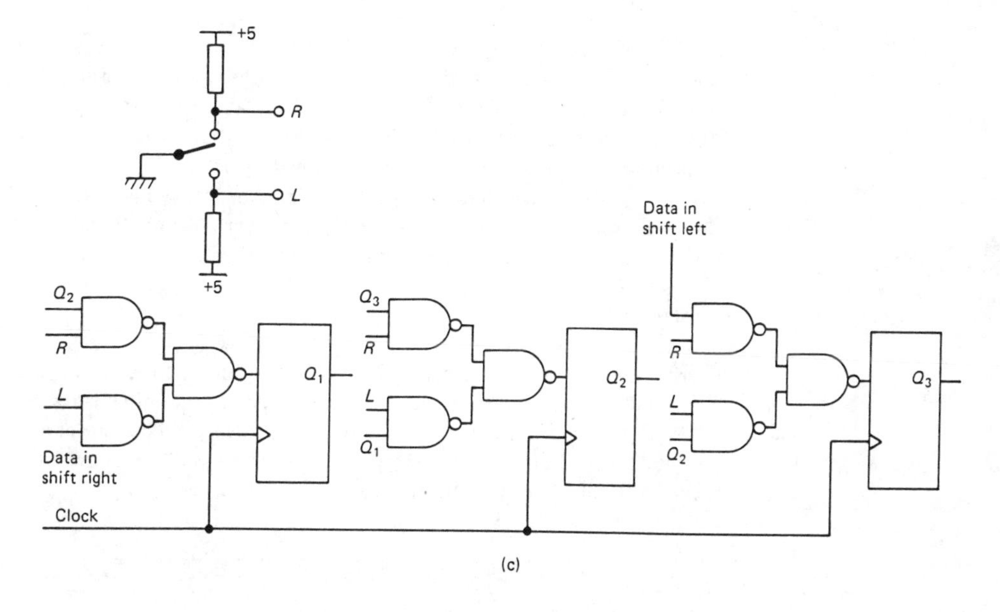

(c)

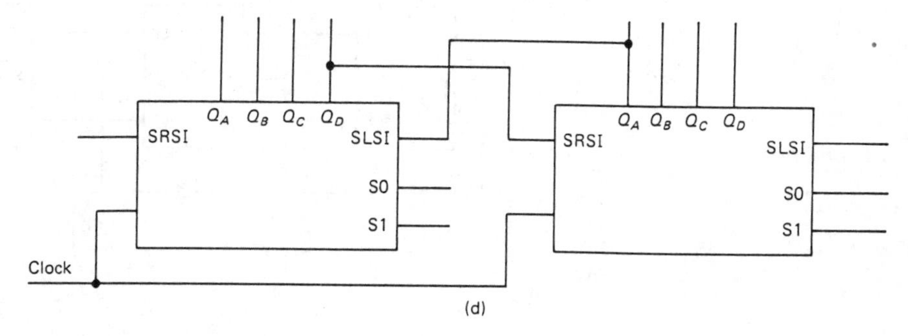

(d)

SO	S1	Function
0	0	Do nothing
0	1	Shift data left
1	0	Shift data right
1	1	Parallel load

Solution 14.6

The shift register using *D*-type flip-flops is shown in diagram (b).

The register may be made reversible by the addition of steering gates as shown in diagram (c).

Two 74LS194s may be connected as shown in diagram (d) to make an 8-bit reversible shift register.

14.6 Unworked Problems

Problem 14.1

Explain the difference between master–slave and edge-triggered *JK* flip-flops. Why is it normally undesirable to use both types in the same circuit?

Show all the stages in the design procedure (e.g. state tables, Karnaugh maps) for a divide-by-five counter using master–slave *JK* flip-flops and NAND gates, in which the states are the first five states of a binary count. Ensure that a state of 'all zeros' is obtained after one clock pulse if the initial state when the circuit is switched on is an unused one. (CEI Part 2)

Problem 14.2

List the different methods available for constructing logic elements and provide brief notes on their performance with regard to speed of operation, power consumption and noise immunity.

Draw a circuit for a simple NOR gate and explain its operation giving typical input and output voltages corresponding to 0 and 1 in the binary scale. Comment briefly on the shortcomings of the circuit and state how they are overcome in practice.

Show by a block diagram how NOR gates may be interconnected to form a memory element that performs the function $Q_{n+1} = S + Q_n\bar{R}$, when the inputs R and S will never both be unity. Give a truth table for this element and mention its uses. (CEI Part 2)

Problem 14.3

Describe the operation of a twisted-ring (or Johnson) counter, and the decoding scheme used.

A modified ring counter is shown in the diagram. The initial state of the counter is $A = 1$, $B = 0$, $C = 1$. Show that it will operate as a divide-by-N counter, and determine N. Does the value of N depend on the initial state?

Draw the complete state diagram for this counter. (EC Part 2)

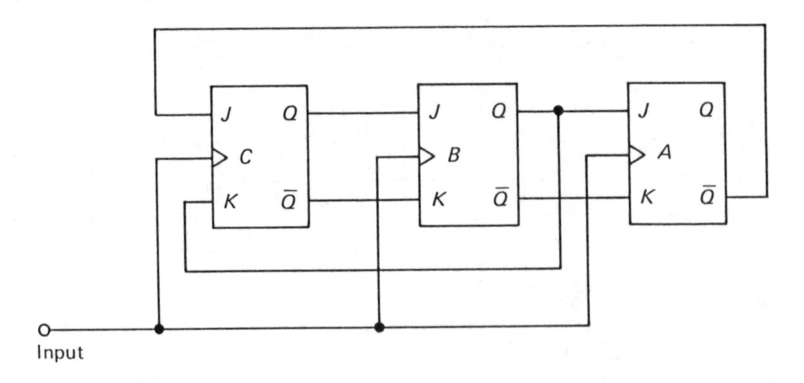

Problem 14.4

Distinguish between synchronous and asynchronous digital counters, giving examples with circuits of each type. Explain briefly the difference between series and parallel carry-in counters and compare the characteristics of these two types of circuit.

Describe the operation of the programmable counter shown in the diagram and list the sequence of states when $Pr_0 = 0$, $Pr_1 = 1$, $Pr_2 = 1$, $Pr_3 = 0$ and all $J = K = 1$. Hence determine the values of Pr_0 to Pr_3 required in order to give a count sequence of 11 states. Comment on the practical difficulties inherent in this simple circuit. (CEI Part 2)

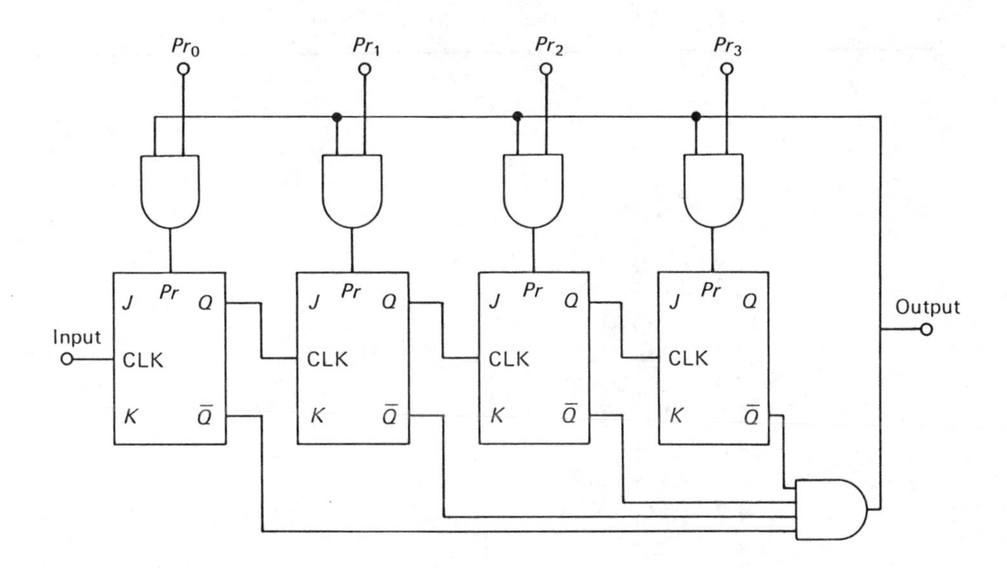

Answers to Unworked Problems

1.1 5.06 A.
1.2 5.4 Ω; 78%; 3.9 W.
1.3 36 μF; 146 V; 0.85 A.
2.1 133 Ω; 34.9.
2.2 480 Ω.
3.1 6 kΩ; –25 V; 47 kΩ; 200 kΩ; 4.7 kΩ.
3.2 20 kΩ; 6 kΩ; –10.
3.3 18.2 kΩ; (a) 184 kHz; (b) 1.3 MHz.
3.4 1.7 nF.
4.1 8 kΩ; –48.
4.2 1.35.
4.3 3.15; 12.6; 25.2.
5.1 750 kΩ.
5.2 4.5; –0.24.
5.3 Overall $A_v = 125$; $R_i = 2$ kΩ; $R_o = 140$ Ω.
6.1 9×10^{-3}; 55.96 dB; 67.13 dB; 2 MHz; $19.6 \angle -78.7°$.
6.2 339; 140 kHz; 434 kHz.
6.3 $86.8 \angle -7°$.
6.4 0.03; 18.4 Ω.
6.5 24.2 dB; 97.4 kHz; 103 Hz; 228 kHz; 44 Hz.
7.1 50.97; 22 MΩ.
7.2 0.999.
7.3 $R_1 = 5$ kΩ; R_2 = say 10 MΩ.
8.1 134.2 pF; 8.8 V.
8.2 5 kΩ; 724 Hz.
8.3 804 kHz; 883 kHz.
9.1 $V_{IH} = 7.30$ V; $V_{IL} = 4.49$ V; $V_{OH} = 12$ V; $V_{OL} = 5.5$ V; $\theta_1 = 15.1°$; $\theta_2 = 197.6°$.
9.2 1 kΩ; 99 kΩ; 1:2.
9.4 971 Hz.
10.2 23.1 W; 56.2%; 71.7%.
10.3 1.46 K/W; 0.14 K/W.
11.1 $7.9 < E < 9.3$; 1 kΩ.
11.2 $R_4 = R_5 = 10$ kΩ.
11.3 2.3 K/W.
12.3 35.75 cm^2; 67.97 W; 57.66 °C.

Index